Systems Analysis and Design

edited by
Don Yeates
Maura Shields
David Helmy

PITMAN PUBLISHING
128 Long Acre, London WC2E 9AN

A Division of Pearson Professional Limited

© Longman Group Limited 1994

First published in Great Britain 1994

British Library Cataloguing in Publication Data
A CIP catalogue record for this book can be obtained from the British Library.

ISBN 0 273 60066 4

All rights reserved; no part of this publication may be reproduced, stored
in a retrieval system, or transmitted in any form or by any means, electronic,
mechanical, photocopying, recording, or otherwise without either the prior
written permission of the Publishers or a licence permitting restricted copying
in the United Kingdom issued by the Copyright Licensing Agency Ltd,
90 Tottenham Court Road, London W1P 9HE. This book may not be lent,
resold, hired out or otherwise disposed of by way of trade in any form
of binding or cover other than that in which it is published, without the
prior consent of the Publishers.

10 9 8 7 6 5 4 3

Typeset by Tek Art, Croydon, Surrey
Printed in England by Clays Ltd, St Ives plc

The Publishers' policy is to use paper manufactured from sustainable forests.

CONTENTS

Preface		xi
1	**The Joy of Analysis and Design**	**1**
	1.1 Introduction	1
	1.2 Successful systems	2
	1.3 Systems development	6
	1.4 The role of the analyst and designer	8
	1.5 Better systems development	10
	1.6 Summary	12
	Case study: System Telecom	13
2	**Approaches to Analysis and Design**	**21**
	2.1 Introduction	21
	2.2 Traditional approaches	21
	2.3 Structured approaches	24
	2.4 Yourdon	29
	2.5 Jackson	31
	2.6 Information Engineering	32
	2.7 SSADM	34
	2.8 Merise	37
	2.9 Euromethod	39
	2.10 Object Oriented design	39
	2.11 Conclusion	41
	Case Study and Exercises	41
3	**Communicating with People**	**42**
	3.1 Introduction	42
	3.2 Types of communication	43
	3.3 Barriers to communication	45
	3.4 Improving your skills	48
	3.4.1 Getting information	48
	3.4.2 Giving information	49
	3.4.3 Meetings	51
	3.4.4 Presentations	53
	3.5 Summary	59
	Case Study and Exercises	60
4	**Building Better Systems**	**61**
	4.1 Introduction	61

	4.2	Quality concepts	61
	4.3	Quality gurus	62
	4.4	The cost of poor quality	65
	4.5	Quality management	68
	4.6	ISO 9000	72
	4.7	Quality in the structured life cycle	74
		4.7.1 Structured walkthroughs	74
		4.7.2 Fagan inspections	76
	4.8	Summary	77
		Case Study and Exercises	81
5	**Understanding the Business**	**82**
	5.1	Introduction	82
	5.2	Business analysis	82
		5.2.1 Levels of understanding	82
		5.2.2 Linkage of IS to business objectives	83
	5.3	Constraints	84
		5.3.1 The user's organisation	84
		5.3.2 Working practices	85
		5.3.3 Financial control procedures	85
		5.3.4 Security and privacy	86
		5.3.5 Legal considerations	86
		5.3.6 Audit requirements	87
		5.3.7 Fallback and recovery	87
	5.4	IT for competitive advantge	87
	5.5	Summary	89
		Case Study and Exercises	90
6	**Project Management**	**91**
	6.1	Introduction	91
	6.2	Stages of system development	91
		6.2.1 Before analysis and design	91
		6.2.2 Analysis and design	92
		6.2.3 After analysis and design	93
	6.3	Project planning	94
		6.3.1 Stages in planning	95
		6.3.2 Planning for quality	99
	6.4	Estimating	100
		6.4.1 Estimating for analysis and design work	100
		6.4.2 Advantages of the structured approach	101
	6.5	Project monitoring and control	102
		6.5.1 The control of quality	103
		6.5.2 Documentation control	104
		6.5.3 Change control	104
		6.5.4 Configuration management	105
	6.6	Summary	105
		Case Study and Exercises	105

7 Systems Analysis: Concepts — 107

- 7.1 Introduction — 107
- 7.2 What is systems analysis? — 107
- 7.3 A structured approach — 111
  - 7.3.1 Structured systems analysis — 112
- 7.4 The PARIS model — 113
- 7.5 Summary — 114

8 Systems Analysis: Planning the Approach — 115

- 8.1 Introduction — 115
- 8.2 Objectives and terms of reference — 116
- 8.3 Constraints — 118
- 8.4 Preparing for detailed analysis — 121
- 8.5 The feasibility study — 123
- 8.6 Summary — 125
- *Case Study and Exercises* — 126

9 Systems Analysis: Asking Questions and Collecting Data — 127

- 9.1 Introduction — 127
- 9.2 Fact-finding interviews — 128
  - 9.2.1 Planning the interview — 128
  - 9.2.2 Conducting the interview — 133
- 9.3 Questionnaires — 141
- 9.4 Observation — 143
- 9.5 Record searching — 143
- 9.6 Document analysis — 145
- 9.7 Summary — 145
- *Case Study and Exercises* — 147

10 Systems Analysis: Recording the Information — 148

- 10.1 Introduction — 148
- 10.2 Data dictionaries and CASE tools — 149
- 10.3 Data flow diagrams — 149
  - 10.3.1 DFD components — 150
  - 10.3.2 DFD hierarchies — 151
- 10.4 Modelling current physical processing — 154
- 10.5 Entity models — 156
  - 10.5.1 The logical data structuring technique — 156
  - 10.5.2 The logical data model — 161
- 10.6 Modelling current data — 163
- 10.7 The data catalogue — 164
- 10.8 Recording the requirements — 165
- 10.9 Summary — 165
- *Case Study and Exercises* — 166

11 Systems Analysis: Interpreting the Information Collected — 167

- 11.1 Introduction — 167
- 11.2 Creating a logical model of current processing — 167
- 11.3 Modelling the required system — 169
- 11.4 Adding the time dimension — 169
- 11.5 Modelling the effects of system events — 171
- 11.6 Entity life histories (ELHs) — 172
- 11.7 Producing entity life histories — 175
- 11.8 Effect correspondence diagrams (ECDs) — 177
- 11.9 Producing effect correspondence diagrams — 178
- 11.10 Modelling enquiries — 179
- 11.11 Defining the user view of processing — 180
- 11.12 Modelling input and output data — 182
- 11.13 Size and frequency statistics — 183
- 11.14 Summary — 183
- Case Study and Exercises — 183

12 Systems Analysis: Specifying the Requirements — 185

- 12.1 Introduction — 185
- 12.2 Agreeing the options — 186
  - 12.2.1 Identifying options — 186
  - 12.2.2 Choosing between the options — 187
  - 12.2.3 The use of prototyping — 188
  - 12.2.4 Quantification of options — 189
- 12.3 Identifying benefits — 191
- 12.4 Presenting the requirement — 192
- 12.5 Writing the functional specification — 194
- 12.6 Summary — 195
- Case Study and Exercises — 195

13 From Analysis to Design — 196

- 13.1 Introduction — 196
- 13.2 Bridging the gap — 197
- 13.3 Design objectives and constraints — 198
- 13.4 An overview of systems design — 200
- 13.5 Summary — 201

14 Systems Design: Protecting the System — 203

- 14.1 Introduction — 203
- 14.2 Damage to the physical installation — 204
- 14.3 Damage to the software system — 205
- 14.4 Damage to the data areas — 210
- 14.5 Damage to the client's business — 211
- 14.6 Discussions with the client — 213
- 14.7 Building in protection — 214

		14.7.1 Using software only	215
		14.7.2 Using a combination of software and hardware	219
		14.7.3 Using hardware only	220
	14.8	Formulating a protection policy	220
	14.9	Summary	222
		Case Study and Exercises	222

15 Systems Design: Human–Computer Interface — 223

15.1	Introduction	223
15.2	Agreeing the system boundary	223
15.3	Output design	224
	15.3.1 Output technology	226
	15.3.2 Presenting information	227
	15.3.3 The use of tables and graphics	229
	15.3.4 Specifying outputs	230
15.4	Input design	233
	15.4.1 Keyboard transcription from clerical documents	234
	15.4.2 Direct input onto the computer system via a peripheral device	235
	15.4.3 Direct entry through intelligent terminals	236
	15.4.4 Input by speech	236
15.5	Dialogue design	237
	15.5.1 Screen design	238
	15.5.2 Dialogue types	239
	15.5.3 WIMP interfaces	243
	15.5.4 User support	246
15.6	Ergonomics and interface design	246
15.7	Summary	249

16 Systems Design: System Interfaces — 251

16.1	Interfaces defined	251
16.2	Analysing interfaces	254
16.3	Physical forms of interfaces	258
16.4	Interfaces to peripherals	262
16.5	Summary	264

17 Systems Design: Logical Data Design — 265

17.1	Introduction	265
17.2	The top-down view: entity modelling	265
	17.2.1 The entity relationship matrix	267
	17.2.2 Summary of entity modelling	268
17.3	The bottom-up view: third normal form analysis	268
17.4	Merging the data models	273
17.5	Testing the data model	274
17.6	The data dictionary	274

		17.6.1 Advanced features of a data dictionary	277
	17.7	Summary	277
		Case Study and Exercises	278

18 Systems Design: Files — 279

	18.1	Introduction	279
	18.2	Types of file	279
	18.3	Storage media	281
		18.3.1 Magnetic disk	281
		18.3.2 Magnetic tape	283
		18.3.3 Other storage devices	284
	18.4	File organisation	284
		18.4.1 Serial organisation	285
		18.4.2 Sequential organisation	285
		18.4.3 Indexed sequential organisation	286
		18.4.4 Random file organisation	287
		18.4.5 Full index organisation	288
		18.4.6 Chained files	289
	18.5	Access methods	291
	18.6	Factors influencing file design	292
	18.7	Specifying files	294
	18.8	Summary	302
		Case Study and Exercises	302

19 Systems Design: Databases — 303

	19.1	Introduction	303
	19.2	Database concepts	303
	19.3	Database models	304
	19.4	File management systems (FMS)	308
	19.5	Hierarchical database systems (HDS)	308
	19.6	Network database systems (NDS)	310
	19.7	Relational database systems (RDBMS)	312
		19.7.1 Data structure	312
		19.7.2 Data manipulation	313
	19.8	RDBMS design	316
	19.9	Futures	318

20 Systems Design: Physical Data Design — 320

	20.1	Introduction	320
	20.2	Quantifying the data storage requirements	321
	20.3	Assessing the required system performance	322
		20.3.1 Factors affecting system performance	323
		20.3.2 Overheads that adversely affect system performance	324
	20.4	Investigating the chosen hardware/software platform	325
		20.4.1 Data storage	325
		20.4.2 Data transfer	327

	20.4.3	The programming language used	327
20.5	Moving from logical to physical data design		328
	20.5.1	Creating a physical data design	328
	20.5.2	Test against client requirements	330
	20.5.3	Refining the physical data design	332
20.6	Summary		333
	Case Study and Exercises		334

21 Systems Design: Program Design — 335

21.1	Introduction	335
21.2	What is meant by program design?	336
21.3	Why break the system down into programs?	337
21.4	What's in a good program specification?	338
	21.4.1 Suggested program specification contents	339
21.5	Software controls	340
21.6	Data action diagrams (DADs)	341
	21.6.1 Levels of data action diagrams	345
21.7	Interaction relationship charts (IRCs)	346
21.8	State transition diagrams (STDs)	348
21.9	Summary	352

22 Systems Design: Choosing Hardware — 354

22.1	Introduction	354
22.2	The evolution of computer systems	354
22.3	Microprocessors	356
	22.3.1 Intel	356
	22.3.2 Motorola	357
	22.3.3 RISC	357
22.4	Processing speed	358
22.5	Processor technology	359
22.6	The impact of open systems	363
22.7	Summary	363

23 Systems Design: Data Communications — 364

23.1	Introduction	364
23.2	Basic concepts	364
23.3	The use and provision of networks	367
23.4	Carrying information across networks	368
	23.4.1 Local area networks	368
	23.4.2 Wide area networks	369
23.5	Standards and standards-making bodies	370
	23.5.1 The OSI reference model	370
	23.5.2 The upper layers	371
	23.5.3 The lower layers	372
	23.5.4 The transport layer	373
	23.5.5 The X and V series recommendations	373
	23.5.6 TCP/IP	374

	23.6	Designing a network	374
		23.6.1 Wide area networks	375
		23.6.2 Local area networks	375
	23.7	Summary	376
		Case Study and Exercises	377

24 Systems Design: Systems Implementation 379

	24.1	Introduction	379
	24.2	Coding and unit test	379
		24.2.1 Employing programmers to write code	380
		24.2.2 Using code generators	381
	24.3	Testing: ensuring the quality	382
	24.4	Data takeon and conversion	386
	24.5	User training	387
	24.6	Going live	389
	24.7	The maintenance cycle	393
	24.8	Summary	394

25 Change Management 395

	25.1	Introduction	395
	25.2	Information technology and people	395
		25.2.1 The role of analysts and designers	397
	25.3	Change management	398
		25.3.1 Unfreezing, moving and refreezing	400
	25.4	The people project	401
		25.4.1 Creating involvement	401
		25.4.2 Building commitment	403
		25.4.3 Providing skills	404
		25.4.4 Managing the benefits	405
	25.5	The change management payoff	405
	25.6	Summary	406

26 What Next? 408

	26.1	Introduction	408
	26.2	How did we get here?	408
	26.3	What's happening to work?	409
	26.4	How will we survive?	410
	26.5	Business process reengineering	412
	26.6	Conclusion	414

Index 416

PREFACE

This book has been written for people who are working in systems analysis and design already or who plan to make a career in it. It has been written by a team of colleagues all of whom are involved in the development of new systems or the development of new analysts for the Sema Group. Many of the team are actively involved in both.

We hope that you will find the book practical, easy to read and easy to use. It can't tell you everything there is to know about analysis and design. No book can ever do that – however, there is no reason why you should not benefit from the experience of others.

The experience in this book is offered to you by Jim Cadle, Alex Cameron, Christine Donaldson, Debbie Paul, Mike Randall, Danielle Sanderson, Tim Walters, Steve Wells, Gerald Wilson, and ourselves the three editors. In spite of what we may have said to them during the project, in the glow of post-implementation euphoria we feel inclined to be more polite and offer just a hint of thanks for all their efforts. The wordprocessing and desktop publishing were done by Elvina Culver and Nicky Wancio; the index was created by Sarah Helmy; and the long-suffering publisher at Pitman was John Cushion.

Don Yeates, Maura Shields, David Helmy
March 1994

CHAPTER 1
The Joy of Analysis and Design

1.1 INTRODUCTION

Not everyone connected with this book is happy about the title of this chapter. At the time of writing, in the UK, there is a widely publicised book called *The Joy of Sex*. Some of our publisher's advisers are unhappy about the association between sex and systems analysis and design. I hope that this apparent association won't give you a problem because we've chosen the chapter title quite deliberately. We want to convey through the title some important aspects of analysis and design.

- Systems analysis and design involves people. Certainly it involves technology, often technology that we don't really understand and that we rely on other people to manage for us, but the best designed systems in the world succeed because the people who use them can do their jobs better. This is because their information systems enable them to achieve goals they wouldn't otherwise reach. There are just too many people on trains and planes using laptop machines to believe that we can ever in the future manage without computers. The importance of this 'people aspect' is emphasised in chapter 26 where we consider change management.
- Systems analysis and design is fun. The only retired systems analysts are people who strayed into it from other trades, or who seized the opportunity to do something new and exciting instead of their old and decaying existing jobs. There are no retired systems analysts who started out their working lives as analysts. If you were in your early 20s in the 1960s and just starting out in systems analysis and design then several things could have happened to you. Either you're still working at it and perhaps getting more of it right than you used to do, or you've left it because it doesn't suit you, because you've made too many mistakes, or because you've been promoted. You certainly haven't retired.
- Systems analysts and designers change the world. This is a bold statement and in one sense, everyone changes the world to some extent just by their very existence. The role of the systems analyst, however we may describe it in detail, is to be a change agent. Unless we are interested in changing the way organisations work we have no need for systems analysts. Unless you are interested in changing the way organisations work, you probably have no need to read this book.

So, who is this book written for? The whole book is about systems analysis and design and it's been written for people who are analysts or designers already or people who are thinking about making a career in analysis and design. This book doesn't cover all you need to know about the job. No book can ever do that: there will always be something you'd wish you'd known, and there will always be something you could have done better. Life will present you with new problems every day. However, there's no reason why you should make all of the same mistakes that other analysts and designers have already made. In this book many good and experienced analysts, designers, consultants and trainers offer you their experience in the hope that you will build on it, benefit from it and be better at your job.

The book is organised in a project-chronological sequence. This means that it begins with analysis and ends with implementation. There are some chapters however that don't fit neatly into this sequence, so they have been put in at the beginning – like this one – or right at the end – like Change Management. They deal with the environment within which analysis and design takes place and are concerned with business, people, management and quality. There's also a basic assumption running through the book. It is that analysts and designers work for customers. We believe that the word 'customer' is very important. Somebody pays for what analysts and designers deliver. New systems have to be justified by the benefits that they deliver. It is easy to use terms like 'the users' and 'user management' – they're used in this book – and forget that they are substitutes for 'the customer'. All of the contributors to this book work for a company whose very existence depends on its ability to build and deliver new computer-based systems and where customer focus comes first, last and everywhere in between. This isn't to say that only those analysts and designers who work for service companies have a customer focus. Indeed, the chapter on Building Better Systems extends the scope of the word 'customer' well beyond its everyday usage and beyond the meaning here in this chapter. We do, however, want you to hold on to the important concept of 'customer' as you read this book.

1.2 SUCCESSFUL SYSTEMS

When the new system you've worked on is implemented and running regularly, and you're assigned to another project, how will you know if you did a good job? How will you know if you've helped to produce a successful system? A typical question often used in analysis and design examinations asks:

> "You're called in to evaluate the effectiveness of a recently implemented system. What criteria would you use?"

Leaving aside project management considerations such as implementation to time, cost and quality you could ask the following questions:

- Does the system achieve the goals set for it? Some of these will be operational running goals concerned with performance, some will be system goals concerned with the production of outputs and some will be business goals addressing the purpose of the system development.
- How well does the system fit the structure of the business for which it was developed? The new system will no doubt have been developed based on an understanding of the then present structure of the organisation and some appreciation of how it might change in the future. However it must not be an 'albatross system' that hangs around the organisation's neck limiting its movement and freedom to reorganise. Systems should be designed in a flexible way so that they can be changed to meet changing business conditions.
- Is the new system accurate, secure and reliable? There will be basic requirements for

financial control and auditing but the system should also be robust so as to continue in operation with degraded performance during partial failure. Security from unauthorised access has also now become increasingly important with the growth in the development of tactical and strategic information systems.
- Is the system well documented and easy to understand? Increasingly large proportions of the budget of system development departments are being used in the maintenance and updating of existing systems. The biggest single way of limiting these expenses in the future is to take account of it when we design today the systems of tomorrow.

This 'single-system' view whilst helping to identify the characteristics of successful systems doesn't however give a sufficiently wide framework for our analysis. We need to begin with an overview of the organisation as a whole. We can, for example, see the organisation in systems terms, operating within its environment and made up of a series of subsystems. A representation of this that has been widely used is shown in figure 1.1.

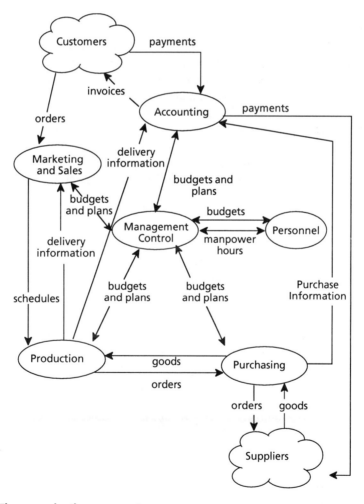

Fig. 1.1 The organisation as a system

It shows an industrial organisation with subsystems for:

- marketing and purchasing; these are the main links with the environment as represented by customers and suppliers. It's important to recognise however that the environment also interacts with the organisation through legislation, social pressures, competitive forces, the education system and political decisions.
- the production system: this is concerned with transforming raw materials into finished products. It applies just as much in service organisations as in traditional manufacturing industry: an architectural drawing office is the equivalent of a motor engine assembly line.
- support systems: these are shown as the accounting, personnel and management control subsystems.

For this organisation to work effectively it has to make good use of information so the need arises for information systems that collect, store, transform and display information about the business. We can represent this information systems structure in two ways: either in a non-hierarchical way showing each subsystem on the same level, or in an hierarchical way where some systems sit on top of others. This multilevel view is often more helpful as it shows the different levels of control, the different data requirements and a different view of the organisation of each system. A typical way of representing this structure is as if the systems are arranged in a pyramid as shown in figure 1.2.

At the top level are strategic systems and decision support systems that informate the organisation. Informate is a term which, in brief, means what happens when automated processes yield information that enables new competitive advantage to be gained. At this level systems contribute towards the formulation of corporate policy and direction. These are concerned almost entirely with financial information and other data that shows the health of the organisation. Strategic systems use information from lower-level internal systems and externally obtained information about markets, social trends, and competitor behaviour.

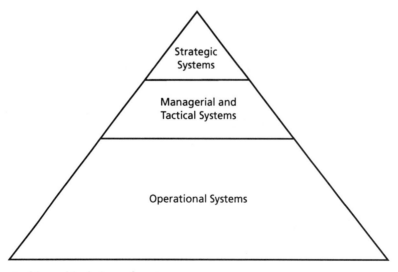

Fig. 1.2 An hierarchical view of systems

Underneath strategic systems lie managerial or tactical systems which are concerned with the monitoring and control of business functions. There is therefore a regular supply of data from the day-to-day operational systems which are manipulated to provide the management information which these systems typically produce. Systems requirements here are for timely, useful and effectively presented reports that enable middle managers to run their departments and divisions effectively. Here also we see the need for systems that can respond to a changing array of ad hoc queries about why the results look the way they do.

The operational systems level is concerned with the routine processing of transactions such as orders, invoices, schedules, statements etc. They help the organisation to 'do what it does' – make parts, distribute products, manage property. They are not concerned with changing the way the organisation works.

We can now see that the organisation can be viewed as a system and that the information systems that support it can be strategic, tactical or operational. This is summarised in figure 1.3.

There is one more view that we must see before we can leave this topic. It is concerned with the evolution of information systems. A useful model here and one widely used is the Gibson/Nolan four-stage model of:

- initiation
- expansion
- formalisation
- maturity

Let's consider an organisation moving through this model for the first time. During the *initiation* phase the first computer-based systems to be developed are those best suited

Strategic Systems	Provide information to managers to enable them to make better informed decisions. Support decision makers in situations that are poorly structured. Used to establish plans for the introduction of new business lines or their closure. Need greater flexibility to be able to respond to constantly changing requirements.
Tactical Systems	Use stored data from operational systems. System outputs are well defined as managers can generally identify the factors influencing decisions that they will have to make. Usually concerned with the management and control of departments or functions.
Operational Systems	Transaction based. Handle the routine business activities of organisations. Often the first systems to be automated as they provide the raw data for higher-level systems.

Fig. 1.3 The three-level systems summary

to the new technology. These projects are almost always at the bottom of the systems pyramid and involve the repetitive processing of large volumes of transactions. They often begin with accounting systems. Following experience here the organisation enters the *expansion* stage and seeks to apply the new technology to as many applications as possible. This is the honeymoon period for the IS department until one day a halt is called to the ever-growing IS budget and the introduction of development planning and controls signals the start of the *formalisation* stage. It is here that the need for information surpasses the need for data and where the organisation begins to plan its way from a mixture of separate data processing systems towards a more co-ordinated and integrated approach. Corporate recognition of the need for integrated systems is the characteristic of the *maturity* stage. Here we see the use of open system architectures, database environments and comprehensive systems planning.

There is a final complication. Organisations don't go through this model once. There is no final nirvana of maturity, of fully integrated systems all talking to each other and producing exactly the strategic information needed by top management. The organisation or its component parts can be at different stages in the model at the same time. Having arrived at the *formalisation* stage with large mainframe computers, many organisations were plunged back into *initiation* and *expansion* with the arrival of personal computers. New technology, the development of new applications software packages or significant price reductions in either hardware or software all throw organisations back into the initiation stage.

1.3 SYSTEMS DEVELOPMENT

The traditional stages of a systems development project begin with feasibility and end with system maintenance in the sequence shown in figure 1.4.

Fig. 1.4 Systems development stages

Systems development projects are initiated for a variety of reasons as we see in chapter 5, but once the process starts we begin by considering the feasibility of a proposed new system. We can expect the initiator of the proposal to be enthusiastic in its support but a balanced view is clearly required before a major investment is made in its development. The feasibility of new system developments is often considered under three headings:

- financial
- technical
- social

While the overall assessment will include an evaluation under each of these headings, it is usual to find that financial feasibility far outweighs the other two considerations. Financial feasibility includes an assessment of the one-time costs of hardware and software of course, but more importantly it addresses the impact the final system may have on the business. The justification for the new system must be that it will increase the profit of the enterprise, improve the quality of service or products which are the business of the enterprise, reduce expenditure, or otherwise contribute towards the achievement of the purpose of the enterprise.

The assessment of technical feasibility is based on system design ideas relating to what can be accomplished with existing or imminently available technology. At times of rapidly changing technological change this is clearly a difficult evaluation to make and it has been estimated that about 35% of average system developments involve new hardware, software or system ideas.

The assessment of social feasibility is now assuming greater importance. This is due to the inroads that new systems are making on the work practices of users and the realisation that the secret of effective new systems lies in their full and comprehensive use by those for whom they have been designed. This issue of managing the implementation of change is discussed in more detail in chapter 26.

The analysis stage is concerned with two activities: firstly, the collection of information about the operation of existing systems and the identification of difficulties, problems and bottlenecks with those systems, and secondly with the specification of the requirements which the newly designed system will have to fulfil. In the terminology of SSADM (Structured Systems Analysis and Design Method) this would be Requirements Analysis and Requirements Specification. In this book, systems analysis is covered in chapters 7 to 12.

The end of the analysis stage will have seen the production of the functional requirement specification which defines the new system in terms of its business requirements. The purpose of the system design stage is to specify the new computer-based system in terms of its technical content. The output will be a design specification. Chapter 13 deals with the transition from analysis to design, and chapters 14 to 24 deal with different aspects of the design stage.

In implementation, the objectives are to produce actual computer programs that process the system's data, and to install the hardware. The new computer-based system is installed. In the maintenance stage – often referred to as the operational stage – the system is in full operation and supports the business. It is called the maintenance stage when viewed through the developers' eyes as the system needs to be maintained to keep it up to date with changing requirements and to put right the system errors found. In terms of the total development budget of an IT department it has been estimated that up to 70% of departmental costs are incurred maintaining existing operational systems.

This life cycle approach is effectively the industry standard but its usefulness is now being questioned following the publication in the early 1990s by Dr Kit Grindley of the London School of Economics of a research report based on work done by Price Waterhouse Management Consultancy.

In 1979, Price Waterhouse established a world-wide panel of about 5000 IT executives who have since then highlighted the problems involved in developing and implementing IT systems. For most of that time 'meeting project deadlines' has been the number one priority. More recently however there has been concern about the way traditional management approaches have been unsuccessful in the management of systems development. It is further suggested that the traditional life cycle approach has failed in that it has left behind too many computer systems which cannot now be altered or extended. With system maintenance occupying up to three quarters of the worlds analysts and programmers, a time may be approaching when new development grinds to a halt. This book won't solve this issue. A variety of solutions have been offered including outsourcing, writing replacement systems based on open system principles, decentralising the responsibility for application systems' development and adopting minimal maintenance policies. One thing you can be sure of however is that the problem must be solved soon and that its solution will almost certainly mean radical changes to the way in which systems are developed.

1.4 THE ROLE OF THE ANALYST AND DESIGNER

Earlier in this chapter we said that systems analysts and designers change the world. In discussing the role of the analyst and designer we should therefore begin with a consideration of this change-making initiative. We have a problem however in that analysts and designers are not always the same person. This is illustrated by a diagram produced by G.M. Weinburg (figure 1.5).

It shows very clearly down the centre what he regarded as the analyst's role. The designers and implementers do quite different things which are not specified, and the analyst is the only link with the users. We also see the difference in recruitment specifications where those for analysts ask for applications and business experience and those for designers specify software environment and hardware architectures. Life is complicated further by the many sorts of analyst and designer titles such as business analyst, applications analyst, database designer, network designer, database administrator, infrastructure manager, systems manager. They all draw on the skills that were once the exclusive territory of the systems analyst. Let us therefore take a step up from this detailed picture that prevents us from seeing 'the wood for the trees'. We can recognise a set of attributes at a general level that all analysts or designers should possess, whatever their job title. As a minimum we should expect our analysts or designers:

- to uncover the fundamental issues of a problem. These might be the bottlenecks in a business system or the logic of a file processing module.
- to be able to prepare sound plans and appreciate the effect that new data will have on them, and replan appropriately.
- to be perceptive but not jump to conclusions, to be persistent to overcome difficulties and obstacles and maintain a planned course of action to achieve results.
- to exhibit stamina, strength of character and a sense of purpose essential in a professional specialist.
- have a broad flexible outlook, an orderly mind, and a disciplined approach as the job will frequently require working without direct supervision.
- to possess higher-than-average social skills so as to work well with others and the ability to express thoughts, ideas, suggestions and proposals clearly, both orally and in writing.

Many designers with very special technical skills and knowledge have in the past sought refuge in their specialisms from the need to exhibit many of the skills and attributes in this list but, in our view, the complete list has validity for all analysts and designers of

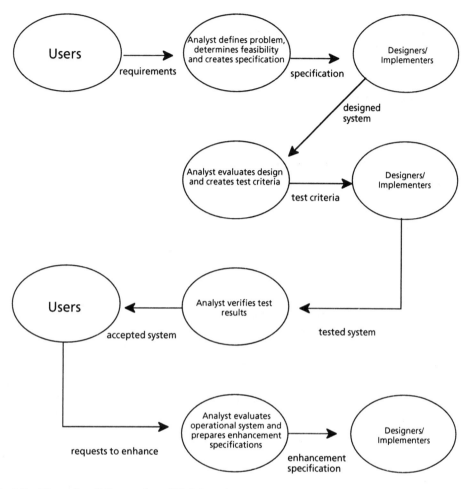

Fig.1.5 The role of the analyst (Weinburg)

whatever job title.

Finally, in this section are two more light-hearted views about the role of the systems analyst. The first is an American view based on Modell's *A Professional's Guide to Systems Analysis* in which he lists the many roles of the analyst, some of which are:

- Detective. A detective, whether private or official, is one whose primary task is to uncover the facts of an event and to determine responsibility for the event.
- Puzzle solver. The puzzle solver is one who either puts things together from component pieces or determines solutions from clues and hints.
- Indian Scout. An Indian Scout is one who is usually the first on the scene and who looks for hidden dangers or for the correct path through the wilderness (of the corporate environment). The Indian Scout may also be the first one to find hidden dangers and may draw the first fire.

Other roles he proposes include artist, sculptor, diagnostician and reporter. Perhaps he was intending to entertain as well as inform his students when he prepared his list. Someone who clearly believes in doing both is Roy Tallis the MD of CMS Ltd and previously a trainer who used to offer his students some Alexander Pope (1688–1744) with their systems analysis when he described the analyst as 'Correct with spirit, eloquent with ease, intent to reason, or polite to please'.

1.5 BETTER SYSTEMS DEVELOPMENT

Developing new systems is a demanding job. The problem is often ill defined. The solution will use new hardware and/or software and change the way people work. The investment has to be cost justified. The people doing the work will include people who have never done this kind of work before, have no body of knowledge or experience to support them and have received inadequate training. In spite of this, brilliant work has sometimes been produced. Because of it, much mundane, poor quality work has been produced. To reduce the amount of poor quality work, the traditional methods of carrying out systems development have been changed. These changes began in a programming context.

In the 1960s in the United States a number of surveys showed what most data processing managers had believed for a long time – that there was substantial variation in programmer abilities and that too much time was spent on debugging programs and on maintenance activities. The full surveys generated much controversy, but the effect they had was dramatic. Suddenly everyone was concerned with programmer productivity and began to examine the way in which programmers programmed. In 1965, Professor Dijkstra of Eindhoven University in Holland presented a paper at the IFIP Congress in New York suggesting that the GOTO statement should be eliminated from programming languages altogether, since program quality was inversely proportional to the number of GOTO statements in a program. In the following year, Böhm and Jacopini showed that any program with single entry and exit points could be expressed in terms of three basic constructs:

(1) sequence
(2) iteration or looping
(3) selection or decision taking

In familiar format these are shown in the diagrams opposite.

From these beginnings was structured programming born. This dramatically improved the quality of programming and of programmer productivity.

There is no doubt that structured programming has been successful, but it doesn't solve all our problems. Poorly constructed system designs can still negate the benefits provided by structured programming. Not surprisingly then, similar principles were applied to the tasks of analysis and design and a full range of structured methods came into being.

To produce well-structured systems designs, the analyst needs to define accurately the outputs from the system, its inputs, the data structures and the processing. To help in this, information is gathered from a variety of sources in several different ways. Further, the analyst checks back with many different people in different circumstances to ensure that what is being done is right. The output from this process is a systems specification which the user is expected to evaluate and approve. Unless we do use structured analysis

- Sequence

- Iteration

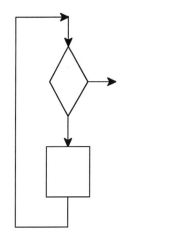

- Selection

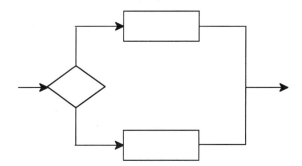

and design methods, we get long, complex narrative descriptions which users have neither the time nor the ability to comprehend. The difficulties with specifications of this kind are:

(1) *They are too big and complex.* It is quite unreasonable to expect users to search through long and complex systems proposals in order to verify the details of the systems proposed.
(2) *They are difficult to maintain and modify.* Simple changes to requirements often cause ripples throughout the specification and in consequence specifications are often not kept fully up to date and the implications of changes not fully understood until programming is reached.
(3) *They often describe the system in physical rather than logical terms.* This means that physical files and actual hardware is described together with descriptions of *how* the system will work. This muddles the picture which the user wants to see and is interested in: *what* the system will do.

Structured analysis aims to overcome these difficulties through involving the user more in the specification of the problem and the development of the solution and through carrying out the analysis and presenting the results in a clearer and more formal way. The following guidelines have been developed to illustrate this process.

(1) Involve the users as much as possible in the development of the new system and in the evaluation of all plans and proposals.
(2) Communicate clearly, bearing in mind the technical competence of the user to handle the material you give. In particular, use graphic documentation as much as you can. One of the features of structured analysis is the use of graphic tools to produce more easily readable and understood system proposals.
(3) Concentrate on building logical systems solutions before becoming concerned with the physical aspects of the design.
(4) Keep the analysis manageable. This means doing two things: firstly, take a top-down approach to analysis, breaking down major systems into smaller subsystems. Secondly, resolve major potential problems at the top as they occur. They cannot be resolved from the bottom up.

Later in this book you will see how to put this into practice and to do 'better systems development'. The tools available to help you will include dataflow diagrams, data structure diagrams, data dictionaries and structured text. A variety of methods have been devised to offer comprehensive approaches to systems development. These are reviewed in the next chapter. Where we've decided to use a specific approach, SSADM has been adopted.

1.6 SUMMARY

In this chapter we've tried to establish some foundations for the rest of the book. These can be summarised as follows:

- Systems analysis and design is an exciting, challenging, difficult and rewarding trade. It is constantly changing; new solutions are discovered every day. If we had the opportunity to rework systems produced three, five, ten years ago, not only would the solutions be very different but so would the problems.
- The customer is very important. Make sure that you know the customer, and what kind of solution is needed.
- There is no such thing in practice as a perfect system. There are successful systems. Understand the customer's criteria for a successful system. What will produce that wonderful phrase 'well done, you've done a good job'?
- Systems development is difficult. Take all the help you can in finding better ways to do things. Today's panacea is the use of structured methods. They work and they improve the quality of the delivered system. These methods will change. They may be replaced.

Finally, enjoy this book. All of it works some of the time. That's the best you can expect. Life is imperfect. You have to work with the book's content, apply it and make it work for you. Good luck.

CASE STUDY: SYSTEM TELECOM

Introduction

Throughout this book, we have included examples of how principles are put into practice. The applications used to illustrate the practice include hospital systems, customer systems of various kinds, a library system, student enrolment in a college and others. This diversity also serves to emphasise the wide application knowledge often required of systems analysts. As a contrast we have included a case example so that there can be a more comprehensive treatment of one application.

The example we have chosen is the customer services function from a European telecommunications company. It's not a real company of course, but is rooted in practicality with data and ideas drawn from several telecommunications companies and the experience of analysts working in this business area. The material here is some narrative describing the company, some of the dataflow diagrams for the required system and a data model. There are exercises based on this case study at the end of most of the chapters.

Company background

System Telecom is a large European telecommunications business with a turnover in excess of £4000m making profits before taxation of just over 20%. It has maintained a record of continuously rising profits since it was established. Its latest annual report says that System Telecom is positioned to achieve above-average levels of growth. It is one of the worlds fastest growing industries where the pace of change is likely to accelerate. It believes that it is now where the motor industry was in the 1930s with many decades of growth and opportunity to come. To perform strongly in the future however it will need to implement a continuing programme of new business development and from the range of opportunities presented to it, it must select those that give it the optimum return. System Telecom is in a capital intensive business and always has to balance the income from existing businesses with investment in new businesses. Reducing the investment in new business enables profit to be maximised in the short term but to the detriment of future growth. Over investment will reduce immediate profits and cash flow for the benefit of the long term. The shareholders in System Telecom pay particular attention to this balance and to investment in research and development, and new technologies. The company spends between 2% and 3% of its turnover on R&D. It believes that the introduction of new services based on technological as well as market research gives it a competitive edge. It views new systems in the same way and expects new computer-based information systems to support business goals and generate competitive advantage.

The shareholders in System Telecom are unusual. Unlike most European telecommunications operations which are state run, recently privatised or on the way to privatisation, System Telecom was set up deliberately to take advantage of the opening up of the European telecommunications market. This followed highly controversial and bitterly contested legislation introduced by the Commission for the European Community aimed at breaking down the protective and nationalistic stances of many EC members. System Telecom's shareholders have pledged long-term funding to establish an aggressive and highly competitive technologically based business which will generate a growing profits stream as their own original businesses plateau or decline.

Strategically, System Telecom intends to be the European leader in three core sectors:

premium and business services, mobile communications and basic telecommunications. Central to its premium and business services is EUROCAB, a transnational data highway linking together Europe's major business centres. Services offered through EUROCAB include Bandswitch, the world's first variable bandwidth data transmission network. This allows companies to transmit high volumes of data at great speed to many distant sites without the cost of permanently leasing large amounts of transmission capacity. Also offered to major companies is the equivalent of a private network and, for individual customers, an international charge card that allows customers to make domestic and international calls from anywhere on the System Telecom network and have the cost of the call billed to a single account.

Future products under consideration include personal numbering, a service already available in parts of the USA, that allows users to redirect their calls from home to office or to wherever they are at particular times of the day.

System Telecom's innovative and entrepreneurial approach has led to a number of highly publicised and commercially successful ventures outside the European Community. It has joined with four substantial Japanese corporations to offer premium services in Japan with the intention of linking these to Hong Kong and Singapore. In the former Soviet Union it has set up several joint venture companies including one with Intertelecom the leading local carrier to provide international telecommunications as the countries of the former Eastern bloc move towards commercial, market-led economies.

The mobile communications business is still largely confined to the UK and Germany where it is experiencing rapid growth in subscribers and number of calls. In the UK the service operates mainly within the M25 ring. A new service has just been launched in Germany in the industrialised areas along the Rhine corridor.

Like most 'hi-tech' companies, System Telecom relies enormously on the skills of its people and places great emphasis on motivating, training and developing people in an international context. It has founded System Telecom University to be the focal point for development and training in technological and business skills.

Although the company has a hard and aggressive business profile it prides itself on being a responsible corporate citizen and it supports a wide range of charitable, community and cultural projects across Europe.

Systems background

The System Telecom Board recognises that the success of its business strategy is largely dependent upon its procedures and systems. The effective use of IT systems will be critical in giving the company a competitive edge. A strategic study was commissioned, and the study report identified specific functional areas where automated systems are essential to the efficient running of the organisations. These functional areas are :

- customer services
- management information
- personnel
- payroll/pensions

The Customer Services function is the area to be considered here. This function provides the administrative basis for the services offered by the company and will be fundamental in achieving business success. The main task areas within this function are :

- customer registration
- call logging and charging
- billing
- payment recording
- number maintenance
- pricing policy.

Customer registration Customers may be from the commercial or domestic sectors and are accepted subject to various credit checks. Customer identification details are held and maintained by the system.

Once customers are registered they may subscribe to services, e.g. fax, telephone, and may have numbers allocated for their use of those services. They then become responsible for payment for all calls logged to those numbers. They may request new numbers or cancel existing numbers as required. Customers who have cancelled all of their numbers continue to be maintained on the system for two years. This widens the customer base for mailshots regarding new services, promotions, etc.

Call logging and charging Calls are logged by three factors: duration, distance and timing. Various combinations of bands within those categories are used to calculate the cost of each call. System Telecom operates a number of tariff and subscription schemes which provide financially advantageous facilities to both commercial and domestic customers. The applicable tariff plans are taken into account when calculating charges. System Telecom aims to provide an efficient service at the lowest possible cost to the consumer and therefore customers are constantly reviewed for inclusion in suitable pricing schemes.

Billing Bills are issued on a periodic basis. This is usually quarterly but may be monthly, half-yearly or yearly, subject to negotiation with the customer. The bills itemise all charges, both for service subscriptions and call charges. Tariff plan discounts are also shown. Reminder bills are issued a fortnight after billing and payment must be made within one month of the issue of this bill or the services are automatically disconnected following issue of a disconnection notice. Payment must be made in full, plus the reconnection charge, prior to the resumption of services. Customers are not eligible for reconnection following a second disconnection. Bad debts are passed to the debt collection department on issue of the disconnection notice.

There is also a set payment scheme where customers make regular fixed payments which are offset against their bills. These payments are shown on the bills. Any credit/debit is carried forward to the year-end reconciliation when a final bill is issued.

Payment recording Payments may be made in various ways:

- by cash/cheque at specified outlets
- by standing order
- by automatic deduction from credit/debit cards
- by direct debit.

Payments are made subsequent to bills being issued. The method is recorded for each customer.

Number maintenance Numbers are held on the system and are either allocated to customers or available for allocation. New numbers are specified periodically and set up so that they may be allocated when required.

Pricing policy The company has several approaches to pricing and uses a number of tariff plans. These are regularly reviewed and updated in order to ensure their continuing competitiveness. Customers subscribe to services via tariff plan agreements.

Additional information about System Telecom is shown on the following charts and diagrams and in the accompanying narrative.

Dataflow diagrams (DFDs)

The Level 1 DFD shows the required system, that is the areas of processing that are intended to be automated. The system will support the Customer Services function within System Telecom and the DFD identifies six distinct areas of processing within this function. These areas are defined as the following :

1. Maintain Customers
2. Record Calls
3. Issue Bills
4. Record Payments
5. Maintain Rates
6. Maintain Customer Facilities.

Process 5 Maintain Rates is shown with an * in the bottom right-hand corner; this indicates that there is no decomposition of this process into smaller areas of processing. Processes marked in this way are explained instead by an Elementary Process Description which is a textual explanation of the process. Every other process on the Level 1 DFD would be further decomposed by a Level 2 DFD and are therefore not marked with an asterisk. Examples of Level 2 DFDs are provided for processes 1 and 4.

There are seven sequentially-numbered datastores. Each datastore represents a grouping of stored data items. The same datastore may be shown more than once on the diagram if doing this makes the DFD easier to read. An additional line at the left-hand side of the datastore box indicates where this has occurred and that the datastore appears more than once.

There are nine External Entities shown on the diagram; each one represents a source and/or destination of data. An additional line across the top left-hand corner of the external entity box shows that they have been included on the diagram more than once, just as with datastores.

The arrows on the diagram represent the flow of data. Data may be

- entered by an external entity - a Customer provides Payment/Bank details to Process 4;
- read from a datastore by a process - Process 3 reads Tariff amounts from datastore d3;
- written to a datastore - Process 1 writes customer details to datastore d1;
- output to an external entity - Process 3 informs the Debt Collection Department of bad debtors;
- output to one process from another - the Level 2 DFD for Maintain Customers shows Process 1.1 sending details of an accepted customer to Process 1.2.

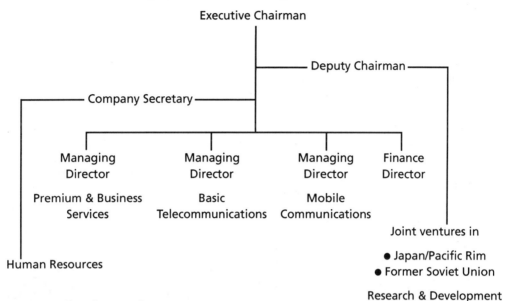

The Level 2 DFDs are enclosed by a boundary box. This separates the processes from the interfaces to the external entities and the logical data stores.

The DFD defines the processing that the system will carry out but not how the system will work. There is no sequence implied in the numbers allocated to the processes or the datastores, they are merely labels. In essence the DFD shows the means of triggering an area of processing and the data that is input, updated and output whilst the process is carried out.

18 Systems Analysis and Design

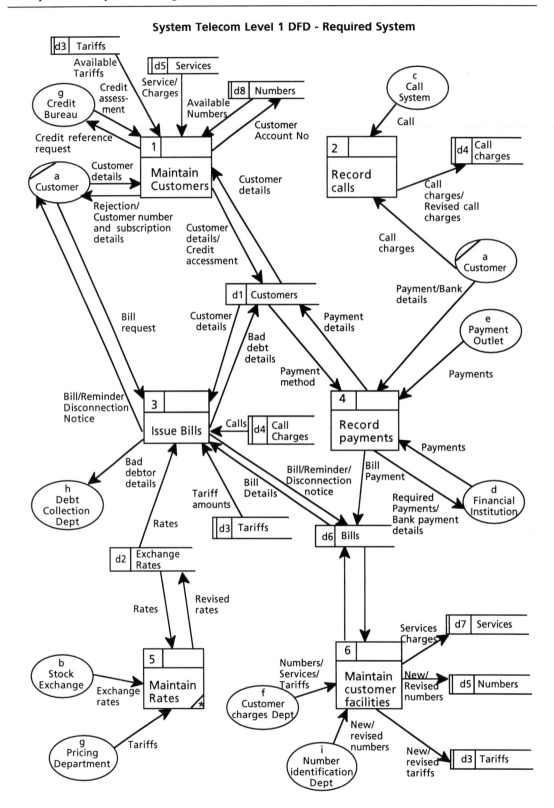

Level 2 DFD - Maintain Customers

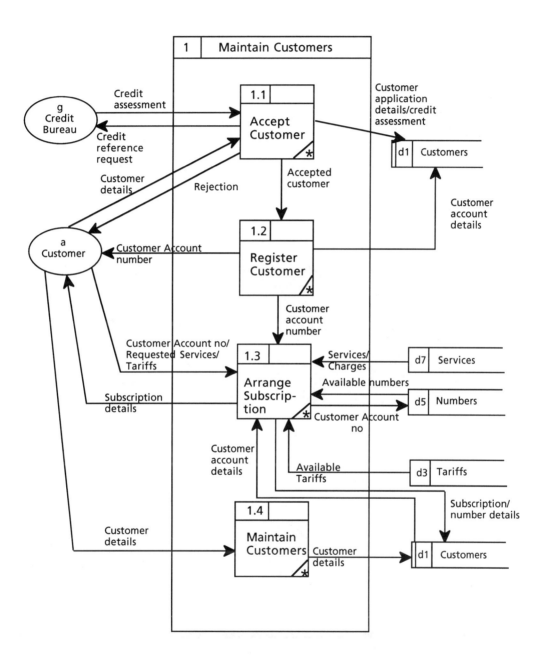

Systems Analysis and Design

Level 2 DFD - Record Payments

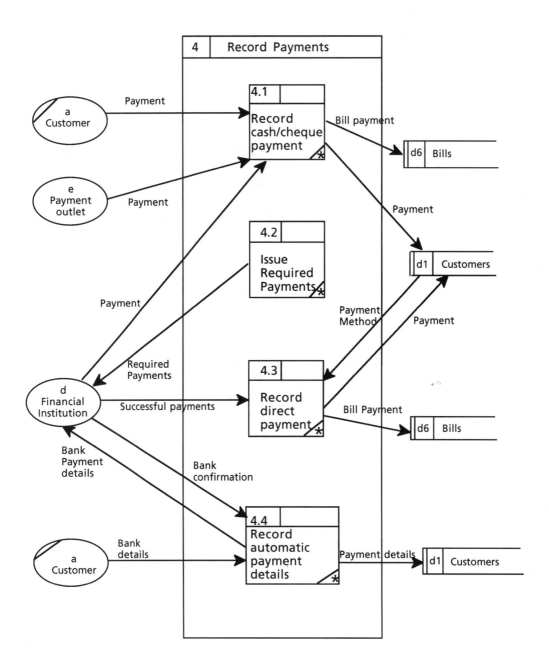

CHAPTER 2
Approaches to Analysis and Design

2.1 INTRODUCTION

In this chapter, we consider alternative approaches to the analysis and design of information systems. In particular, we look at some of the major 'structured' methods and examine their advantages. A complete survey of all structured methods is, of course, beyond the scope of this book so we have selected a number of methods that are either widely used in the UK (for example SSADM) or which UK-based practitioners will encounter sooner or later (like Merise and Euromethod). We also look briefly at the idea of object-oriented design.

2.2 TRADITIONAL APPROACHES

Generally, a traditional approach will involve the application of analysis and design skills by competent practitioners. They will use their experience and knowledge of the business being studied and of the technical environment, to devise and propose a system which meets the users' needs as they understand them. Usually, the consideration of business issues – what the system is supposed to do – will go side by side with the technical evaluation – how the system might work – with no explicit boundary between the two.

By the very nature of the traditional approach, it is not possible to define a sequence which will summarise its use in all projects. However, the following sequence is reasonably typical and it can be contrasted with that for the structured approach which follows later. Refer to figure 2.1.

(1) Analysis of requirements
The analysts examine the current system, if there is one, and discuss with the users the problems with that system and the requirements for a new one. Generally, no attempt is made to separate functional requirements, like provision of on-line transactions, from non-functional requirements, like system response times. Usually, the analysis documentation will not be delivered to the users and will only be used by the analysts in devising their specifications. Therefore, the users will not necessarily review analysis documentation unless specifically asked, for example, to check the completeness of interview notes.

(2) Specification of requirements
The analysts now sift through their documentation and produce a specification of requirements for the new system. This will include functional and non-functional requirements and will often include a description of the proposed hardware and software as well as the users' business requirements. Very often, layouts for on-line screens and

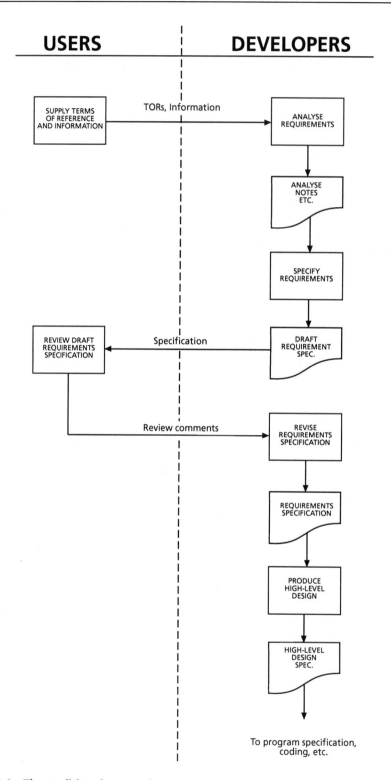

Fig. 2.1 The traditional approach

printed reports will be included in the specification. The specification must usually be approved by the users before the project continues but the degree to which users examine this document carefully varies greatly, depending on the knowledge and interest of individual users, how much time they have available and on the importance they attach to the project. It is also very likely that the specification will be mainly in text, perhaps illustrated with some flowcharts or other diagrams. For a big system, this could result in a very large quantity of text which, again, will act as a barrier to the users reviewing the specification properly.

(3) High-level design
Once the requirements specification has been approved, the designers take over and produce a high-level design for the system. This will include the database design and the general structure of menus, on-line enquiries and reports. Individual program specifications will usually be produced later, during detailed design.

Now it will be seen from this outline that the traditional approach does not *mandate* the involvement of the system's users to any great degree. Of course, good analysts have always worked hard to involve their users, to get their views on ideas as they emerge and to review important analysis and design documents, like interview notes for example. But the users only *have* to be involved when they are presented with the system specification to review and approve. The result is that analysis and design come to be seen as the province of the technician – of the analyst and designer – rather than as a partnership between developers and users.

The most obvious advantage of the traditional approach is, of course, that it is quite familiar to analysts and that the general methods of working are well understood. There is a lot of experience of traditional methods and most of those concerned – analysts, project managers and also users — know roughly how the project will proceed. As a result, there is generally no specific, and possibly expensive, training required, although there has always been a need for analysts to be trained in the basic fact-finding techniques, such as interviewing. The documentation resulting from traditional methods is a big, mainly textual, document. As we will suggest later, there are several problems with this but at least those who need to review the documentation, either IT professionals or users, will possess the basic skill – reading – even if they lack time to examine everything that is presented to them.

In considering the problems inherent in traditional analysis and design techniques, we need to face up to the fact that the history of IT development is littered with a large number of projects that were either started and never finished or which, even if completed, failed to provide the users with what they wanted or needed. In addition, developers have often laboured to provide areas of functionality which, for one reason or another, the users have never got round to using. In this respect, a large proportion of the money spent on information systems must be regarded as having been wasted. There is more about this in chapter 4. In considering this depressing state of affairs, it became apparent that problems were generally created during the analysis phase. It was concluded that traditional methods had the following defects:

- Large quantities of written documentation act as a barrier, rather than as an aid, to

communication between users and developers.
- There is a lack of continuity between the various stages of analysis and between analysis and design, so that requirements get lost in the process.
- There was no way to ensure that the analysis and design work was complete because it was difficult to cross-check the findings from the analysis stage.
- Systems developed traditionally lack flexibility and are therefore difficult, and also expensive, to operate, maintain and adapt to changing circumstances.
- Traditional development methods tend to assume the use of a particular hardware and/or software platform; this constrains the design so that the users may not get what their business really needs – and the project can be thrown badly off course if the organization changes platforms during development.

In other words, traditional methods have signally failed to deliver the goods in terms of robust and flexible systems which meet the needs of their users.

2.3 STRUCTURED APPROACHES

In the late 1970s, a number of people in the IT industry began to consider why so many IS developments had gone wrong and failed to live up to their promises and, mostly, they reached a similar conclusion: projects went wrong initially during the analysis phase and efforts to improve the situation later were usually a waste of time and even more money. Most of these authorities agreed that there was a need for new methods of analysis and design which would offer:

- Greater formality of approach that would bring systems development nearer to the scientific method or to an engineering discipline than had been common in IS projects.
- More clarity of stated requirements by using graphical representation as well as text.
- Less scope for ambiguity and misunderstanding.
- A greater focus on identifying and then satisfying business needs.
- More traceability, to enable a business requirement to be followed through from initial analysis, into the business-level specification and finally into the technical design.
- More flexible designs of system, not unduly tied to specific technical platforms.
- Much more user involvement at all stages of the development.

All of the structured methods described later in this chapter result from attempts to meet these requirements in one way or another. As a result, there are some features that are common to all structured methods and to the structured approach in general.

Focus on data structures
Most of the methods concentrate heavily on a thorough examination of the data requirements of the proposed information system. The reasons for this concentration on data are twofold:

- It is a fact that, whereas the processing requirements of organisations can change often and may change significantly, the underlying data is relatively stable; therefore, the data provides the soundest base for the development process.
- An inflexible data structure can act as a severe constraint on an organisation wishing

to change its systems to match changing requirements. Evolving a flexible data structure early during development yields many benefits later on in the life of the system.

With a sound and flexible data structure in place, it is possible to adapt and amend the processing to meet the changing needs of the organisation.

Use of diagrams and structured English
Another common feature of structured methods is their heavy reliance on diagrams to convey information. The reasoning here is that:

- Diagrams are generally easier for people to assimilate than large quantities of text, thereby providing an easier means of communication between users and developers
- It is less easy to commit sins of omission or ambiguity with diagrams; for instance, failure to mention an important dataflow, or an incorrect statement about the direction of flow, may not be noticed in a long written specification but it will be immediately obvious on a dataflow diagram.

Together with diagrams, many methods make use of 'structured English'. This aims to reduce the complexity of the language and reduce the written component of the documentation to brief, terse, unambiguous statements. Alongside structured English goes the extensive use of decision tables or similar to show the processing logic.

Concentration on business requirements
Another important feature of structured methods is the separation of the logical and physical aspects of the analysis and design process. This is done so that both analysts and developers focus on the business requirements of the proposed information system, rather than considering too soon the technical details of its implementation. So, with a structured method, the developers will focus for much of the time on the business requirements of the proposed system. They will look at the data needed to support the system and at the kind of processing which the users will want to support their business. Only once these things have been clearly established will they consider how these features may be implemented.

It may be objected that leaving physical implementation considerations so late in the analysis and design process is rather foolish since, if it then proves impossible to provide the services the users require, the only result can be dissatisfaction and disappointment. Whilst this is a possibility, the increasing power, flexibility and availability of the range of hardware/software platforms makes this outcome less likely. On the other hand, focusing on the business requirements, rather than on technical considerations, avoids the possibility of the users having to accept what the developers want to deliver, rather than what they really need. Also, if the full business requirements cannot be met because of some technical impediment, it may be better to abandon the project rather than to press on and deliver some limited functionality which may not assist the users much in their everyday work.

We can contrast the traditional approach which was described earlier with the following sequence of analysis and design which reflects the general structured approach. Refer to figure 2.2.

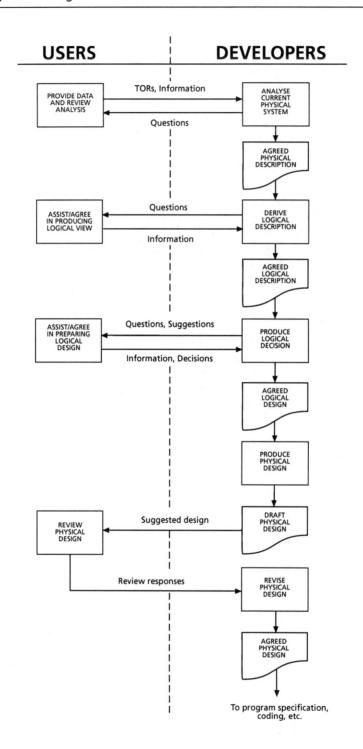

Fig. 2.2 The structured approach

(1) Analysis of current physical system
The existing system, if there is one, is studied and documented 'warts and all'. This means that the models will cover all sorts of peculiarities (like sorting and re-sorting of data) which are purely the result of *how* the system is implemented as distinct from what it is supposed to achieve. The users are required to review and agree the analysis results.

(2) Derivation of current logical system
By stripping away the physical aspects of the current system, like the data sorts already mentioned, the analysts form a picture of *what* the system does. This can then be used as a sound basis for the specification of a new and improved system. Again, the users are asked to review and agree the logical system description.

(3) Specification of required logical system
Further detailed analysis is now carried out to ensure that the requirements of the new system are fully understood. This is then designed, initially at the logical level. Again, this provides a very clear statement of *what* the system is supposed to provide, in business terms uncomplicated by technical implementation considerations. The users are involved throughout this work and must approve the finished specification before the project can continue.

(4) Specification of required physical system
Only when all the business requirements have been specified is the design converted into a physical design for implementation on a specific hardware and software platform. As the issues dealt with here are mainly technical, user involvement will be less than during earlier stages, but user approval must still be obtained to the final specification.

In general, the use of structured methods brings the following benefits:

- We have said already that structured methods concentrate on systems' data requirements. Since these are more stable than the processing requirements, a system built around a flexible data structure will prove amenable to change and have a longer life.
- The use of diagrams and structured English improves communications between users and developers and increases the chances of the former getting the system they really need.
- Because of the rigour and cross-checking inherent in structured methods, it is relatively easy to spot errors and omissions in the analysis which could otherwise lead to problems later in the development.
- Systems built using structured methods are based on business requirements and should deliver business benefits. Technology is not used for its own sake but in support of business objectives.
- Because technical issues are addressed relatively late in structured development projects, much of the design will be suitable for implementation in a variety of environments. Should the users' hardware or software policy change, it should be possible to go back to the logical documentation and re-develop from there for the new platform.

- A system built using structured methods will have complete, rigorous and consistent documentation which will greatly help in maintaining and enhancing the system over time. Also, because of the business-level documentation represented in the logical specification, it is possible to evaluate accurately not only the technical, but also the business, consequences of proposed changes to the system.
- It is possible to define precisely the training requirements for the analysts and designers, and for users too. With some methods, qualifications are available which provide some level of confidence in the basic ability of the people carrying out the work.
- Finally, using a recognised structured method means that the users are not tied to any one developer. This means that one could, for instance, commission one firm to carry out the analysis and another to produce the design.

Together with these advantages, however, there are some problems with the use of structured methods which developers and users need to consider if they are to achieve real success with them. Chief among them are the following:

- Structured approaches tend to shift the balance of effort in an IS project. Traditionally, most of the work went into coding and testing the actual software, with analysis and design forming a lesser component. In part, this resulted from much of the detailed analysis – of processing logic for instance – having to be done by the programmers. With a structured method, however, the balance of work shifts noticeably and more of the effort is now expended earlier in the project, during analysis and design. There is a commensurate decrease in the programming effort, though not testing effort, and indeed, with some modern development tools, programming may become a relatively trivial task. An obvious problem with this is that the users seem to have to wait for a long time before they actually see any concrete results from the development such as actual screens or reports. A special responsibility therefore devolves upon the manager of a structured development project to ensure that the users understand that increased work at the beginning of the project will be compensated for later on, and the higher-quality system that will result will incur lower future maintenance costs.

    It is worth noting that the specifications resulting from many of the structured methods can be fed directly into a code generator for the production of the required programs. This process, in effect, removes the programming phase from the development and means that the actual system can be delivered to the users quite soon after the completion of analysis and design.
- The success of an IS project has always – whether conducted traditionally or otherwise – depended substantially on the degree of user involvement. However, with traditional methods, the lack of sufficient user involvement may not become obvious until the finished system is being demonstrated and it becomes apparent that their requirements have not been met. With a structured method, user involvement is usually explicitly set out and the users will be asked to contribute to reviews and approvals at various stages. Once again, the result should be the production of higher-quality systems which better meet their users' needs. The amount of user involvement must, however, be carefully explained at the beginning of the project and the necessary commitment, to the time and effort involved, must be obtained. In addition, it may be necessary to provide suitable training so that key users can understand fully the documentation being presented to them.

- Some structured methods remain the intellectual property of their designers and developers who will provide advice, training, consultancy and sometimes support tools for their use. The dangers in this, of course, are that one can get 'locked-in' to the single source of supply and that the high prices charged for the various services may add a considerable extra cost to the development budget.
- CASE (computer-aided software engineering) tools are becoming more widely used in systems developments and some of the structured methods become very difficult to use indeed without some form of computerised support. At least one – Information Engineering – is predicated on the basis of an Integrated CASE (I-CASE) being available. The reasons for this are the sheer volume of documentation produced, particularly that to support the data modelling techniques, and the need for cross-checking which computer systems do so well but humans find tedious and time-consuming. The development of CASE tools has not – except where they have been built to support specific proprietary methods – quite kept up with the development of the structured methods themselves. The situation is improving, although setting up the necessary CASE infrastructure can, once again, add to the development budget.

2.4 YOURDON

Edward Yourdon is one of the pioneers of the structured approach to systems development and his ideas have had a major impact on thinking in this area. Yourdon's approach (figure 2.3) has evolved over the years, most fundamentally to downplay the need to model the user's current system during the analysis work. The reasoning is that this has proved wasteful of time and effort and has often proved to be very unpopular with users – 'why waste time modelling the current system, when what we want is a new one?' Thus, the developer is encouraged to use analysis to build what Yourdon calls the *essential model*, that is a logical model of the *required* system. This has two components:

- *The environmental model.* This shows the boundary of the required system in the form of a context diagram, an event list and brief description of the reason for having the system.
- *The behavioural model.* This illustrates how the system will work within its boundary, using various diagrams – dataflow diagrams, entity relationship diagrams and state transition diagrams – supported by process descriptions and a data dictionary. Dataflow diagrams show the processes which the system carries out, the various stores of data and the flows of data between these two; entity relationship diagrams illustrate the data items (entities) which the system uses and the relationships between them; and state transition diagrams model the time-dependent behaviour of the system and show the states which the system can be in and the valid connections between different states.

These two component models are *balanced* against each other so that together they form a complete and consistent picture of what the required system is to do. Of course, in developing the essential model, developers may well have to study the current system; but this is regarded as a means to an end and not as an end in itself.

The Yourdon approach is not particularly prescriptive in that it does not mandate that

30 Systems Analysis and Design

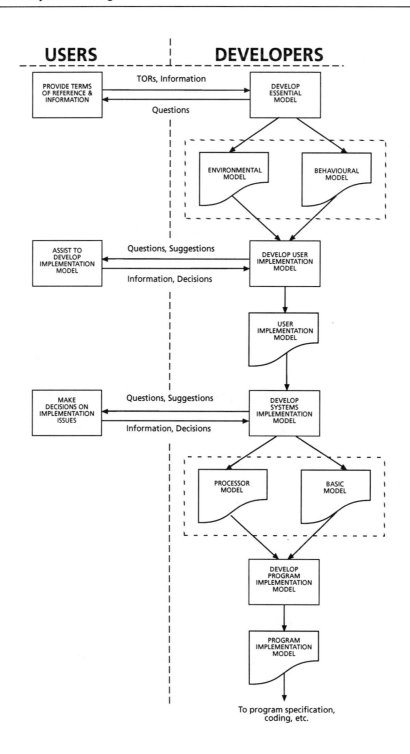

Fig. 2.3   The Yourdon approach

the developer must use all of the techniques described. Nor is the use of other techniques excluded if they can contribute to the completeness of the analysis. The emphasis is on choosing and using tools and techniques which are appropriate to the requirements of the individual project.

Once the essential model has been built, it forms the basis for the *user implementation model*. This differs from the essential model in that it takes into account how the system is to be implemented. Options considered here could include, for example, adopting a client-server architecture, using a fourth-generation language of some sort, or seeking a wholly (or partially) packaged solution.

So, major issues considered at this point include:

- The boundary between the computer system and the manual processes.
- The nature and type of the human–computer interface.
- Other operational constraints such as response times or system security.

The user implementation model then forms the input into the design process, which Yourdon conceives as the development of further models:

- The *system implementation model* which shows how the required processing will be allocated to the selected hardware. It has two sub-models concerned with the processor and the tasks to be performed.
- The *program implementation model* which maps how the individual tasks which make up the system will be developed as program modules.

Yourdon's approach is well-proven and based on practical experience. It is suitable for most types of application, real-time as well as commercial applications, and it uses techniques, like dataflow diagramming, which have now become widely understood. It is, perhaps, less prescriptive than some other methods, which makes it less easy for novices to grasp and apply but more flexible for the experienced developer.

2.5 JACKSON

Jackson System Development (JSD) was developed by Michael Jackson and John Cameron. It is fairly widely used in the UK, especially where organisations have already adopted the complementary Jackson Structured Programming (JSP). JSD covers the development life cycle from analysis to maintenance and works by the composition of processes upwards from an atomic level. It is very concerned with the time dimension of the system and thus, unlike some of the other structured methods, JSD is suitable for the development of real-time systems.

JSD proceeds through a sequence of six steps, as follows:

- *Entity action step*. Here, the developer examines the real world which the IS will model and describes it in terms of the entities involved and the actions which they will perform or have performed on them.
- *Entity structure step*. This involves an examination of the time component of the IS. The actions performed by or on each entity are arranged in their time sequence.
- *Initial model step*. So far, the method has described the real world. Now, a process model is built which simulates, as the IS will simulate, these real world entities and actions.

- *Function step.* Functions are now specified which will produce the outputs required from the system. If necessary, additional processes may be identified at this point.
- *System timing step.* This involves the consideration of process scheduling to ensure that the system's functional outputs are correct and are produced at the right time.
- *Implementation step.* Finally, the hardware and software required to implement the IS are identified and the model is transformed into a form suitable for operating in that environment.

It can be seen that JSD relies heavily on the use of models and this is, in fact, one of the claimed advantages of the method. By concentrating on the identification of entities, and of the actions which they affect or which are affected by them, the developer avoids building a system which simply provides for existing levels of functionality. Instead, what results from JSD is a very flexible basis for later development and the resultant system should prove amenable to enhancement and amendment to meet changing business requirements.

Another advantage of JSD is that it does deal very satisfactorily with the time component of systems. Some structured methods are very good at handling commercial applications – payroll for example – but are less successful when applied in a real-time situation, like a factory process-control system. JSD, however, is very much concerned with time and the sequencing of events and so can be used for the development of systems where process is more significant than data.

The major problem with JSD is that it is initially quite difficult to grasp some of its concepts and then to apply them in a real project situation. This, however, is to some extent true of any method and proper training is, as always, the answer.

2.6 INFORMATION ENGINEERING

Information Engineering (IE) was originally developed by I. R. Palmer and his colleagues in the mid-1970s. It has been commercialised by James Martin Associates as the Information Engineering Method and several proprietary variants are now available. IE is rather more than a technical method for the analysis and design of information systems. It is, in effect, a philosophy for information systems management which rests on a number of basic premises:

- Information systems are, or will be, central to the survival and growth of businesses and, therefore, their use should be included in the planning of the business at the highest level.
- The piecemeal approach to systems development hitherto used has resulted in a plethora of non-integrated systems which are hard to control and manage, expensive to maintain and not flexible enough to respond to rapidly changing business circumstances.
- The key to flexibility and responsiveness is the development of corporate data models which can support a whole range of business systems.
- The application of engineering rigour to systems analysis and design will result in systems which are better designed, better meet the users' needs, and are robustly constructed.

- Integrated CASE (I-CASE) tools are essential to support the complex data management required in IE and to control – through the imposition of logical cross-checks – the analysis and design process.
- I-CASE tools will generate the program code more or less automatically once the design has been properly completed.

IE thus provides an overall framework for the development of information systems and also defines a tightly-integrated toolset which will support the analysis, design and development processes.

In its analysis and design components, the method uses many of the diagrammatic techniques encountered elsewhere, including process decomposition diagrams, dataflow diagrams, entity-relationship diagrams, process-entity matrices and action diagrams. A key feature of IE, however, is the fact that these various diagrammatic viewpoints are coordinated and cross-checked through the notion of an I-CASE 'encyclopedia', so that an individual system component, like a process, can be modelled in a number of ways, rather like the various angles presented in a three-view engineering drawing.

The greatest strength of IE is also, to some extent, its greatest weakness, in that it is rather an all-or-nothing approach. The greatest benefits of IE will only be attained if it is adopted in its entirety and this involves:

- High-level (probably Chief Executive Officer, CEO) commitment to its use.
- A strategic IS plan, linked to the business plan, within which sit individual IS projects.
- The implementation of the I-CASE environment.
- Strong centralised control of the development infrastructure including tools, methods and, of course, the corporate database.

The fact that the development environment is centrally controlled does not mean, however, that individual projects need to be managed from the centre. Indeed, one of the benefits of IE is that individual departments can safely be allowed to produce systems which meet their individual needs provided only that the integrity of the corporate database is protected and that the approved toolsets are used. Of course, applying the IE rigour to the development of individual systems should still provide benefits. The system will have been rigorously analysed and designed and the database and program code will have been logically derived from this analysis. The system should also prove to be much more maintainable since changes in requirements are met by changing the business model – that is, the design – and then regenerating the code as necessary. However, the full payoff from information systems will, in future, only come when a business's systems are fully integrated and this does require that IE be applied throughout the organisation.

The greatest problem, though, is that the initial implementation of IE is likely to be very expensive and will not yield immediate financial returns. Unless a CEO is very alive to IS issues, and understands the technicalities involved, it may prove difficult to justify the necessary investment against a payoff which will only occur several years into the future.

2.7 SSADM

SSADM is the Structured Systems Analysis and Design Method and is the preferred method to be used in projects for the UK Government. It was devised first in the early 1980s as the CCTA (the Government agency responsible for advising on computer and telecommunications matters) sought to find a way of getting greater efficiency and effectiveness from the Government's very large IT projects. The method was initially developed for CCTA by Learmonth and Burchett Management Systems. However, each subsequent version has introduced new concepts and techniques and the current version was produced by a consortium working for the CCTA.

The main user of the method is, of course, the UK Government and its use is mandatory on many public-sector projects. However, the fact that it is an established and open method, and that SSADM skills are widely available, means that it is becoming a *de facto* standard method in the wider marketplace. It is also used outside the UK. The Information Systems Examinations Board (ISEB), a subsidiary company of the British Computer Society, provides a recognised qualification in the use of SSADM and accredits training organisations to run courses leading to its certificate. A tool conformance scheme is in operation for CASE tools which support the method and there is a large and active international user group.

The method presents a 'three-views' model of an information system. The method views:

- The data in the system
- The events to which the system must respond
- The functions in the system, as perceived by its users.

Extensive cross-checking between these three views provides a high degree of rigour in the analysis and design process.

SSADM is documented in a set of definitive manuals which describes:

- The structure of an IS project using the method, in terms of the modules, stages, steps and tasks by which the work is tackled
- A set of analysis and design techniques, to be applied at various stages of the project
- A series of product definitions, including the quality-control criteria to be applied at each stage
- 'Hooks' so that the method can be coupled with structured approaches to project management, particularly to CCTA's PRINCE method, and to programming, mainly to Jackson Structured Programming or JSP

The 'life cycle' used in SSADM is illustrated in figure 2.4.

It should be noted that SSADM has nothing to say about the earlier stages of an IS project – a strategy study, for instance – or about the actual development, implementation and maintenance of the system. The outputs from SSADM, however, do fit quite well with, for example, JSP.

The major techniques employed are:

- Requirements analysis
- Dataflow modelling
- Logical data modelling

Approaches to Analysis and Design 35

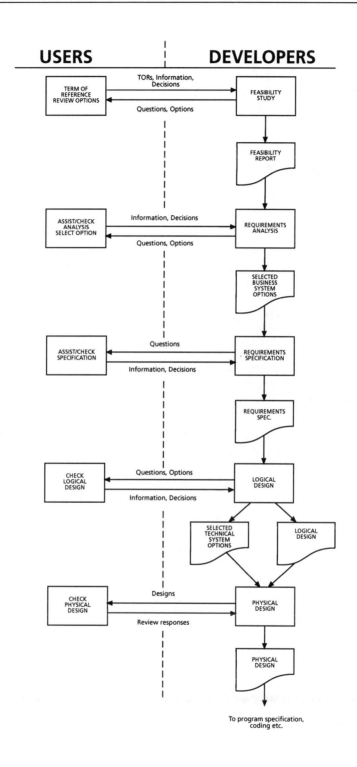

Fig. 2.4 The SSADM life cycle

- User/user role modelling
- Function definition
- Entity/event modelling
- Relational data analysis
- Logical database process design
- Logical dialogue design.

Prototyping can be used, but SSADM confines it to confirming and clarifying requirements, rather than the prototypes being carried forward into the design. In version 4 of SSADM, much use is made of Jackson-like structure diagrams. Some of these – entity life histories, for example – form part of the analysis process but the Update and Enquiry Process Models, in particular, can be implemented directly using some non-procedural programming languages.

As one would expect from an established and widely used method, SSADM has a number of points in its favour. It is a mature method which has undergone several evolutions to reach its present stage so it can be used with confidence as a reliable and stable platform for development. Partly because of the support of the UK Government, but also because of the decision to make SSADM 'open' – that is, no licence is needed to adopt and use it – the method has become very popular and is used in a wide range of businesses and industries. Skilled staff, therefore, whilst not exactly plentiful are at least available in reasonable numbers. Because the method is open, many firms offer consultancy, training and CASE tools. This provides a competitive situation unlike any other comparable method and makes the cost of adopting SSADM relatively lower. There are also several books available, explaining the method from various perspectives – manager, practitioner and user. Most of the SSADM techniques are developments of methods widely practised elsewhere, so that these are well-proven and generally familiar: included here are dataflow diagramming, entity life history analysis, logical data structuring and relational data analysis. The process modelling techniques are less familiar, but do mesh well with Jackson Structured Programming methods. Finally, the method follows the well-established structured route of moving from current physical, through current logical and required logical, to required physical systems specification. There are major user decision points where business systems options (the *what* of the system) and technical system options (the *how*) are considered. User involvement is mandatory and extensive.

There are some disadvantages to the use of SSADM which must be considered however. The first problem is that of learning the method in the first place. Although most of the SSADM techniques are well established, some are not and in any case the sheer number of them can pose problems for the student practitioner. The structure is logical, but with five modules, seven stages and dozens of steps and tasks, there is a lot to learn. The ISEB mandates that approved courses should include at least 80 hours of instruction but experience shows that it takes much longer than this for someone to gain even a basic grasp of SSADM. Many developers feel that SSADM is overly complicated and places too much emphasis on the earlier stages of a project but this criticism can also be levelled at other structured methods. However, with version 4, it is certainly assumed that the developer should tailor the application of the method for each project rather than apply it slavishly whatever the circumstances and this approach may in time help to defuse this criticism.

One problem that does arise with SSADM is that of controlling the large amount of documentation produced. There are difficulties in coordination if large teams are working on a project and the configuration management of documents becomes a major issue. The quality of CASE tools to support SSADM projects has not been too high until recently. Today, there are several competent diagram editors available; many tools provide for various degrees of cross-checking and some of the better tools offer semi-automatic generation of some SSADM deliverables. Still, there is no fully comprehensive tool on the market. Finally, SSADM becomes rather vague as it moves into the areas of physical design. To some extent, this is understandable as – with the vast range of possible implementation environments which are available – it is impossible to offer detailed prescriptive advice as can be done for the analysis process. However, some developers find this aspect of the method rather unsatisfactory and not really in accordance with the rigour one might expect from a structured method.

2.8   MERISE

Merise was designed in France at the beginning of the 1980s and is used widely there and also in Belgium, Spain and Italy. Its use is recommended in many public-sector projects and it is widely employed for commercial projects as well. A survey in 1989 revealed that more than half the people in France who were using a structured method were using Merise. The control of the method is rather less formalised than, say, for SSADM. There are separate user groups representing the French Computer Society, various companies which use the method and academics working in the systems field and these originate and define amendments and enhancements which may be incorporated into the method by the authors of the reference manuals.

Along with SSADM, Merise is one of the major methods used in Europe which have given rise to the idea of the 'Euromethod', which is examined later.

Merise focuses on the development process required to develop information systems and addresses the standard life cycle of requirements analysis, specification, design, code production, implementation and maintenance. Like Information Engineering but unlike, for example, SSADM, the Merise Development Process Model covers the entire range of the development process, that is:

- Master plan – of a whole organisation or a major part of it
- Preliminary study – of one or several business areas
- Detailed study – producing a logical model of one or several systems
- Technical study – detailed design of one or several systems
- Code production – for the system/s designed
- Implementation – of the developed system/s
- Maintenance – of the delivered system/s.

It provides techniques to support the earlier phases of this life cycle – analysis and high-level design – but assumes the use of other methods or techniques – like Jackson Structured Programming – for the later phases.

Merise's development process model results from three cycles:

- The life cycle: this is the conventional idea of representing the creation, life and

decommissioning of the system
- The decision cycle: this represents the series of decisions which must be made during the system project
- The abstraction cycle: this is the series of models which are developed to express and document the system.

These three cycles operate in parallel and are interdependent and, between them, determine the shape of the IS project.

Merise produces a number of models of the information system including:

- A conceptual data model, which shows the data entities, their relationships, their attributes and so on. This model is built using a kind of third normal form data analysis.
- A conceptual process model, which shows the interaction of events, external or internal to the system, and the operations which the system performs in response to them. This model also incorporates the rules which the system must follow in responding to events.
- A logical data model, which can be a representation either of a CODASYL model or of a relational model; there are rules for mapping the conceptual data model onto this logical data model.
- An organisational process model, which builds on the conceptual process model but adds detail on where, and how, the various operations take place.
- A physical data model, expressed in the data description language of the chosen DBMS.
- An operational process model, consisting of the hardware/software architecture, flowcharts or chain diagrams for batch procedures and some sort of description, for example a program specification, of how the processing will be implemented.

It can be seen that the development of these models is a sequence moving from conceptual, through logical, to physical implementation.

An obvious strength of Merise is that the method deals with the whole of the systems development life cycle, including construction, implementation and maintenance. Thus unlike, say, SSADM it does not need to be coupled with other methods outside of the scope of systems analysis and design. Having said that, Merise is more conceptual than procedural when it comes to the later stages of development.

The conceptual and organisational process models provide a powerful means of modelling the dynamic behaviour of the system. They are probably more accessible to users than the combination of dataflow diagrams and entity life histories which would be required in an SSADM project. The control of the method is rather informal and not as tightly defined as, for example, the Design Authority Board for SSADM. Although all methods tend to be interpreted by individual users, this makes it harder to say what is 'complete' Merise and what is not, although the manuals are generally regarded as the 'orthodox' version of the method. On the other hand, this does make the method more flexible in allowing the immediate introduction of new tools and techniques.

For the English-language practitioner, however, the main problem is that there exists no suitable documentation of the method in English. Apart from the 'official' manuals, there are a number of books available which could be used by someone with a good understanding of technical French.

2.9 EUROMETHOD

Euromethod is an initiative of the European Community (EC). Among the EC's objectives which are relevant to Information Technology are:

- The creation of a large European market of a similar size to that enjoyed by the USA or under Japanese influence
- Encouraging open competition within this market
- Removing technical obstacles to such competition
- Facilitating the mobility of labour within the Community.

With regard to information systems, the proliferation of structured methods in the EC countries acts as an obstacle to achieving these objectives. The market is fragmented, competition is discouraged because the competitors are offering different, and incompatible, approaches and people trained in one method have difficulty in switching to a different environment. The EC has therefore set up a project to see if it is possible to achieve some harmonisation of methods within the Community.

Three possible approaches to Euromethod were identified initially:

- To select one of the existing methods. This was rejected as offering an unfair advantage to the country or company owning that method.
- To create a brand new method. This has been temporarily set aside as it would involve all sorts of migration difficulties from existing methods and, in any case, is probably too ambitious an idea to have any prospect of early implementation.
- To harmonise existing methods and to guide them towards convergence in a suitable framework. This would involve devising a suitable structural model within which the existing methods could be used, and the agreement of some common terminology.

The Eurogroup Consortium, a body representing major European systems companies, has been established with a remit to pursue the third approach.

Euromethod is, in effect, a meta-method (a tool representing methods) built on the theory of process modelling. Thus, the systems development process will be represented as a set of activities – stages, steps, tasks and so on – and also as a set of deliverable products. The model will also incorporate the decisions which will be needed to trigger each of the activities. The structural model is tailorable according to a set of rules so that it can be adapted to various sorts of project by size, risk, type and so on. As well as the structural model itself, there is a common and multilingual terminology, so that the equivalence of different terms can be understood. Finally, there is a procurement framework which will facilitate the assessment of bids made using different methods.

Once the structural model is in place, it is intended that Euromethod should facilitate the harmonisation of the various methods, either by specifying a European approach to systems development, or by assuring the convergence of existing methods within the Euromethod framework.

2.10 OBJECT-ORIENTED DESIGN (OOD)

The idea of object-oriented design has been gaining momentum in the last few years. The basic principles of object-oriented systems are simple enough but the problems arise from how such systems are to be developed, and the methods that will be needed to build them.

An *object* is a 'thing' of interest to the system and consists of both data and the logic about itself. Thus it is rather different from a conventional entity which contains only data. An object belongs to a *class* and, for each class, there is a set of *methods* that defines the behaviour of objects of that class. The behaviour is invoked when the object receives a valid message; this message could come from outside the system, triggered by a real-world business event, or from another object. Finally, objects *inherit* data and methods from their classes and can pass these on to their sub-objects; the subordinate object or sub-object can either behave as would its superior or it can have different methods defined for it. Thus, an object-oriented system is a network of interconnected objects. Each object is self-contained, with its own data and logic, and the total system is the total of all objects. When it comes to physical implementation, the objects can be stored as tables of information, rather like a relational database. It is common to read about object-oriented databases and object-oriented programming but, in a true object-oriented system, the two would be one and the same thing.

So, how does working in an object-oriented environment affect the analyst/designer, especially within the context of structured methods? So far, no methods have emerged which embrace the object-oriented approach completely (except for HOOD Hierarchical Object Oriented Design, developed for the European Space Agency and targeted towards systems to be built using ADA). Because of this, analysts and designers find themselves trying to adapt existing methods to object-oriented principles, with greater or lesser success. Since most of the structured methods use some sort of data modelling technique, this is the obvious place to start with object-orientation. Most of the data structuring rules are still valid and the normalisation rules can be extended to embrace logic as well. Specifying processing is more tricky as, in the object-oriented world, the fundamental units of processing are smaller and the need to catalogue more processing triggers – or events – is greater. However, what is probably needed is to continue the decomposition of processing to lower levels than has been necessary previously. The documentation adopted would have to be altered radically, since it is inappropriate, in object-oriented systems, to document data and logic separately. Practitioners will have to devise a scheme that fits in with their chosen method but this will probably be based upon the logical data model, with details of the objects – data and logic – held in some sort of objects database.

Two major advantages spring to mind when considering the object-oriented approach. Firstly, it is easier in an object-oriented design to cope with the complexity of business requirements. Instead of having logic built into a number of separate programs – all of which might need changing if the business requirements change – this logic is held at the lowest possible level: the object. So, if there is a fundamental change, it is made to the logic of the object, which is then reflected wherever that object is involved in the overall system. This obviously simplifies that bugbear of all system developers, configuration management. And, of course, it is easier to evaluate the possible effects of a change, since one is examining those effects from the atomic level upwards.

Secondly, the possibility of software re-use is greatly enhanced. Since object-oriented systems are built on the atomic principle, it follows that the data and logic contained in a single object are likely to be usable in several places within the system, and even in other systems. So, for instance, if there were a 'Customer' object, it is very likely that this could be pertinent in sales, credit-control and marketing applications.

The main problem with object-oriented development seems to be that it is a very difficult concept to get into one's head and it is even harder to see how the concept will be implemented in practice.

2.11 CONCLUSION

In this chapter we've examined some of the different approaches to analysis and design methods. Why are there these different approaches; if there is a best way of doing things why doesn't everyone just follow the same method? Firstly we have to recognise that different methods have been created by different people based on their views about system development and on their experience of the then problems of developing systems. Over time therefore different methods – Jackson, Yourdon, and Information Engineering – have been developed. Similarly, new methodologies have been developed to meet specific organisational requirements such as those of the UK government in calling for the development of SSADM and of national bodies in France that led to the development of Merise. In the final analysis the choice about what method to use depends on each installation selecting the method that will work best in the particular circumstances that apply.

For the individual analyst, the important features are that the method should be fully documented and be supported by adequate training and CASE tools. There should preferably be practical experience available from more experienced people and a sufficiently widespread use to ensure the long-term life of the method.

CHAPTER 2   CASE STUDY AND EXERCISES

Q1 System Telecom has chosen to use SSADM, and the data flow diagrams, data models and so on are drawn using SSADM conventions. If you had been responsible for choosing a structured method for System Telecom, which would you have chosen? What would be the reasons for making your choice; there is no reason to assume that just because SSADM has been chosen in the book, it is necessarily the method that a European-based telecommunications company would have chosen. A survey of European methodologies shows that there are several available and widely used methods.

Q2 People are sometimes critical of structured development methods because they say that these methods increase the time and cost of the early stages of development. Do you think that this is true compared with traditional development methods?

Equally, it is asserted that structured methods reduce the costs of implementation and maintenance so that system lifetime costs are reduced overall. Do you agree with this? What examples could you suggest to show how structured methods have the potential to reduce system lifetime costs?

Q3 Structured methods are said to improve communications between system developers and their customers (the users). Do you think that this is the case? Can you suggest ways in which this improvement is achieved?

CHAPTER 3
Communicating with People

3.1 INTRODUCTION

In this chapter, we shall be looking at how people communicate – that is, at how they convey facts, ideas and feelings to one another. This is an important skill for all analysts. Much of your time will be spent trying to identify the needs, wants and expectations of your clients. Much of your success as an analyst will depend on how well you understand those around you and on how well they understand you. To be an effective analyst, you will need to know exactly what you want to achieve when you communicate with your customers. It may be that you want to know how or why customers carry out a particular task in a particular way, or it may be that you want to influence them to accept your idea of how the system needs to be organised. But, whatever your purpose, you need to have it clear in your mind at the start. Ask yourself these two questions when you communicate:

- How are you trying to affect the people who are on the receiving end of your communication?
- What words or actions on their part will convince you that you have communicated successfully?

If you don't know the effect you want to have on the other person, and you have no clear idea of what you intend as an end result, you won't know when you have been successful.

Effective communication is more than just telling someone something and noticing whether or not they seem to be listening. It happens when facts, ideas and feelings are conveyed accurately and understood accurately but it's not complete until we know that this has happened. We need to see some confirmation that our attempt at communication has been successful.

When we tell someone something, the communication model looks like this:

You may think that another word for Target here could be Receiver, but that assumes too much. After all, the target of your message may not even notice it, let alone receive it or respond to it. As a responsible sender, however, you will want to do what you can to turn your target into a receiver. For example, you will take into account what you know about the target, so that if your target speaks French only, you won't speak in English. And if you are to know whether your target has understood what you are saying and how he or she responds to it, you will need some feedback. With feedback, our model looks as follows. Notice that the target is now the receiver: because we get feedback, we know that the message has been received.

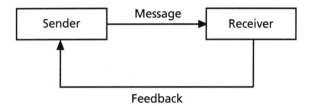

For an analyst, though, this is not a complete model of communication. It is not enough to be able to tell your client what you think and get some feedback about their views, it is even more important that you both understand each other. In other words, you will each be both sender and receiver. A full model of effective communication will therefore look like this:

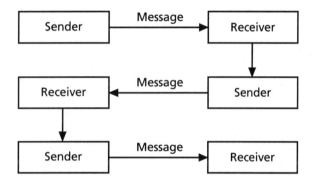

In this model, both you and the person you are communicating to are Senders and Receivers and will alternate roles. As soon as you have sent, you are ready to receive; having received, it is again your turn to send. Neither of you is merely acknowledging the other's ideas or feelings but is using them as a cue to express your own thoughts, feelings and understanding of each other's point of view.

3.2 TYPES OF COMMUNICATION

Our communication model works in a number of situations. Although we may think of face-to-face communication in meetings and interviews, there are other forms of communication and in this section we distinguish between written and spoken communication and consider each from the point of view of both the sender and the receiver. Since it is important that appropriate body language backs up the verbal message, we also consider non-verbal communication.

The main forms of verbal communication include meetings both formal and informal, interviews, presentations, and telephone conversations. Written communications include letters, memos, meeting minutes and reports. Some communication is even in a mixture of two forms, as in meetings where agendas are sent out first and minutes taken during the meeting. Each of these types of communication has its own perils and pitfalls, but in all of them we are trying to communicate with the receiver in a form that can be readily understood and accepted.

When choosing the most appropriate form of communication for your purpose, some questions you should ask yourself are:

- Which one of these types of communication would best enable me to get the information I need?
- Which one of these situations would best enable me to influence the potential receiver(s)?
- In which one of these situations would I get the best feedback?

To answer these questions, there are three things to consider: the purpose of the communication, the target audience and their level of understanding in the area.

- *Purpose*. In thinking about types of communication, remember that you are looking for the best type of communication to suit your purpose. If your purpose is to tell 30 people some information which you know will not need any discussion, it would not be sensible to arrange to meet each one of them separately. A memo, report or presentation would probably be a good choice here.
- *Target audience*. Here you need to take account of the needs, expectations and feelings of the target audience. If you know that the customer expects you to meet each senior manager at least once, then you will have to have some compelling reason not to do so, but if one particular person is very difficult to meet, you may have to resort to written communication, on the grounds that a memo is better than nothing.
- *Level of understanding*. It is very important to gauge the level of understanding of your intended receivers. If you don't do this, you run the risk of using over-simplified explanations and perhaps being seen as patronising or, alternatively, of using complex technical jargon with someone who doesn't understand it. It is clearly important to communicate at the right level in order to create your intended impression, but how can you do this? The answer is thorough preparation. When preparing to communicate, whether for an interview, a meeting, a presentation or in writing, you must always prepare your ground, to ensure that you are not wasting your time.

Having chosen the most appropriate form of communication, it is useful then to consider good ways of ensuring that you achieve your purpose. Let's look at the advantages of spoken, written and non-verbal communication, from the point of view of the sender first and then from the point of view of the receiver.

One of the strongest reasons for an analyst to use spoken communication is that people tend to feel more personally involved in the work of analysis if they have participated in an active way. Another major advantage is that it allows you to tailor your information, remarks or questions to the responses of the person or people you are talking to. This will mean that you will be able to get your points across more effectively and understand theirs better. If there are strong feelings about the subject under discussion, you are much more likely to find this out in a meeting than at any other time, because people are more likely to talk of their feelings as well as hard, plain facts. Finally, it will enable you to use non-verbal communication. When we communicate with people who are physically present, we read their body language constantly, looking at their expression and their gestures, and evaluating their tone of voice. Since we do this constantly, it is possible to take for granted the skill that we all have in reading others. If communication breaks down, however, we are most likely to notice this through body language. For example, if someone tries to avoid talking about something with you, you are most likely to notice that something is going on through their body language first, then through their tone of voice, and lastly, a long way behind the other two, through the actual words they speak.

Many of the same advantages apply to the receiver as to the sender. When you're a receiver, however, the main advantage is that you can tell the sender straightaway if there is something in what is being said that you don't like. So if you don't understand something, you can ask for further details, and if the sender has got something wrong or is missing something out, you can say so straightaway. This means that any of your concerns can be dealt with directly. Being able to change role between sender and receiver is the greatest benefit for both sides as it enables greater clarity and agreement in the communication.

When writing, you will be better able to choose your words more carefully than when speaking. This means that you will be able to change and edit what you write until you are satisfied that it is accurate. Your writing can be used to support your point of view. A piece of writing can prove whether you gave information or your opinion, asked questions of and received answers from a particular person on a particular date. Only something written down can easily prove this. This means that accuracy in written communications cannot be overemphasised. Communicating in writing can also be cheaper. You can circulate a large number of people with a memo much more quickly than you could hope to talk to them all, and this will save everyone's working time.

There is, however, a major problem about communicating in writing: the sender rarely gets immediate feedback from the receiver. This can be for a number of reasons. It could be that the letter, memo or report never reached its intended receiver, who is still waiting patiently for it. It could be that the receiver was too busy to read it. It may be that the receiver read the communication hurriedly and failed to see its point. All this time you will have no idea whether you have said the right thing, whether you have written too much or too little, or whether you have put your point clearly. You must bear this potential problem in mind when choosing how to communicate.

One of the greatest benefits of written communication for the receiver is that you can deal with it in your own time. This means that you need not be rushed and can give due time for thought before giving a response. Another benefit is that you receive someone's best thinking on the subject, the clearest communication they can make on this subject, so when you keep it, you have the best possible record of their thoughts at that time. This does not mean that their thoughts will always stay the same, but it does mean that you will know if they change in the future.

3.3 BARRIERS TO COMMUNICATION

You already know that communication isn't always successful. Let's look at some of the barriers that can prevent you from communicating as effectively as you would like.

Sometimes you may not come across clearly. For example, you may not really understand the information you are trying to communicate or you may not have a clear purpose for your communication. You may not have decided what effect you want to have on your receiver or you may not know enough about your receiver to be able to judge the impact of your communication. You may not have thought of convincing examples and analogies that would help to back up your case. All of these would indicate that you have not prepared thoroughly enough in advance.

If you are in a meeting, there is a strong probability that you will be able to overcome this problem if you listen carefully. Once you start talking, your receiver might help you by asking for examples or by asking you to expand on a particular topic. But it is unwise to rely on your receiver's skill to make up for your deficiencies. Your preparation should always be as thorough as you can manage. You may find it useful to make a list of topics you want to cover, points you want to make, questions to which you need answers. You can then refer to this list in the meeting. There is no harm in letting your receiver know that you have a list to help you – it indicates that you have prepared, not that you don't know what you are doing! Before important presentations and even some meetings, rehearsing exactly how you can make your main points is always useful. Some people fear that rehearsing can make them look artificial on the day, because it takes away the spontaneity of the moment. While it is true that spontaneity is reduced, you will probably feel more relaxed and better able to take advantage of the new ideas that may come to you as you are speaking. It is always better to look practised and competent than it is to appear spontaneously unprepared!

So far, we have been looking mainly at the message we intend to communicate, but there are other messages we both send and receive over which we don't have so much control. The importance of non-verbal communication has already been mentioned. We can learn a great deal about someone, especially about their feelings, by observing them closely. People are often unaware of what their body language is saying to others and we often respond on an emotional level without really being aware of what someone else has done that makes us feel the way we do. For example, you may feel alienated by a limp handshake or by someone who does not look you in the eye or else seems determined to stare at you all the time. If someone stands very close too you and follows you when you try to move away, you may feel that they are trying to dominate you. Because of this, you may make assumptions about what they are like. You may see the person as 'wet' or a bully, but you may be wrong in your assumption. So you will always find it useful to be aware of how you interpret other people's body language. Whether you know it or not, you will be making judgements about them, and to be of real use to you, those judgements should be explicit, so that you can challenge them, check them out, and even, if appropriate, change them.

Words convey only a small part of the communication on their own; tone of voice backs up the message you are putting across. Suppose you are telling a customer that the changes you suggest will save the company a lot of money. If you say this in an uncertain tone of voice, you may not be believed; but if you sound absolutely confident, there is more chance your suggestion will be taken seriously.

Accent is one of those things that should never be allowed to influence how seriously we treat someone's communication. Unfortunately, many people are affected by another's accent. A strong accent, regardless of where it comes from, can sometimes prevent someone's message being taken seriously if they are working outside of their own region. Similarly, a 'posh' accent can sometimes result in someone being taken more seriously than their communication would deserve. If you have a strong accent, you may not be aware of it. Listening to your voice on a tape will let you know how strong your accent is. You may not be able to change the accent, and you may well not want to, but if you are speaking to someone who is not local to you, avoid any dialect words or unusual local phrases, and perhaps speak more slowly.

The pitch of your voice is often more significant than your accent, because you are even less likely to be aware of it. Youthful voices tend to be higher in pitch than older ones. This can mean that those men with naturally light voices may be seen as less serious and less important than those with naturally deeper voices. Deeper pitches seem to carry more conviction and authority. This seems to be also true for women: Margaret Thatcher, the former British Prime Minister, was trained to lower the pitch of her voice to improve her communication.

If you are communicating in writing, you may be able to spend more time carefully crafting your message to make it wholly complete and utterly clear, but you will lack the feedback that would enable you to rephrase parts that your receiver found difficult or omit parts that your receiver clearly already knew. There are some things you can do which might help to overcome this problem:

- As you write, think all the time of the person who will be reading the communication. It can sometimes help to imagine them peering over your shoulder as you write, and interrupting now and then to ask for clarification of this point or expansion of that point.
- Ask someone else to review your first draft. If possible, make sure that this person is similar to your intended reader. If your piece of writing is to be read by a large number of people, it may be worth asking one or two of them to comment on it before you send it out to everyone.
- After writing, put what you have written away for a time. Even a break of a day will enable you to look at it critically. As you review it, imagine that it was written by someone else. What advice would you give that person in order to make a good piece of writing even better?

It is possible to spoil the whole impact of your communication by poor presentation. If, for example, you wrote a memo because it was a clear, quick, accurate form of communication, but your receiver was upset because he considered it to be cold, impersonal and indicating that you were not interested in his views, then you have presented your communication poorly, regardless of how good the memo was. Even when we choose the correct way of communicating, we may present the information poorly. When speaking, either in a meeting or when giving a presentation, it is important not to speak too quickly. It will feel unnatural and uncomfortable at first, but it will be a lot easier for your listeners. Your attention must be on making life easier for those who are listening to you, not on making life comfortable for you. When writing, it is essential to make sure that the written output is clear. Unless your handwriting is exceptionally clear, you should always type any important communication. Many people cannot read others' handwriting easily, and so they may not read it at all. For both written and spoken communication, you must organise your information before you start. Even if that means saying to someone, 'I need to get my thoughts in order before I can answer you', it is worth taking a bit of time to know what you mean to say. When answering questions, you will usually find that you need a little time for thought, a breathing space.

Communication sometimes breaks down because the sender and the receiver have different assumptions and are not aware of this. For example, a word or phrase may mean completely different things to each of them. Your customer may ask you to send her a report to present at a meeting on Monday the 18th. You assume that this means you can give her the report on the Monday morning when you have arranged to meet, but on the previous Tuesday she rings you up because you have not yet sent the report. She needs the report on the previous Tuesday because papers presented at the meeting are always sent out three days in advance. Each of you made an assumption, and as a result of that unstated assumption, effective communication broke down.

It is sometimes hard to accept, but there will be times when someone will decide that it is not in their best interests to tell you all you need to know and they do not communicate fully with you. They may omit something, or embellish something or they may express something ambiguously, hoping that you will misinterpret it. They may even lie, hoping that you will not be able to do anything even if they are found out. Sometimes this will happen because people want to appear more important than they really are, saying something like 'I control everything around here'. Sometimes they do not want others to know what really goes on, saying things like 'We always process every order within one working day'. Because this possibility is always there, never rely on just one person to give you important information. If two people tell you exactly the same, you can be fairly certain that what you now know is accurate, but if they differ in a number of ways, then you know that you have to keep looking for the accurate answer.

3.4 IMPROVING YOUR SKILLS

It is difficult to become a successful systems analyst without having good communication skills. However, no-one is so good that they don't need to improve! In this section you will find ideas to help you improve your communication skills. Of course, ideas written in a book are all very well: to improve what you actually do, you will need to go out and try things in a new way. We'll look firstly at getting information, then at giving information, dealing with meetings and giving presentations.

3.4.1 Getting information

There are a number of ways in which you can get information. Some information will come from reading, some from listening to people, and some from non-verbal communications.

Every systems analyst has too much to read. You may find that there are reports about what you are doing and about what you propose to do. You will be expected to have some familiarity with them all. There are numbers of books on reading efficiently, and you may well find it useful to go to a bookshop and browse until you find one that you feel comfortable with. In order to use your reading time effectively, you may find it useful to develop a systematic approach, using your concentration well. In order to make best use of your time, you need to give your task your undivided attention. You will know when you have lost concentration when you notice that you have read a sentence or a paragraph or a even a page and have taken nothing in. As soon as this happens, take a break. Few people can concentrate deeply for more than twenty minutes without a brief break, so the odds are that you will need a break every twenty minutes or so.

Analysts spend a large part of their time listening, but listening is not a simple activity:

there are many ways of listening. You can listen *for* something, like when you are waiting for someone to arrive by car, you listen for the sound of that particular car engine, or the clunk of the car door closing. This is a little like a lawyer in court listening to evidence so that he can spot contradictions. Here, the listening is selective, concentrating on one part of the message only. This is part of the listening that you will need to develop as an analyst, and it involves listening only for the essentials, refusing to let yourself be distracted by side issues, even if they sound interesting. Another way of listening is to listen *to* something, when you are actively trying to grasp what it is the other person is trying to communicate, often helping the other person to formulate their views and express them clearly. This is another part of the skill you will need to learn as an analyst, and this time it involves listening to the whole message, not just those parts which you have decided in advance are the ones to concentrate upon.

Fortunately, both of these types of listening can be put together and called *active listening*. We call it 'active' because it is a very long way from settling back and letting someone's words wash over you. This kind of listening only works if the listener takes an active part in the process of communication. It has been said 'The speaker dominates the conversation: the listener controls the conversation'. Let us explore how a listener can control the flow of conversation.

Active listening starts with the listener's genuine desire to know what the other person is thinking and feeling about something. As an analyst, you must be prepared to listen to people's feelings as well as their thoughts: every new system engages people's emotions as well as their minds. What you do when you listen to someone actively is get inside that person, begin to see things from their point of view, even if it is the opposite from your own. And, more than that, you have to convey that you are seeing things from their point of view. When someone speaks, there are normally two parts to what is said: the *content* and the *feelings behind the content*. Put together, these give us the meaning of what they are saying. If you are listening to someone talking about how incoming mail is dealt with in a complaints department, they might say, 'The man responsible for the mail leaves it in a large box on the table by the window'. This is fairly clear – you get an idea of who brings in the mail and where it is left. If your speaker had said, 'The man responsible for the mail eventually brings it up and then dumps it in a huge box on the table by the window', then the content of the message is the same but the meaning of the message is different.

To be an active listener, you would want to respond to the content, perhaps by indicating the table and checking that you had correctly identified it, but you also need to respond to the feeling behind the content, perhaps by saying something like 'You don't sound too happy about the mail. Is it a problem for your department?' In this way, you indicate both that you have heard the message and that you are prepared to probe to find out the cause of any problem. To be effective as a listener, you need to be able to convey to the other person that you have a sincere interest in them and their views. Your listener will pick up any pretence, consciously or unconsciously, and will no longer speak freely.

3.4.2 Giving information

Analysts have to communicate ideas, instructions, information and enquiries in writing in an effective way, accurately, briefly and clearly. Your writing style must conform to the normal rules of spelling and punctuation, you need a style that is neither too terse nor too wordy, and you have to organise your ideas in a way that makes them clear to the reader. Let's look at some guidelines for good writing.

Many people feel that they should be able to sit at a keyboard or a piece of paper and produce expert, clear prose. When they can't do this, they feel that there is some skill which others have which eludes them. The truth is that few people write well at first attempt – the best writers make many attempts until they are satisfied. It is useful to think of structuring your time and effort into three parts:

- Prewriting, when you prepare to write
- Writing, when you do the 'real' work
- Rewriting, when you edit what you have written into something that others will enjoy reading.

The first stage is to find out why you are doing the piece of writing. Ask yourself why you have been asked to write something and what you expect to gain as a result of writing it. To write something that is effective, you need to know what you want your reader to do as a result of reading it. It's not enough that the reader now knows something new, or is made aware of something new – what do you want the reader to do as a result of this new awareness or knowledge? Different people react differently. In order to be able to identify what we want our reader to do, we must have a clear idea of what our reader is like. Will the readers have specialist understanding of the main subject of the report? Will you have to remind or tell them why the report was commissioned? Is the subject already important to them? And, if it is, do they know that it is? What do they expect the report to say? What do they want to read – as distinct from what they expect to read? Having considered both your potential reader and your purpose in writing, you then need to gather all the relevant information and to organise it to suit the reader and achieve the purpose.

Having done all the prewriting, we are now at a good place to go on to write whatever it is we need to write. What we want to produce is writing that is clear, concise, correct and appealing to read. Accurate spelling and punctuation add clarity as well as correctness to a report.

So far, when looking at prewriting, we have been looking from the perspective of the writer. Let's look at writing from the other perspective. As a reader, you are more likely to read something if it looks interesting. Also, many people who have to read a lot of reports only read parts of them.

Most reports will contain a title page, a management summary, a conclusions/ recommendations section, a main body and perhaps some appendices. Of these, the most important parts to make interesting and clear are the summary and the conclusions/recommendations. Many senior managers read only these parts consistently and rely on their staff to know the detail. Clear interesting writing usually has few cliches, little jargon and is active, not passive. Most readers prefer such writing, but many writers rely on jargon, cliches and the passive, mistakenly believing that they are writing the correct, accepted style. Nothing could be further from the truth.

Finally, there is the rewriting stage when you polish what you have already written. At this stage you need to know the mistakes you commonly make, so that you can correct them. You will probably have to write a lot of reports in your job as an analyst. If you follow the *prewrite, write, rewrite* plan, you will produce an accurate, clear document. Let's look at one or two factors which can help to make that document more readable.

The order of sentences in a paragraph of a report can make a considerable difference to the reader's understanding of that document. Paragraphs are easiest to follow if they start with a flag sentence which tells the reader what the topic of the paragraph will be. This paragraph, for example, deals with the order of information in paragraphs. Since you are told this in the first sentence, fast readers will be able to scan the topic of the paragraph quickly and then decide whether to read the rest or not. All paragraphs should start with the key idea first, then with a logical sequence of sentences all relating to that key idea.

Word order in sentences should follow the same principle. The most important word or words should come first, like a flag to attract the reader's attention. For example, key words and phrases like 'in conclusion' and 'to summarise' should always be at the start of the sentence, not hidden in the middle where they could be easily missed.

If you want people to agree with your conclusions as well as follow your argument, it is a good idea to start with the weakest argument and progress to the strongest. So, if there are a number of possibilities for a new system, all of which have some merit, your report should start with the weakest possibility, explaining why this is the weakest, and progressing on through the others with explanations as to why they should not be chosen, ending with the strongest, together with reasons as to why this one should be chosen.

3.4.3 Meetings

If you ask people to describe what makes them effective at meetings, the chances are they will list:

- Careful preparation
- Well-structured information
- Clear presentation of the ideas.

It follows then that if you are to be effective in meetings, you should consider these three topics.

Most meetings circulate something to the participants in advance. This may include an agenda, minutes of the last meeting, briefing papers, proposals, reports. As a participant, you have to decide what is important in this mass of paper. A properly prepared agenda will help you to do this. It will give you titles of the topics to be discussed and an idea of what will be required of the participants – a decision, a briefing, for information only, and so on. You should also be given an idea of how long the topic is expected to last. That way you can devote your preparation to those topics which require some action by the group and which are expected to take some time. Beware of the meeting where nothing is circulated before it – it probably means that most people will do no preparation and that the meeting will be ineffective.

In order to be able to contribute well in the meeting, you will want to ensure that you know what you intend to say. Some written notes can be valuable, particularly if you take time to think of what could persuade others to your point of view. You need to ensure your notes contain information that is accurate and not open to argument. If you have one previously undetected error in what you say, you may well find that people pick on that error and so lose the whole point of your argument. Also, if you have a good clear case with one or two supporting examples, you can make your point. If you were to bring in other arguments in support of your position, you could fail. Some listeners pick on the weakest of the arguments and dismiss it, and with it the whole position. People are more interested in people and specific cases than in abstractions, so use examples, illustrations and case histories, especially if you can use information known to your listeners where you can be certain they will agree with your conclusions. Write your notes in a way you can easily understand and then refer to them when you need to. Many people don't use notes for meetings, and you may feel that you will look inexperienced if you do so, but it is better to look inexperienced and thoroughly prepared than to look inexperienced and totally confused! Even though you have your notes, listen to what others are saying, and be prepared to modify your views. Finally, don't over-prepare. Come with an open mind, not just a series of answers and conclusions that you came to without hearing the other points of view.

You will have noticed that a large part of careful preparation is about ensuring that the information you present at the meeting is well-structured. There are, however, some other things you can do to ensure that you make a good impression at meetings. If you will be making a proposal at the meeting, one good thing to do is to lobby other attendees beforehand. It is especially useful to see the most influential people before the meeting and try the proposal out on them. If they agree with you, you will have an influential ally; if they disagree with you, you can find out why and still have time to do something about it before the meeting. You could adjust your proposal so that you do have support, you could marshall your arguments so that you can convince others or you could drop the proposal, at least for the moment, if it is clear you will not succeed. But whatever happens, you will have lost nothing.

As an analyst, one of the areas you will be talking about in meetings is change. If you expect opposition to what you will be saying, the best plan is to anticipate what others may say, and then prepare counter-arguments. You may not know the details of what the others will say, but if you are proposing a change, the two most common counter-arguments are that it's too expensive and that it won't work. If you can anticipate these attacks and have your position already prepared (and backed up with facts and figures), you are more likely to have your point of view accepted.

If you disagree with what someone says in a meeting, there are some things you can do to make your point have more chance of being accepted.

- Having established that it is the idea you are criticising, not the person proposing it, hesitate before you disagree. There is a fair chance you have not understood clearly what the person meant. It is often a good idea to restate succinctly in your own words what it is you think you heard. In this way, the speaker can clarify for you what was meant. Either there will be no disagreement, or the disagreement will become more sharply focused.
- If the person putting forward the idea is speaking in the abstract, it is usually a good idea to ask for a specific example. Someone may say, 'All that senior management is interested in is cutting costs, not in making this a better system'. This may or may not be true, but if you disagree with the abstract statement, you could generate conflict.

A better approach would be to ask for supporting evidence or to wait until the person gives the evidence. Then you have something concrete to work with.
- Agree with anything you can of the other person's point of view, making statements like, 'I agree that there will be a significant cost saving if we adopt this plan, but there will also be the possibility of generating more business, and that possibility is what I think we should be discussing here.'
- One of the best ways to avoid an argument is to ask questions rather than attack someone's views. So you could say, 'Geoff, you said a number of times now that you think this proposal would never work here. What makes you think this?' rather than 'Of course this proposal would work here'.
- Once you know why the person holds a particular view in opposition to yours, you may find that you both have the same objectives and only differ in your preferred ways of achieving them. In this case, you should concentrate on the similarities between your views before you go on to disagree. In this way, you have taken some of the sting out of the opposition, and stand more chance of winning the other person round to your viewpoint.

3.4.4 Presentations

Most analysts will have to speak in public. You may sometimes look at good public speakers and envy them their natural ability, but while it is true that some people do have exceptional natural ability, the majority of good speakers work hard to develop their skill. It is important to be as thorough as you can about preparation. Doing so allows you to deal with some of the essentials before you stand up to face your audience. If you know you are well prepared, you can then turn your attention to the task in hand – making a good impression on your listeners. When you first hear that you will have to give a presentation, there are four areas you need to concentrate on:

- your audience
- your purpose
- your material
- your intended location.

The first two of these will have an enormous impact on your material and the last one will have an impact on how you will deliver your material.

The first area to look at is your audience. This will come first because the first questions you need to ask are about the audience. As when you write, you have to tailor everything about your presentation to your audience. Even if you think you know nothing about them, there are bound to be things that you can find out. You can call their secretaries, you can talk to other people who have experience of speaking to such an audience so that you can have a clear idea of what is expected from you. You will need to know, in particular, how much they already know about the subject you'll be presenting. In addition to this, there are always a number of things that you already know about every audience you will be presenting to. First of all, in some ways, each member of that audience will be a bit like you. So if you consider when you have felt uncomfortable during others' presentations, you will have some ideas about what to avoid during your presentations. Also, if you remember what you have enjoyed about others' presentations, you can use this as a guide for what could be effective for you.

Once you have decided what your audience is likely to need from you, you can then turn your attention to the material. This time the first question to be answered is: 'Why have they chosen me to make this presentation?' You need to be clear in your own mind about why you have been chosen. Do you have a particular skill or ability that others lack? Or were you the one member of the team with adequate time to prepare? Whatever the reason, knowing why you have been chosen will help you to meet the purpose of the presentation.

The next question to be answered is 'Why am I giving a talk?' People are often asked to make presentations without knowing the reason. If you have no idea why a talk is needed, rather than a letter, report, article in a journal or a series of telephone conversations, then you will not be able to do a good job.

Lastly you need to find out the answer to the question 'What is my purpose?' When you present, you may be trying to do a number of things, but you need to have the purpose in your mind. Are you trying to persuade? Or to inform? Or to train? Or to sell? Do you want to give background information or detail? Once you know this, you can move on to decide what your material should be and how you can structure your material to achieve your objective.

When preparing your ideas, research your topic and, when you know all you need, write down your ideas in any order. Collect everything you may need for the topic at this stage. You can prune your ideas later. At this stage, collect even information that you find dull or boring – your audience may need it or may find it interesting. And do it as early as you can, as your thinking will mature with time. Then review your collected material after a little time. Decide what to include and what to leave out. Include only material that will have the impact you intend upon your particular audience. Consider a structure that will enable you to get your message across. This will not only help you with your preparation, it will help your audience follow where you are during your talk. Then prepare your notes to help you keep to your prepared structure. These can be as full or minimal as you prefer, but make sure that you pay extra attention to the start and the end of your presentation, as these are the two areas where your audience will particularly notice your performance. Your notes will give you the best chance of remembering what you want to say and help you neither get lost nor dry up.

Very few people present without any notes at all. Even fewer do it well. Some guidelines for notes are:

- Once you have made them, use them to help keep you on track. Beware of the last-minute inspiration as you stand up, as you have no idea of how it will affect your timing or the balance of your talk.
- Write them yourself. A surprising number of people give talks using notes written by someone else who was originally scheduled to give the talk but now can't. At the very least, if you have to use someone else's notes, ensure that you understand them thoroughly and rewrite them for your own use.
- Use only one side of the paper, and number all of your pages.
- Consider using index cards. They are easier to handle and less obtrusive than paper pages.

- Use colour to highlight important points, headings, new topics and optional parts to be included if there is time.

There is a continuum from 'no notes at all' to a 'full script'. You should choose the place on the continuum where you think you will feel the most comfortable and try out your notes when you practise.

Comprehensive notes are suitable for most people most of the time and will be quite full, probably with most of the following:

- A sequence of headings and subheadings, each with the first few sentences
- Links from one section to another
- Summaries and conclusions
- References to visuals
- Prompts to help with any interactive parts
- Examples you intend to use
- Timing notes, to ensure you keep to time.

Simple notes, sometimes called skeletal notes, are much briefer and are most suitable if you are familiar with the subject. They would normally consist of:

- Key headings
- Key words/phrases to jog your memory
- Prompts to sequence the ideas.

When your notes are complete, you will know exactly what you will be doing during your presentation. Now make a rough version of anything you will need for the presentation. You may need to use visuals, show graphs or give out a handout. None of these should be left to the last minute. Finally, have a run-through beforehand. This will not only give you the chance to assess your timing and adjust it if necessary, it will also let you experience what it will be like presenting that material. If there is anything you feel uncomfortable with, you can change it before your audience has to experience it. You may want to get some feedback from colleagues, or from a video camera, so that you can get an idea of your impact.

When listening to you, your listeners will not be able to go back and re-read parts they did not understand the first time. You need to do something to ensure that they are unlikely to lose their place, and, if they do, that they can find it again quickly and easily. The best way to give your audience this help is to have a clear and accurate structure. A well-known and frequently used structure is often called the 'Three T's'. It looks like this:

- Tell them what you are going to tell them
- Tell them
- Tell them what you told them.

To put it another way, you could have an introduction in which you outline the main sections of your talk, a middle where you present each of these topics, and a conclusion where you sum up what you said previously and ensure that your listeners go away with your message ringing in their ears.

In the introduction, you need to make a good impression at the start. You need to attract the interest of your audience and build on their desire to listen. One way to do this is to look comfortable and happy with presenting. It is a paradox that in the first two minutes of your talk the audience will be judging how comfortable you are as a presenter when it is the very time that you are likely to be feeling most nervous. If you have a good introduction that is well rehearsed, you will get through those first few moments more comfortably and make a good impression.

In your introduction, you will need to welcome your audience and to introduce yourself and the topic. Because you have prepared thoroughly, you will know why you are presenting, and it may be appropriate to tell this to the audience, to build up your credibility. It will also be useful to define both the subject you will be talking about, what you will cover and what you will not, any parts that you will concentrate upon in particular, and so on. You also need to describe the procedure, how long you intend to talk, whether you will give a handout, whether it would be useful for them to take notes, whether they should ask questions as you talk or whether you would prefer them to ask questions at the end, and so on. All of this establishes a comfort zone for you and your audience for that presentation. With everyone comfortable and knowing what is expected from them, you can move on to the main body of your talk.

Within the main body, a good general guide is to work from the known to the unknown. You cannot transfer new knowledge or information to someone in an effective way unless you relate it to something they already know. As well as this, people will accept your ideas and conclusions better if you establish a context for them. Also, you will be talking about a number of related topics. Since the ideas are not already linked in the mind of your audience, you need to create this link for them. One way of doing this is to introduce and sum up each topic, ensuring that there are adequate signposts for your audience to be able to see where they are going, where they are and where they have been. Another structure you could use for a persuasive or sales presentation is the 'Four P's'. These stand for:

- Position (the audience should already know this)
- Problem (they should know this too)
- Possibilities
- Proposal (your recommendation).

An example might be that our *position* is that we all want to go out together tonight. The *problem* is that the car we were intending to use has broken down. There are a number of *possibilities*: we could call a rescue organisation, we could take a taxi, we could take a train, we could take a bus. I *propose* that we take a bus to the station then go by train, as that way we will all be able to have a drink. It is a logical structure which can be expanded well beyond a few sentences.

A structure which is rarely extended beyond a few sentences is that of the 'sound bite', the mnemonic for which is PREP. This stands for:

- Position
- Reason
- Example
- Position.

Once you are aware of this structure, you will notice a number of politicians use it in answer to questions. It goes like this:

My position on this matter is ... My reason for this is ... Let me give you an example

. . . So, I firmly believe that (and the position is restated) . . .

It is an ideal structure for sounding expert and authoritative in a few words.

At the end, your audience will probably appreciate being reminded of your main messages. If something you have said was important, consolidate by restating it. This is your last chance to get your message across, your last chance to make a good impression on your audience. Many poor presenters falter into silence, shuffle their notes and then sit down looking apologetic. You do not have to be stunningly original, but you do need to be memorable in those last few moments. Ensure that you finish on a high note, where you convey your enthusiasm for your subject. You will only do this if you plan it well before you stand up.

Many a good presentation has been spoilt because the presenter failed to check the venue in advance. You need to know about the audio-visual equipment. Try to visit the room and use the equipment beforehand. That way you will get a feel for the place which will help you in your final preparation of the material. You will also be able to see if you can operate any audio-visual equipment, like an overhead projector, a 16mm projector or complex video equipment. You will also see how the equipment and furniture are set out. You can then decide whether you like them like this or whether you want to change them.

Having dealt with all the preliminaries beforehand, your time and attention during the presentation is on your delivery. When you present, you want to be your own natural, relaxed self, but just a little larger than life. How can you do this? In order to stand up and be yourself, you need to have controlled your nerves to some extent. You will probably never be free from nerves – many great presenters and actors were racked with stage-fright all through their careers – but you can control nerves with some simple techniques. Many people find when they start to speak in public that they cannot quite breathe properly. Sometimes they run out of breath partway through the sentence, leading to a loud first part where the breath was sufficient and a quiet second half where it was not. Sometimes they need to breathe in the middle of sentences, and so create an unnatural pause. Sometimes they breathe very shallowly and sound rather like they are attempting to run a marathon. You can avoid problems like this, and feel more calm than you did before, by doing some simple breathing exercises.

It is also important to use your body to reinforce your message, not negate it. It will establish you as someone who can be trusted to tell the truth. It will establish you as expert in the subject you are talking about. Because it will be appropriate to the situation, it will vary from time to time, but there are a few things that are constant:

- *Eye contact*. You will only build up rapport with your audience if you look at them. This is particularly true in the first few minutes of your talk. If you avoid their eyes then, you will have to work hard to regain the ground you have lost. Avoid all extremes, so do not stare at anyone either. You need to distribute a calm even gaze throughout your whole audience, particularly to those people who look less than delighted. If someone is smiling and nodding, they are already on your side, so give them just enough attention to keep them happy while spending more attention on converting the doubters.
- *Open posture*. If your body position is closed, with arms folded and shoulders hunched, you will not look relaxed. If you look defensive, which is how a closed body posture

is often interpreted, your audience may believe that you do not know what you are talking about.
- *Avoid mannerisms*. You will have seen presenters with mannerisms: they jingle coins or keys in their pockets, play with pens, stand still like a stuffed dummy or march like a soldier up and down. No doubt you could add other mannerisms to this list. Regardless of what the mannerism is, it conveys to your audience that you are uncomfortable in some way. This discomfort is catching, and they will feel it too. But that is not all: these distractions will stop the audience focusing on the message you are conveying.
- *Be yourself*. However uncomfortable you may feel, you should be presenting yourself as well as the content of what you say. There will be a way you want to present, a way you want to sound and look: whatever it is, it should reflect you the person not someone else. Enhancing your own personal style means that you will be presenting yourself a little larger than life.

One thing that will help you achieve the effect you want is if you focus your attention where it should be during your talk: on your audience, not on your state of mind. Your audience is listening to you because they need something from you. This something they need will depend on what your purpose is: to persuade, sell, inform, and so on. But if you focus on them and their needs, you will take some of your attention away from yourself and your nerves. You will know when you are achieving the effect you want if you read your audience. Look to see what you think their state of mind is. Are they confused? Or sceptical? Or bewildered? Or happy? You will get this information from making eye contact with them. If you also have some brief pauses in what you present, you can allow your audience to catch up with you and will give yourself time to look at them all and see how you think they are feeling. If they ask questions, you can judge their tone of voice as well as the content of the question.

Whatever you think their state of mind to be, do not leave it there. You need to check that you have received the right message from them, so reflect back what you have assessed to see if you are accurate, saying something like:'I seem to have confused you here.' This gives them the chance to reply: 'I am confused. You described what you were going to do and that's not what you've done.' Since the source of confusion is now out in the open, you can clear it up. If you do not clear it up, that person could leave your presentation confused and lacking support for your ideas. By handling the confusion, you have gained an ally.

It is not always appropriate to invite questions, but presenters generally do so. Communication is two-way, so most presenters feel that question-time will allow the audience to have their say. At the same time, many presenters worry a lot about such questions. However you decide to handle questions in your presentation, you must tell your audience what you expect from them. There are a number of ways you can structure the time for questions:

- Spontaneously, ask them as they occur to you
- Interim – you ask 'Are there any questions?' at times when you feel there may be
- At the end of the talk
- Written questions presented in advance, at the end or after a short break.

Whenever they occur, here is a model for answering questions effectively:

- *Listen*. There are a number of presenters, some politicians among them, who do not answer the question they have been asked. Sometimes this is because they prefer to answer another question but sometimes it is because they have not heard the question. If you are not sure you have heard it accurately, ask the speaker to repeat it.
- *Pause*. This gives you a little time for thought and gives the audience time to assimilate the question.
- *Repeat/clarify*. Often, people in the audience do not hear the question the speaker has been asked. By repeating it, they get to hear it for the first time. But this is not all, by repeating it in your own words, you are checking back that you have understood the question, so clarifying your understanding. An extra bonus is that you get an extra moment to think of the answer.
- *Respond*. This may seem too obvious to state, but answer the question. Do not answer one you would have preferred to have been asked, do not waffle around the subject because you cannot answer the question, never lie, do not make this the opportunity to put in those points of your talk that you forgot to give at the time.
- *Check back with the questioner*. It is courteous to check that you have dealt with the topic to the questioner's satisfaction. Doing so will also ensure that you can clear up any lingering queries.

This model will help you deal with almost any type of question and help to keep the audience thinking of you in a positive way. Whatever method you choose, ensure that it is appropriate to you and your talk and that you have told your audience, preferably in your introduction, when and in what format they should ask questions.

3.5 SUMMARY

This chapter has presented a model for effective communication, where the communicators alternate between sending and receiving. Such a model works in a number of situations: face-to-face communication, in meetings and interviews, written communication such as reports, and non-verbal communication. Whatever the situation in which you are communicating, there are three things you always need to know in order to ensure that you are successful: the purpose of your communication, the nature of your intended audience and their expertise in your subject. We looked at barriers to effective communication: not coming across clearly, not listening effectively, poor presentation, different assumptions or someone not communicating fully. There were also guidelines for coming across effectively, for improving your listening skills, your writing skills, your contributions in meetings and your presentation skills.

People spend a lot of their time communicating, sometimes well but sometimes badly. It's worth spending time to improve your skills, as the payoff can be huge, not just in your job but in your life as a whole. Communication is an area where many of us are complacent, so reading this chapter every six months or so is a good way of keeping the ideas in mind and ensuring that you make the most of what you have learned.

CHAPTER 3   CASE STUDY AND EXERCISES

Q1   A particular application in System Telecom is giving cause for concern because of the increasing number of programming errors being found during its early operational life. Herr Norbert Rothaas has asked you to chair a meeting that will clearly state the problem and recommend appropriate action. He then wants you to give him a short verbal summary of the main issues and back this up with a report. How would you plan for this assignment?

Think first of all about the implications of the assignment. You need to

- find out who can contribute towards resolving this issue and bring them together in a constructive environment
- organise an effective meeting
- investigate the problem and identify solutions
- make a brief presentation to him
- write an accurate and comprehensive report.

Each of these is a systems task in itself. Three of them have been covered in this chapter but you will have to be creative to work out how to complete the first task of bringing the appropriate people together.

Q2   Assume now that you have completed the assignment. You've made your presentation to him and submitted your report which he has accepted. You are now required to present to the user department manager and senior staff, and the application development team for this system. Your presentation will be different. What will be the structure of the presentation and the headings you will use? This event is likely to be a more emotionally charged event. How will you handle this?

CHAPTER 4
Building Better Systems

4.1 INTRODUCTION

In March 1988, at an annual meeting in Gleneagles, the Computer Services Association of Great Britain announced the results of a study undertaken by Price Waterhouse which revealed that £500 million a year were wasted by users and suppliers in the United Kingdom as a result of quality defects in software. An earlier United States Government Accounting Office report had investigated a number of federal software projects costing a total of $6.3 million. Out of these, half were delivered but never used; a quarter were paid for but not delivered; and only $300,000 worth was actually used as delivered, or after some changes had been made. The rest was used for a while, then either abandoned or radically reworked. In 1989, the British Computer Society advised the Department of Trade and Industry that losses due to poor quality software were costing the UK £2,000 million per year.

Reports and studies such as those described above have highlighted the problems associated with quality failures in systems development and have led to moves in the 1990s towards building better systems which meet the needs of the user, are cost effective and produced on time and within budget. Developers of software products and information systems are being encouraged to adopt definite policies and practices which are carried through from requirements analysis to maintenance and user support phases. For example, in Britain, the Department of Trade and Industry is sponsoring a quality management certification initiative, called TickIT, the aim of which is to achieve improvements in the quality of software products and information systems throughout the whole field of IT supply, including in-house development work.

In this chapter we wiil turn our attention to this important area of quality, looking firstly at some definitions and then at the contribution made by a number of 'gurus' to our understanding of quality. We will discuss quality management, including the standard ISO 9001, and then go on to describe techniques for building quality into systems analysis and design and into all parts of the software development process.

4.2 QUALITY CONCEPTS

The term 'quality' means different things to different people, depending on their perspective. To some it means 'finding the errors' or 'making sure it's correct' and involves checking the deliverables of a system. For others it is about the process of production and means 'doing it right first time', 'achieving the standard' or 'getting the job done in the best possible way .' Actually these definitions point to the fact that there are a number of dimensions to the concept of quality, which we will examine. In this book, our working definition of quality is:

'conforming to the customer's requirements'

where 'customer' can be either an external customer (a client) or an internal customer (a colleague) and 'requirements' relate to both the product and the service delivered.

Customers The concept of an internal customer, and an internal supplier, is important to an appreciation that our definition of quality goes beyond the interface between the software supplier and the client and includes all our working relationships. The roles of internal customer and internal supplier are constantly changing. For example consider the situation when an analyst hands a report to a secretary to be typed. Before handing it over, the analyst (the supplier) will have checked with the secretary (the customer) the form in which the secretary would like to receive the written draft (for example, with pages numbered, paragraphs clearly marked, written legibly etc.). When handing back the typed document, the secretary becomes the supplier, and will ensure that the requirements of the customer (the analyst) have been met, for example completed on time, in the required format and with spellings checked.

In both internal and external customer–supplier relationships, the supplier must first talk to the customer to ensure they fully understand the customer's requirements if a quality product or service is to be delivered. The requirement will include details of:

- what is required
- the most appropriate way of producing or delivering it
- the involvement and contribution expected from each party during the process.

Product and service quality The deliverable to the customer, the thing they see or experience, is a product or a service or both. Product quality can be defined as the degree to which a product meets the customer's requirements. According to this definition, what the customer thinks about the quality of the product is all that counts. One can only speak of better product quality if the customer *perceives* the product to be better, no matter what the experts may consider to be objective, factual improvements. Service quality can be defined as the degree to which a service meets the recipient's requirements. The quality of service can be described as having two components, *'hard'*, the tangible content of the service such as the number of times the phone rings before it is answered, the user guide, the number of post office counter staff available at lunch time, the comfort of the aircraft seat; and *'soft'*, the emotional content of the service, the friendliness, flexibility, helpfulness of the service provider, the atmosphere of the premises, the treatment of complaints. Studies conducted in the United States show that the human factor has a crucial bearing on the customer's perception of the service quality. 'Soft' service is therefore often more important than 'hard ' service in a customer's perception of quality.

All too often, suppliers concentrate all their effort on investigating the customer's product requirements, but in order to deliver a quality package to internal as well as external customers, their service requirements (hard and soft) must also be investigated.

4.3 QUALITY GURUS

In this section we'll look at the work of three influential 'quality gurus' – W Edwards

Deming, Joseph Juran and Philip Crosby – in order to understand the origin of current approaches to the subject. There have been three 'generations' of quality gurus – firstly the Americans such as Deming who developed the philosophy and took the idea to the Japanese to help them rebuild their economy after the second world war; the second generation were Japanese gurus who built on the original ideas and introduced techniques such as quality circles and fishbone diagrams; and finally came the third wave, typified by Tom Peters and Philip Crosby, who, influenced by the Japanese, have developed new ways of thinking about quality in the West.

W Edwards Deming The origin of the Japanese ideas of Quality Management can be traced back to an American named W. Edwards Deming. Deming found that managers in post-war Japan were much more receptive to his concepts than their Western counterparts – whose perception was that quality increased the cost of the product and adversely affected productivity. Under Deming's guidance Japanese companies took a number of important actions:

1. They invited customers into their organisation and worked with them to improve quality.
2. They removed responsibility for quality from a separate department and ensured all line managers worked to their own clear quality objectives.
3. Responsibility for quality was delegated to all levels of staff within an organisation.

Deming was trained as a statistician and his ideas are based on statistical process control (SPC), a method which focuses on problems of variability in manufacture and their causes, identifying and separating off 'special' causes of production variability from 'common' causes using statistical process control charts. The special causes would then be analysed and problem solving methods applied. He proposed a systematic, rigorous approach to quality and problem solving and encouraged senior managers to become actively involved in their company's quality improvement programmes. Deming's philosophy was described in fourteen points addressed to all levels of management. The key messages were:

- Create a sense of common purpose throughout the organisation
- Build quality into the product
- Buy from the best supplier rather than the cheapest
- Establish a programme of continuous training
- Improve communications within the company and with customers
- Drive out fear – encourage people to work together.

Deming argued that emphasis should be placed on controlling the production process rather than concentrating on the end product, and that a higher quality of product results in reduced costs.

Joseph Juran Joseph Juran, another American writer who was also very influential in Japan in the early 1950s, began his career as an engineer. His message is primarily aimed at management who he claims are responsible for at least 85% of the failures within companies. He believes that quality control should be conducted as an integral part of management control and that quality does not happen by accident but has to be planned. Juran's central idea is that management should adopt a structured approach to company-wide quality planning, and that this should be part of the *quality trilogy* of quality planning, quality control and quality improvement. This would mean:

- identifying customers and their needs,
- establishing optimal quality goals, and
- creating measurements of quality.

in order to produce continuing results in improved market share, premium prices and a reduction in error rates.

Philip Crosby   Philip Crosby is another influential writer and speaker on quality whose book, *Quality is Free*, published in 1979, became a best-seller. He developed the concepts of 'Do It Right First Time' and 'Zero Defects'. Recognising the role management and employees play in the framework of an appropriate quality culture, he defined the four absolutes of quality management:

- Quality is defined as conformance to the customer's requirements, not as 'goodness' or 'elegance'.
- The system for implementing quality is prevention not appraisal.
- The performance standard must be Zero Defects, not 'that's close enough'.
- The measurement of quality is the price of non-conformance.

According to Crosby, the price of non-conformance – the cost of doing it wrong and then having to put it right or do it again – is about 20% of revenue for manufacturing companies and up to 35% for service companies.

All three of the gurus described above agree that quality is a continuous process for an organisation; that building quality into the production process – prevention – is more effective than just testing or checking the product at the end of the process – inspection; that management is the agent of change; and that training and education should be continuous processes at all levels.

What are the implications of these ideas for software developers? Four main points seem to result from these ideas about quality:

- There needs to be a greater investment of time and effort in the early stages of system development, if the cost of correcting errors at the end is to be reduced. This is consistent with the message that prevention is more effective than inspection. Research conducted by IBM has shown that 75% of software development costs are associated with testing, debugging and maintaining software, and if this figure is analysed, over 80% of the testing, debugging and maintenance cost can be traced back to problems introduced during analysis. A more systematic approach to analysis could make a dramatic difference to these figures, and this is the purpose of the structured methods described later in this book.
- The cost of correcting an error during implementation is many times greater than the cost of putting it right during analysis. For every hour spent tracing a bug at the analysis stage, it takes a hundred hours to find it at the testing stage. The use of reviews and walkthroughs at every stage of development would help with the early detection of errors. This supports the idea, put forward by the 'gurus', that quality is a continuous process. There is sometimes a fear that quality assurance will only delay a project, but

if managers don't invest the necessary time and resources to build reviews into the development process, they are putting themselves in a similar position to a frog placed in boiling water. If you put a frog in cold water and gradually turn the heat up, the frog will not notice that the water is boiling until it is too late. If problems are not detected early, during formal reviews, management may not be aware of them until too late, when the cost of putting these problems right can be huge.
- Clients who have been questioned about their satisfaction with the product and service delivered by software suppliers have identified that one cause of dissatisfaction is their discovery that project personnel do not have the level of experience or expertise which they were led to expect. By investing in the training and development of their staff, not only will a company be supporting one of the main recommendations of the gurus for ensuring quality, but they will be raising their chances of doing the job right first time and also building the client's confidence. Philip Crosby has said that before you can expect people to do it right first time, you need to tell them what 'it' is, and show them what doing it right looks like. Crosby describes this as investing in prevention, an idea we will explore further in the next section.
- The gurus agree that management is the agent of change. Managers on software projects must take the lead in making quality an intrinsic part of the development process. Often software development project managers don't give the projects a chance to achieve this because they only focus on two issues: 'Is the project running according to budget?' and 'Is it on schedule?', rather than first asking the key question, 'Are we meeting the customer's needs?'. Once this has been addressed, the other questions about schedule and budget can then be asked.

4.4 THE COST OF POOR QUALITY

It is sometimes claimed that there is no real way of measuring quality. This is not true however. It can be measured by calculating the total costs incurred by a company not doing the job right first time. This measure is called the cost of poor quality, or the price of non-conformance, and includes the following types of cost:-

- Prevention
- Appraisal
- Internal failure
- External failure

Prevention The cost of prevention is the amount spent to ensure that the work will be done correctly. It includes risk reduction and error prevention, for example ensuring the design is right before beginning production. Examples of prevention costs are supplier evaluations and the training and development of staff.

Appraisal Appraisal costs are those associated with inspection and testing of both the company's own products and products received from suppliers. On software projects, appraisal costs might be associated with testing, walkthroughs, Fagan inspections and design reviews. The idea of appraisal is to spot the defects as early as possible in the process so they can be fixed, saving additional costs later. Figure 4.1 shows a comparison

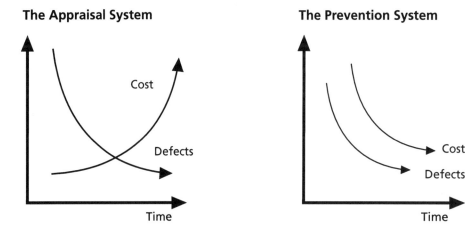

Fig. 4.1 The costs of appraisal and prevention

between appraisal and prevention costs. As the cost of appraisal within a company rises, because of more resources being devoted to inspecting and testing, the number of product defects is reduced significantly. For example, the more design reviews carried out during system development, which involves an increased investment in appraisal, the greater the chance of spotting and correcting defects.

With prevention, on the other hand, the costs fall as the number of defects fall. If, for example, everyone on a software project is trained in the use of a structured development method, the only additional training required would be for new people joining the project and so the cost of training – the prevention cost – would fall.

Internal failure The cost of internal failure is the cost of rectifying everything that is discovered to be wrong while the product or service is still in the company's possession or under its control. An example would be part of a software system having to be scrapped, and then designed and coded again, as a result of a major problem being unearthed during integration testing. In other words, thinking back to the discussion about customers earlier in this chapter, internal failure costs are incurred in putting right problems with the product or service before it is delivered to the external customer because the requirements of the internal customers have not been met.

Some companies spend a lot of time inspecting their products and fixing defects, without attempting to prevent the problem which originally caused the defects. This has been described as scraping burnt toast, based on the premise that if your electric toaster is burning the toast, you can either fix the problem, the toaster, or deal with the symptoms of the problem – scrape the burnt parts off the toast. This phenomenon is illustrated in figure 4.2, where an automated system has been developed to enable the symptoms of the problem to be put right, while the real problem remains unresolved.

External failure External failure costs are incurred by a company because defects are not detected before the product or service is delivered to the external customer. Examples of these are the cost of fixing software or hardware problems during the warranty period and the cost of handling customer complaints. It also includes the cost to the business of

Fig. 4.2 'Scraping burnt toast'
(Reprinted with permission from *The Team Handbook* by Peter Scholtes)

an external customer cancelling or withholding repeat business as a result of poor service.

If quality costs within an organisation are examined, internal and external failures will typically account for a third to a half of the total cost of poor quality. A similar proportion of the cost of poor quality will relate to appraisal and only a small percentage to prevention.

This is shown in the figure 4.3 which represents the effect on a company's quality costs, expressed as a percentage of total operating cost, of implementing a quality improvement program over a 5 to 7 year period. The total cost of quality which initially is 25-40% of total cost drops to a figure between 5 and 15%. Of the final figure over 50% is spent on prevention measures rather than on appraisal or on correcting failures. The margin between the initial and final cost of quality is measured in savings to the company.

In addition to the direct, measurable costs, there are the indirect costs of poor quality

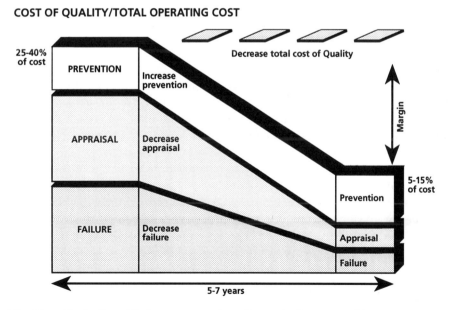

Fig. 4.3 The effect of quality improvement on the cost of poor quality

which are very difficult to measure. The following story illustrates this point: A manufacturer produces bicycles for children. Stabilising wheels are provided as an optional extra and are held on by the use of nuts and bolts which are an integral part of the bicycle. The stabilisers are produced separately to the bicycles. The manufacturer is made aware of a problem when bicycles start coming back in under guarantee because the stabilisers break off due to the weakness of the bolt.

This is an example of external failure. The direct cost of this quality failure to the manufacturer is incurred in supplying new, stronger bolts and distributing these to retailers and to customers who have complained. However, consider the child who is given the bicycle as a present on Christmas Day, and because of the broken bolt, can't ride it. The indirect cost of poor quality will include the petrol used in taking the stabilising wheels and the broken bolt back to the shop where the bicycle was purchased, the time wasted by the child's parents in sorting out the problem, not to mention the fact that their Christmas has been spoilt because the child was so unhappy about the broken bike. Then there is the loss of repeat business from the customer and the loss of potential future business when the parents tell their friends and colleagues about the problem. On average it costs five times more to win a new customer than to keep an existing one, and according to one survey, if you buy a product and are pleased with the purchase, you tell, on average, eight other people about it and how happy you are with it. But, if you are unhappy with the product you have bought you will make your dissatisfaction known to 22 other people. Add all this together and you begin to get an idea of how great the indirect costs of quality failure are – and much of this will be invisible to the manufacturer.

So how can a software developer ensure that quality is taken seriously so that better systems can be built? One answer to this question is to measure the cost of poor quality in their organisation, by developing metrics which give them numbers which indicate where problems exist, and which enable improvement to be monitored. Another answer is to introduce a quality management system in which measurement may play a part, and this is the subject of the next section.

4.5 QUALITY MANAGEMENT

To ensure that quality is maintained when developing software so that products and services are delivered which conform to the customer's requirement, three concepts are important – quality control, quality assurance and quality management.

Quality Control is the task of ensuring that a product has been developed correctly – to requirements and to standard – and that the procedure identified for its development is effective and has been followed. Quality control is done best by the person or team who did the work, but it should include an independent contribution from a peer – someone who could have done the work, but who didn't. This might be provided by an equivalent member of a different team. For instance, a completed system design document should be reviewed not just by the person who has written it, but also by an independent person. It is surprising how many previously unseen errors and problems can come to light when the author 'walks through' a document with someone else, or examines a product against its requirements and relevant standards. Quality control also covers the procedures and methods used for the work. These must be identified beforehand, even if they are the usual ones, in a *quality plan*, and any deviation from them must be explained and assessed. It is important to maintain records to show that quality control has been carried out, and to indicate on the product or, if this is not possible, on an associated record, that it has been reviewed successfully. Quality control is the responsibility of everyone in the organisation.

Quality Assurance is the responsibility of a smaller group of people. Someone independent of the work area or project checks that quality control has been performed, that it has been effective, and that the products are complete and suitable for delivery or for further use by someone else within the project. A formal audit of a software project is an example of quality assurance in action. The principal aim of quality assurance is to achieve confidence that the job or product will be acceptable to the external customer or to those involved in the next stage of development – the internal customers. Usually, this is done by the supplier – preferably by someone in an independent quality assurance role – and the evidence recorded and made available to the customer. In effect, quality assurance is a check on quality practice in terms of the performance and effectiveness of the quality plan.

Increasingly, there is recognition that in addition to quality control and quality assurance, a further level of monitoring is necessary which can be described as *Quality Management*. This describes the establishment and maintenance of a quality system within the organisation, the company, the division or the project and is usually the responsibility of senior people in that area. The hierarchical relationship between quality control, quality assurance and quality management is illustrated in figure 4.4, the quality pyramid.

The foundation of an organisation's quality management system (QMS) is a statement of its objectives and policy for quality, which should of course correspond to the type and scope of product or service being offered. There must be a description of the responsibilities and the internal organisation for the QMS, to ensure that quality control and quality assurance practices are understood and are operated effectively. A major reason for doing this is to allow an external customer to assess the supplier's attitude and approach to quality, both before work is placed with the supplier, and throughout the

Fig. 4.4 The quality pyramid

progress of the work. The QMS is a company's framework, within which all work is performed, using only procedures and methods which are defined, checked and visible.

Standards A QMS will specify the standards to be used for tasks carried out within the organisation. Standards are needed to help people to do the job right first time, as a means of communication, making it easier for teams to work together and ensuring that products developed will be compatible and contribute to producing consistent maintainable systems.

As an encapsulation of expert knowledge, standards can help newcomers and save time by preventing the need for different individuals and teams to have to 're-invent the wheel'. They play an important part in quality control, enabling individuals to check that they have done what they should have done and usually cover technical, administrative and managerial procedures within a company. Although standards are supposed to make life easier, their effectiveness is reduced if they are seen as inappropriate, out of date, not helpful, ambiguous, long winded, too detailed or too vague. In order to deliver the benefits described above and be seen as important and useful on projects, they must be living documents. In other words, they must be owned by somebody. The owner must be willing to listen to feedback, both positive and negative, about the standard and be prepared to make changes – or recommend that a new standard is created to fill the gap. To enable this to happen, a QMS must include:

- Procedures for the regular – at least annual – review of every standard
- Mechanisms to allow users of the document to direct comments about the standard to the owner
- A catalogue listing all the standards in the system and a central keeper of the latest version of each standard.

It is also the responsibility of each individual in the organisation to make the standards system work by giving feedback to the owner of the standard and by ensuring at the start of a piece of work that the standard being used is the latest version.

A lot of time and effort is usually expended in introducing a Quality Management System to an organisation. Is all the work justified? In a report published in 1991, which contains the findings of an NCC survey of IT companies who had introduced a formal QMS, the benefits to the organisations concerned were summarised by responses such as these:

- "It helps to identify the company."
- "It helps to ensure repeat business."
- "It makes our practices coherent" (gives a 'house style').
- "It helps new joiners settle in more easily'.
- "You realise that all processes are QA" – everything contributes to Quality.
- "Now we know what we do and how we do it."
- "It has brought greater confidence in our ability to deliver."
- "Our products now have the stamp of quality."
- "It has brought us closer to the customer."

The survey also pointed to the fact that businesses became more professional in their approach, that a sense of ownership was created within the organisations and that a quality culture attracted quality people to work for the company.

A *quality manager* will usually have a day-to-day responsibility for the QMS, although this may not need to be a full-time role once the quality system is established and running effectively. The QMS is documented in a *quality manual*, which also includes or refers to descriptions of the methods and procedures used on work tasks. This manual also becomes a valuable marketing and selling aid, as it provides evidence to the outside world of the means by which a supplier achieves quality of work and product, and is part of the basis on which the quality system can be assessed.

The current approach in forward-looking companies is for quality managers to assist, advise and support quality initiatives but not to take direct responsibility for them. This allows for the development of TQM (*Total Quality Management*). TQM can be defined as 'implementing a cost effective system for integrating the continuous improvement efforts of people at all levels in an organisation to deliver products and services which ensure customer satisfaction'. TQM requires a stable and defined QMS and a company-wide commitment to continuous improvement, which involves everyone in the organisation working together towards a common goal. In addition, to be effective, TQM needs two other important components to work together:

- Tools and Techniques: To enable every employee to be involved in continuous improvement, common tools and techniques must be adopted throughout the organisation, and individuals and groups should all be trained in the use of these tools.
- Human Factors: It is vital that everyone is motivated to take part in the process and for their contribution to be recognised. The culture of the organisation must encourage co-operation and team work. Leadership styles must be appropriate to enable this to happen.

TQM represents a fundamental shift from what has gone before. Quality control and quality assurance remain important, but the focus is on a process of habitual improvement, where control is embedded within and is driven by the culture of the organisation. Senior management's role is to provide leadership and support, the main drive for improvement coming from those people engaged in product and service delivery. The task of implementing TQM can be so difficult that many organisations never get started. This has been called TQP – total quality paralysis! Understanding and commitment are vital first steps which form the foundation of the whole TQM structure and these must be converted into plans and actions. Implementation begins with the drawing up of a quality policy statement, and the establishment of the appropriate organisational structure, both for managing and encouraging involvement in quality. Collecting information about the operation of the business, including the costs of poor quality, helps to identify those areas in which improvements will have the greatest impact, after which the planning stage begins. Once the plans have been put into place, the need for continued education, training, and communication becomes paramount.

In summary

Quality Control (QC) involves ensuring that a task has been done correctly.

Quality Assurance (QA) is the process of checking that QC has been carried out satisfactorily so that products are complete and suitable for delivery to the customer.

A *Quality Management System* (QMS) is concerned with implementing the quality policy of an organisation. It includes the management of QC and QA, the keeping of records, the maintenance of standards and the identification of individuals responsible for the various tasks.

The *Quality Manual* documents the Quality Management System adopted within an organisation.

Total Quality Management (TQM) goes beyond the QMS and refers to a scenario in which quality lies at the centre of an organisation's business, permeates every area of activity and involves everyone in the company in a process of continuous improvement.

Clients are becoming more assertive in demanding that software suppliers can demonstrate that quality underpins the development of computer systems. A formal way of demonstrating that a working quality system is in place is certification to the international standard for quality management, ISO 9000 – also known as BS 5750 or EN 29000 – and clients are now making certification to this standard a prerequisite for suppliers wishing to bid for their work. Because of the importance of ISO 9000 to software developers, the next section of this chapter summarises the requirements of the standard.

4.6 ISO 9000

The British standard for quality management, *BS 5750*, defines a minimum set of activities that manufacturers should engage in to give the customer confidence in the quality of their products. BS 5750 concentrates on the *organisation* and the *process* rather than on the actual goods produced and lays down the requirements for cost effective quality systems, concentrating on how they should be established, documented and maintained. The British standard has since been adopted by the International Standards Organisation as the international standard for quality management systems, *ISO 9000*, and has also been adopted within the European Community as *EN 29000*. Despite having different covers, the content of these three standards is identical.

The International standard ISO 9000 consists of the following parts :

ISO 9000   Quality management and quality assurance standards – guidelines for selection and use.

ISO 9001   Quality systems: model for quality assurance in design/development, production, installation and servicing.

ISO 9002   Quality systems: model for quality assurance in production and installation.

ISO 9003   Quality systems: model for quality assurance in final inspection and test.

ISO 9004   Quality management and quality systems elements – guidelines.

As the names of the documents suggest, ISO 9000 provides guidance on the tailoring of the standard to specific situations and on the selection of the appropriate quality assurance model – ISO 9001, ISO 9002 or ISO 9003. The choice of which of the three ISO quality management standards is appropriate will depend on which of the main stages in the production or manufacturing process – design and development, production, installation and servicing – are part of an individual company's business. If a company is involved in all four stages then ISO 9001 is appropriate; if it is only involved in production and installation, the appropriate standard is ISO 9002 ; while if the company is not involved in production, but only in installing the product, ISO 9003 is appropriate. Part 4 of the standard provides guidance on how to develop and implement a quality management system (QMS).

The principal concepts of ISO 9000 can be summarised as follows:

- *The organisation should provide confidence to the purchaser that the intended quality is being, or will be, achieved in the delivered product or service provided.*
- *The organisation should provide confidence to its own management that the intended quality is being achieved and sustained.*
- *The organisation should achieve and sustain the quality of the product or service produced so as to continually meet the purchaser's stated or implied needs.*

Applying ISO 9001 to software development A software company developing information systems for clients will be involved in all four stages of the production process – design and development, production, installation and servicing – and therefore the standard ISO 9001 is appropriate. ISO 9001 requires a company to do three things with regard to their Quality system:

- Say clearly what they will do
- Do what they say they will do
- Prove it by keeping careful records.

However, because of the use of manufacturing terminology in the standard, its applicability to software development has been overlooked. But there are four main reasons why a systems development company would adopt the standard:

1. Customers expect it.
2. It is a condition of tendering for some business.
3. The work involved in becoming registered and the discipline of following the standard provide a good platform for general improvement.
4. Competitors are, or are becoming, certificated to the standard.

To help with its interpretation, the International Standards Organisation has published a guide to explain how ISO 9001 can be applied in a software environment. This guide, *ISO 9000-3: Quality management and quality assurance standards – Part 3 : guidelines for the application of ISO 9001 to the development, supply and maintenance of software*, describes the different aspects of the quality management system under the headings *framework* dealing with management responsibility and the quality system, *life cycle activities* covering all stages of software development from contract review to maintenance activities and *supporting activities* describing activities important at every stage of software production.

The TickIT initiative In Britain, the Department of Trade and Industry (DTI) has sponsored a QMS certification initiative, called *TickIT*, to achieve improvements in the quality of software products and information systems. TickIT provides a scheme that meets the special needs of the software industry, enjoys the confidence of professional staff, and commands respect from purchasers and suppliers. The target is to achieve widespread adoption of the minimum best practice requirements set out in the European harmonised standard EN 29001, which is identical to the ISO 9001 standard and British Standard 5750 part 1. As part of the initiative a guide has been prepared by a group of leading quality management and IT professionals co-ordinated by the British Computer Society under contract to the DTI. The contents of the TickIT guide include an overview of ISO 9001, a copy of the document ISO 9000-3, a purchaser's guide, a supplier's guide and an auditor's guide.

4.7 QUALITY IN THE STRUCTURED LIFE CYCLE

In the traditional approach to developing an information system, there was little or no quality checking at each stage of the development process. There was ample testing of programs, interfaces, sub-systems and finally the complete system which ensured that when the system went 'live' it worked. However, the quality system did not ensure that the working system satisfied the requirements of the customer who had asked for it.

If testing is properly designed and planned, it can be very effective at locating defects. However it can never be completely comprehensive. Checking every path through even a simple program can take a large amount of time. It is particularly difficult to trap defects introduced in the earliest stages of analysis and design, or facilities that have been 'lost' between one stage of development and the next. Such defects are often the ones that cause the most concern. Another drawback of relying on testing is that it can only be done in the later stages of development as it actually exercises code. Various studies show that at the testing phase the cost of correction may be sixty times greater than a correction made before the coding begins, and that the correction of a defect in an operational system can be a hundred times greater. As a result, the cost of defect correction for large products can be over half of the total development cost.

With the introduction of structured approaches to system development, each stage of the project became a 'milestone', the deliverables of which had to be signed off by the developer and the customer before the next stage began. This ensures that not only does the system work when it finally goes live, but also that the client is in full agreement with the interpretation of the requirements - from the earliest stage to final implementation. A procedure for agreeing and signing off milestones, which is part of many structured methods, is the *structured walkthrough*, while a more formal technique which is also used on software development projects is the *Fagan inspection*. In this section we will describe these two review techniques.

4.7.1 Structured walkthroughs

A structured walkthrough is the review of products at the end of a stage in the development of a system by a group of relevant and competent persons. The prime objective of the walkthrough is to identify problems and initiate the necessary corrective action. There are two types of walkthrough – formal and informal. A formal walkthrough is a full review of all the work done in one stage of structured development, and involves the client. An informal walkthrough, on the other hand, is internal to the development project and reviews each step of the development within a stage. User involvement is optional in an informal walkthrough. In any walkthrough there are at least three people, but the recommended maximum is seven. The roles played by the attendees of the walkthrough are as follows:

Presenter is the person who has done the work and is now submitting the relevant documentation for quality assurance.

Chairperson is the person responsible for circulating the documentation prior to the meeting, choosing the time and the location, and chairing the meeting. The chairperson must also be familiar with the appropriate standards.

Secretary is the person responsible for documenting the problems raised and then reading them back at the end of the meeting so that priorities can be assigned and follow-up action agreed.

Reviewer is a person from the same project as the presenter.

In formal walkthroughs there will be extra reviewers:

A *user representative* mandatory for the formal agreement and signing-off of a development stage.

An *extra observer* from the same project.

An *unbiased observer* from another project, an optional role which can be useful in providing objectivity.

A walkthrough is divided into three stages: preparation, the walkthrough meeting itself and the follow-up. Preparation takes place at least three days before the walkthrough and involves the *presenter* preparing the documentation on the product to be reviewed and passing this to the *chairperson* who then distributes the documentation and notifies the reviewers of the time and location of the walkthrough meeting. The Walkthrough Meeting should be kept short – between 60 and 90 minutes – and is a meeting in which the *presenter* walks the *reviewers* through the product. The prime objective of this session is to ensure that a product meets the requirement and conforms to the appropriate standards and that any defects are identified. An additional benefit of a walkthrough is the spread of information, knowledge, ideas and new approaches. A number of follow-up actions are available to the walkthrough team:

- Accept and sign-off the product;
- Recommend minor revisions with no need for a further review.
- Recommend major revisions and schedule another walkthrough to review the revised product. In this case the person creating the product will have a written record of the identified problems, produced by the secretary, and the actions required. The necessary corrections are then made for resubmission in the next walkthrough.

Problems associated with walkthroughs The problems with walkthroughs which often arise are: inadequate preparation by the reviewers; too much time spent discussing solutions rather than identifying defects; the author being defensive about his work; the walkthrough rambling on for too long. It is the chairperson's responsibility to avoid these problems by enforcing a time limit, ensuring walkthrough standards are adhered to and reminding attendees, before the meeting, of the purpose of the walkthrough and the importance of preparation.

4.7.2 Fagan inspections

A *Fagan inspection* is a formal review technique developed by Michael Fagan, a British engineer who worked for IBM. It was based on established review methods such as the structured walkthrough, but was designed to eliminate the problems associated with walkthroughs described above. An Inspection can be defined as a formal examination of an item, against a previously produced item, by a group of people led by an independent chairperson, with the objectives of finding and recording defects, using standardised checklists and techniques; initiating rework as necessary; monitoring the rework; accepting the work, based on stated exit criteria; and adding to and utilising a base of historical defect data.

The technique has been continually refined in the light of experience and is applicable to all types of documents, including functional specifications, program designs, code listings and test output. The inspection process involves checking a completed document for conformance to documents at a higher level – parent – and at the same level – sibling – and to relevant standards. Each non-conformance – defect – identified is recorded and used in the rework of the document, and statistics are collected to monitor the efficiency of both the software development and the inspection process. This process is illustrated in figure 4.5.

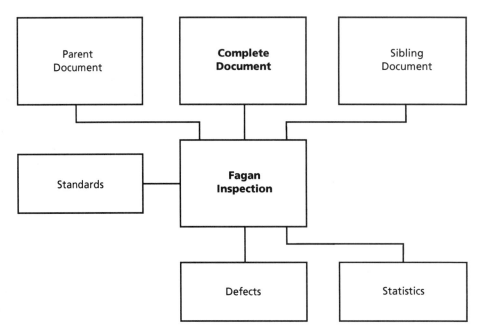

Fig. 4.5 The inspection process

The objectives of Fagan inspections are to identify and correct as many defects as possible early in the development process, so that the next stage can proceed with confidence, and to minimise the number of defects in the final system so that maintenance costs are reduced. The inspection process consists of a number of fixed stages: *Planning* during which the inspection team is appointed and any administration is performed; *Overview*, the purpose of which is to ensure that those inspecting the document understand how it fits into the system as a whole; *Preparation* during which each member of the team takes time to become familiar with the item to be inspected and all related items; *The Inspection Meeting*, which is the most visible part of a Fagan inspection, at which the document is formally examined; *Rework*, the task of correcting all defects found, performed by the author of the inspected item; *Follow up*, checking that the rework has been performed adequately.

The optimum number of attendees at the inspection meeting is between three and six. A number of roles have been defined for attendees, and these are summarised in figure 4.6.

At the meeting the document is paraphrased aloud by the reader in segments agreed with the moderator. To ensure that everybody can hear the reader, the reader sits at the furthest point from the moderator, and if the moderator can hear distinctly, it is assumed that everyone else can. A suggested seating plan is shown in figure 4.7. After each segment has been paraphrased, the moderator asks each inspector in turn for their comments on that part of the document. Comments are expected to be constructive criticisms about aspects of the document. All comments made are termed 'defects' and are recorded by the scribe. Any discussion of comments is restricted to ensuring that the nature of the defect is clearly understood. The meeting is intended only to bring defects to light, not to discuss solutions to any problems raised, nor to criticise the author. At the end of the meeting, the moderator appoints one of the inspectors as reviewer of the rework, and makes a preliminary decision on whether or not the work needs to be reinspected. After a maximum of two hours, the meeting ends. After the meeting, during the so-called 'third hour', possible solutions to issues raised during the inspection meeting may be discussed informally by the attendees.

The moderator ensures, after the meeting, that the rework is completed, reviews the decision about whether another Fagan inspection of the reworked document is required and ensures that all relevant statistics have been recorded.

4.8 SUMMARY

In trying to give you an idea of the scope and diversity of quality issues, we have covered a lot of ground, from quality concepts to methods and techniques for ensuring quality in systems development.

We began by introducing our working definition of quality, 'conforming to the customer's requirements', and the key ideas of product and service requirements, and internal and external customers. The ideas of the quality gurus, Deming, Juran and Crosby (and others) have caused people to question the way quality is managed in organisations, and have led to a move from inspecting deliverables for defects to building quality into the process of production. We have discussed the implications of these ideas for systems

THE MODERATOR

The Moderator is specifically trained for this role and, to provide objectivity, should be drafted from outside the Author's project team.

The Moderator will be familiar with the type of document being inspected and should:

- check that the entry criteria have been met;
- appoint Inspectors, detailing special roles;
- liaise with the Project Manager to set up the Inspection;
- send out details of the meeting, the document to be inspected and other relevant material;
- plan the Inspection process;
- chair the meeting;
- record the effort spent on preparation, inspection, rework and review;
- ensure that the Author understands the rework required;
- follow up the rework;
- decide whether reinspection is necessary;
- collate and distribute statistics.

THE SCRIBE

The Scribe records all defects found. Each defect is normally given a unique number and its location marked on a clean copy of the document.

THE READER

The Reader is responsible for guiding the inspection team through the material during the meeting by paraphrasing the content of the inspected item in portions agreed with the Moderator. The Author may not be the Reader as the object of reading the item is to ensure that the contents are not open to misinterpretation. The Reader should:

- liaise with the Moderator to establish a reading pattern (and NOT ask the Author for clarification);
- prepare the document for paraphrasing;
- read the document as planned (focus and pace are vital, in doing this)
- contribute to the defect detection process.

THE AUTHOR

The Author attends the meeting as a reference source and as a validator. The Author should:

- liaise with the Moderator only;
- supply the Moderator with documents;
- act as a reference source or validator in the meeting, and NOT to volunteer information'
- take copious notes during the meeting so that the reason and source of all logged defects is known at rework time;
- understand the defects found and their knockon effect;
- provide estimates for rework within 24 hours;
- rework the inspected item, assigning a category and severity to each defect.

THE INSPECTOR

Inspectors prepare for the meeting by reading the document, checking for understanding of the document contents and for any specific points requested by the Moderator (e.g. checking against a specific higher level document, looking at a specific interface, checking against project standards). At the meeting, each Inspector should:

- contribute to the defect detection process;
- avoid introducing any distraction to the meeting (e.g. alternative methods of achieving the same result).

Fig. 4.6 Roles in a Fagan Inspection meeting

Fig. 4.7 Seating plan for a Fagan Inspection meeting

developers, and for business in general, describing the difference between the control, assurance and management of quality, and explaining the meaning of the acronyms QC, QA, QMS and TQM. The content of the international standard for quality management, ISO 9000, which has become increasingly important to our industry in recent years, has been described, and we have discussed how it can be applied to software development.

The last section of the chapter has introduced techniques for building quality into systems analysis and design – the structured walkthrough, which is an integral part of a structured approach to systems development, and the more formal Fagan inspection.

We would like to conclude by bringing together a number of ideas presented in this chapter in a model to help you, either as part of an analysis or design team or as an individual, to build quality into your work. This model is represented graphically as figure 4.8.

The job of an individual or a team can be thought of as a series of tasks, each with its own input and output. For example a task could be carrying out a fact-finding interview, presenting system proposals to a client, developing a data model of the system or writing a report. The *outputs* of the task are the products or services supplied to the customer. This could be an internal customer or an external customer. The *inputs* describe everything that is needed to produce the outputs, and the *task* is the process which transforms the inputs. Once the task is completed, an *evaluation* must be carried out to review how well the task was achieved and to provide information for improvements in the future.

Inputs Often the way in which quality is assessed on projects is by looking at the deliverables – the products or services supplied to the customer. However while it might be valid to inspect these deliverables, find any defects and then trace them back to where

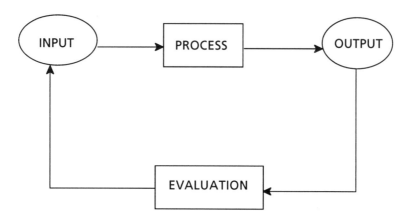

Fig. 4.8 A model for building quality into your work

they occurred, building quality into a task means going even further back in the process and examining the inputs before beginning the work.

There are four inputs to any task :

- a statement of the customer's requirement
- a method for completing the task
- the resources available
- the individual's ability, attitude and working standards.

To build quality into the task it is important to ensure that all of the inputs to the task have been reviewed and are appropriate.

To be clear about exactly what the customer – external or internal – wants, you need a statement of the customer's requirement. This should include an explanation of why the product or service is required, and the detailed requirements as specified by the customer – including their 'soft service' requirements, which will always include the need to feel valued and to be treated with courtesy and respect. The list of requirements should be agreed formally or informally with the customer who should be notified if there are problems in meeting their requirements, for example if you can't complete a report by the agreed time.

It's important to establish early on whether there is an approved method for carrying out the task, and if standards or guidelines exist. As part of the quality improvement process, questions should be asked – is this the best method, the most efficient, most cost-effective, fastest, safest, or could it be improved ?

Before beginning the task, you need to check what resources, materials and people, are available. Equipment should be adequate, appropriate, safe to use, in good working condition and properly maintained. It is not easy to do a task properly if the correct equipment is not provided. Any materials needed to do the task must be correct for the task and of the appropriate quality. Last, but not least, you will need to check who is available to assist with the task, to support, to guide, to provide information or supply specialist skills.

The ability, attitude and working standards of the person or people doing the job are the most important input to consider when building quality into a task. People must have the ability to do the job properly and receive the appropriate training. A positive attitude makes the job more satisfying, and decreases the incidence of human error, while the standard of performance the individuals sets for themselves will be critical to the overall quality of the work done.

The combination of an individual's attitude, ability and working standard can also be described as *personal quality*. Personal quality depends on a number of factors such as self-perception, state of mind, motivation and awareness of the customer's requirements. The scope for improving personal quality can be determined by comparing your actual performance – what you achieve here and now – with your ideal performance – your expectations, requirements and goals. Personal quality is important for the team as well as the individual. The sum of everyone's scope for improvement is a measure of the overall scope for development of a project team, a business group or a department. It can be argued that personal quality is at the root of all other quality within an organisation.

Outputs In addition to the product or hard service that you hand over to the internal or external customer, it is worth emphasising here that the output also includes the soft service. In a meeting with the client, for example, this means making them feel valued, treating them with courtesy, giving them your full attention and not showing any signs of negative opinions you may have formed.

Evaluation Assuming that we have built quality into the task by getting the inputs right, and ensuring that output meets the customer's requirements, a further question needs to be asked: 'How well did we do?' For example, was the system delivered on time and within budget? Was the information provided during fact-finding interviews accurate and complete? Did we treat the customer with courtesy and helpfulness at every stage of the development? Did we, individually and as a team, live up to our own working standards? This process, called evaluation, is often left out on projects but it is important that it is done in order to 'close the loop' and build quality into future tasks. If the answer to any of the questions asked above is 'no', the failure must be analysed (what went wrong? why did it go wrong? which inputs need to be improved?). It is only by the evaluation of success or failure that improvement can be a continuous process.

CHAPTER 4 CASE STUDY AND EXERCISES

Q1 System Telecom has a policy of developing and using computer-based systems to give it commercial advantage. One of the consequences is that new systems need to be 'developed on time, within the budgeted cost, and to quality.' What does 'to quality' mean in this context?

Q2 It is said that the use of structured methods improves the quality of the developed system. Bearing in mind your understanding of 'quality', explain how structured methods contribute towards achieving it.

Q3 Would it be useful for System Telecom to apply for ISO 9000 certification? What are the internal and external benefits that might come from certification, and what would be the cost of a successful application for certification?

Q4 System Telecom has decided to buy an application package for its personnel application. There are several available to choose from and no particular advantage to be gained from developing a new system in house. The most appropriate package appears to be a relatively new one with, as yet, few users. It is therefore particularly important to make a rigorous assessment of the quality of the package. You are assigned to assess the techniques used by the supplier to ensure software quality. What techniques would you expect to find being used?

CHAPTER 5

Understanding the Business

5.1 INTRODUCTION

Information systems are expensive to develop and maintain, so it is clear that businesses do not commission them just for the fun of it. The need for an information system must grow out of some perceived business requirement and the justification for it must be expressed in business terms.

Although this is well enough understood in theory, it is surprising how often IS projects do start without clear links back to business plans and strategies. System developers often take the sketchiest of briefs and start to develop something they think meets the need – and are then unpleasantly surprised when the user or sponsor of the system refuses to accept it because it does not properly meet their specific business requirements.

So, before starting to develop a new IS, before even beginning the analysis for that system, the developers must investigate, document and agree with the users the nature of the business need and the contribution that the IS is to make.

5.2 BUSINESS ANALYSIS

5.2.1 Levels of understanding

A proper analysis of system requirements requires that those carrying out the investigation have a balance of skills and knowledge. We saw in chapter 1 that this includes a good general understanding of the operation of business and of the factors which affect the viability of all businesses. This involves a broad-based knowledge of finance, accounting, marketing, production and distribution, ideally gained in a variety of business environments. There will also be a need for a more specific grasp of the important features of the particular business being studied. So, for example, an analyst working for a high street chain of shops should have knowledge of the retail sector and be able to discuss retailing matters knowledgeably with the customer. Finally of course a broad knowledge of the possibilities and limitations of information technology is also required.

This balance of skills is needed for three reasons. Firstly so that the analysts understand what the users of the proposed system are saying to them and can recognise the implications of the requirements they are capturing. Also because users will not expect analysts to adopt a completely passive role they will want them to challenge assumptions and interject ideas to stimulate them in thinking out their requirements. Without an understanding of the business of the organisation or how other organisations have tackled similar issues, it will not be possible to play this catalytic role. Finally the users will generally expect the analysts to devise and propose technical solutions to their business problems.

Systems analysts, however, are often generalists and, particularly in service companies, are quite likely to be sent to work with a travel agent on one project, a bank on the next and a manufacturing company on the third. Whilst this can facilitate cross-fertilisation of ideas between different sorts of business, the disadvantage is, of course, that they sometimes lack the specialist knowledge to deal with users on an equal basis. So if this is the case, what can be done to bring relevant experience to bear on a particular assignment?

Firstly, and most obviously, the project manager should try to find analysts with previous experience of the business to be studied, or something similar. Someone who has worked on a distribution application, for instance, may have achieved a reasonable understanding of areas such as stock-keeping and just-in-time delivery systems which can be brought into play on a retail assignment. If, however, analysts with relevant backgrounds are not available, additional support will have to be provided by arranging some training for the analysts in the areas of interest in the form of public courses, self-teach packages or by hiring an expert to coach them in the new business areas. Alternatively, suitable background reading can be provided. Another approach is to provide consultancy support, which can be used either to lead the investigation with the customer or provide background advice and guidance for the analysis team.

A particular challenge for analysts, and one which seems more difficult for those from a technical background such as programming, is to keep the business requirements as the focus instead of getting hooked up on technological solutions. It is very easy, for example, to take a messy and cumbersome manual system and produce instead a messy and cumbersome computer system. Analysis should at all times be tightly focused on the business objectives which the proposed system is supposed to fulfil and only when these are thoroughly clear should the identification of technical solutions be attempted. As we have seen in chapter 2, structured methods, particularly SSADM, provide a very good discipline in this regard, since they tend to conduct requirements analysis and requirements specification at a wholly logical level and deliberately exclude the consideration of technical issues until the business requirements have been settled.

5.2.2 Linkage of IS to business objectives

There are many reasons why businesses should want to develop information systems but some of the most common objectives are:

- *To reduce manpower costs*. The introduction of computer-based systems has often enabled work to be done by fewer staff or, more likely nowadays, permitted new tasks to be undertaken without increasing staffing levels. The automation of many banking functions, such as cheque clearance, falls into this category.
- *To improve customer service*. Computer systems can often allow organisations to serve customers more quickly or to provide them with additional services. Supermarket point-of-sale systems producing itemised bills provide an illustration of this.
- *To improve management information*. Management decisions can only be as good as the information on which they are based, so many computer systems have been designed to produce more, or more accurate, or more timely information. With modern database query facilities it is even possible to provide systems which do not require the data retrieval requirements to be defined in advance, thereby enabling managers to institute new types of enquiry when changing business conditions demand new or different information.

- *To secure or defend competitive advantage.* This is becoming a major justification for spending on information systems and is examined in more detail later in this chapter.

Ideally, the analyst should work from a hierarchy of objectives, each one posing a challenge to, and imposing constraints upon, those at a lower level. The lower-level objectives are sometimes referred to as *critical success factors*, that is they are things which must be achieved if the top-level objective is to be met. The critical success factors will become more detailed and tightly focused as one works down the hierarchy, and perhaps this may be best illustrated by an example.

Let us consider a motor-car manufacturing company. It currently has, say, 10% of the market for its products and the board has defined a five-year mission of raising that proportion to 20%. But this is a very broad target and, to achieve it, the organisation will have to define a set of more tangible objectives which will lead to its being met – in other words, the critical success factors. It may be felt that one key to increasing market share is to offer a more frequent choice of new models. So, a lower-level objective for the design team may be to reduce the time to develop a new model from, say, five years to two. And the production department will have to be able to switch over assembly lines in less time, say a reduction from six months to three.

Coming down a stage further still, the designers will want to introduce technology which can produce manufacturing instructions, documentation to support the issuing of tenders to subcontractors, component listings and setup instructions for the assembly lines from their drawings. Now, we can derive some very focused critical success factors for our information system based upon these detailed requirements. It can be seen, then, that business objectives and critical success factors 'cascade' from one level to another and the whole set should form a pyramid supporting the overall aims of the business.

5.3 CONSTRAINTS

The range of potential solutions which may be proposed by the analyst will be limited by constraints imposed by the user and by the nature of the user's business. The analyst should ideally undertand these constraints before analysis begins but certainly as it progresses, and must keep them firmly in mind as the ideas for the proposed system emerge.

5.3.1 The user's organisation

The first thing for the analyst to consider is the structure of the user's organisation. It may, for example, operate in a very centralised fashion with nearly all decisions being made at head office. If this is the case, then clearly information systems must reflect this pattern and be designed so that relevant information can be processed rapidly and presented at the heart of the business. If, on the other hand, the organisation allows considerable autonomy to managers in subsidiary parts of the business, then the systems must be designed to provide these managers with what they need to run the business effectively. In other words the systems must reflect the structure of decision making in the business.

It is important to remember however that organisations are not static and tend to oscillate between centralisation and devolution as circumstances, the business and intellectual climate and the most recent theories of management gurus dictate. Nowadays, the difficulty of altering information systems can prove a major obstacle to reorganisation Whilst some might think this to be a good thing, it is important not to make systems so inflexible that they actually prevent the organisation being operated in the way the management decides is necessary.

5.3.2 Working practices

It is easiest to introduce a new information system if it leaves existing working practices largely undisturbed. Conversely, systems that require a lot of change may prove very difficult to implement.

However, the analyst must not allow a fear of the difficulty of implementation to prevent the best solution – best for the business as a whole, that is – being advocated. The most radical changes will often provide the greatest gains and if that is so then the problems of implementation must be faced and overcome and in chapter 26 some ideas are given that help in this difficult process of managing change. Alternatively, the management of the organisation may want to use the introduction of new technology as a catalyst for change, to shake up a sleepy, backward-looking department for instance.

The key point is that the analyst must make some assessment of the climate prevailing in the organisation. Will its management wholeheartedly push for change, give it lukewarm support or just run away from its implications? Finally, at a more prosaic level, some working practices which may appear over-elaborate and cumbersome will turn out to have evolved for good business reasons – like the maintenance of safety standards on a railway for example. The analyst must make very sure that these business reasons are not ignored in proposing new and more streamlined systems.

5.3.3 Financial control procedures

Various aspects of the way an organisation manages its finances can impact on IT developments. The first is the concept of capital versus revenue expenditure. Most IT developments involve capital expenditure in that they are funded as one-off projects rather than out of continuing expenses. But the payoff for a capital project may be a reduced continuing revenue cost somewhere downstream. Depending on the organisation's rules for the 'payback' on capital projects, a capital project cost of, say, £3 million may not be justifiable even if, over a seven-year system life, it could produce revenue savings of 'only' £1 million per annum.

Also organisations may only have limited funds earmarked for capital projects in a given year but have reasonably generous revenue budgets for ongoing work. In these circumstances, it may be sensible to propose a limited initial capital expenditure for a core system, with enhancements and additions being made gradually as funds permit.

The analyst needs to consider who actually holds the purse strings for a particular development and what it is that will convince that person of the worth of the proposed development. Let us suppose, by way of example, that we are to examine the requirements for a new payroll system in an organisation. The paymaster could be the payroll manager, who wants a fully comprehensive system that will enable him to offer new services and perhaps even take over the functions of the corporate personnel system, or it could be the IT director, who is developing a strategy of packaged systems running on 'open' architectures, or the finance director, who wants a system which will cut the number of staff, and hence the costs, in the payroll department. The objectives of each of these managers is rather different and if the analyst is to get a solution adopted, it must be geared to the needs of the person, or people, who will approve and pay for it.

5.3.4 Security and privacy

The analyst needs to determine fairly early on which sort of security conditions will be required for the proposed information system. These could include:

- Ordinary commercial confidentiality, where the main aim is to ensure that sensitive commercial information like, for instance, the production cost breakdown of products cannot be stolen by the competition,.
- More sensitive systems, like the Police National Computer where special considerations apply to the holding of, and access to, data.
- Very secure systems such as those which support the armed services and government agencies.

Clearly, the need for rigorous security control could impose major constraints and development costs on the project.

5.3.5 Legal considerations

It is becoming increasingly the case that the users of information systems are liable for the consequences of things done, or put in train, by those systems. If there are such liabilities, the analyst must examine them and allow for them in the proposed system. Safety-critical systems are the most obvious example and if one were examining the requirements for, say, an air traffic control system, safety considerations would constitute one of the main constraints of the proposed solution. Other systems have also fallen foul of the law recently and it cannot be too long before, for example, a credit-scoring agency is found to be liable for the consequences of wrongly deciding that someone is a bad risk.

So far, too, there has been little legal exploration of the subject of consequential losses arising from the use of information systems, and developers have been able to hide behind contract clauses which limit their liability to the cost of the systems's development or some arbitrary figure. There must be some possibility that this will change in the future so the analyst, in assessing the risks from some proposed solution, ought at least to think about what might be the consequential losses resulting in a system failure.

The law does already have something to say on the subject of storing information about individuals, in the form of data protection legislation. This has two aspects which particularly concern us. These are ensuring that the data is held only for defined and declared purposes and enabling those with a statutory right to inspect information held about them to do so. So, if the proposed information system may hold information on individuals, the analyst needs to ensure that the requirements of the legislation can be met.

5.3.6 Audit requirements

An organisation's internal and external auditors will want access to systems to ensure that they are working properly and that the financial information they produce can be relied upon. They may also require that certain self-checking mechanisms and authorisation procedures be incorporated into systems. In some types of system – those supporting pension funds or banks are obvious examples – the need to check and extract audited information forms a large part of the requirement itself. It is very much better, not to say easier, if these audit requirements can be taken into account at the specification and design stage rather than added after the system is complete, so the analyst must talk to all the relevant authorities and find out their requirements alongside those of the more obvious users of the system.

5.3.7 Fallback and recovery

Most information systems have some sort of requirement for fallback and recovery. These requirements could include the ability to 'roll-back' the system to some point before failure and then to come forward progressively to bring the information up-to-date, or some back-up means of capturing data while the main system is off-line. Standby systems which normally perform less urgent tasks can take over from 'critical' systems and if necessary full system duplication or even triplication may be provided for critical real-time or command-and-control systems. Provision of back-up, though necessary, is expensive and so the case for the arrangements provided must be examined in strict business terms and the effects on the business of system failure assessed. 'What would be the costs involved' and 'how long could the business go without the system' are two of the more important questions to be answered.

Sometimes the analysis of these consequences can produce truly frightening results. In one case, there was an investigation of a system which supported a major undertaking and which had distributed data-capture and centralised control. It was found that, if the central processors went out of action for more than two days, the backlog of data in the distributed processors would be such that the system could never catch up, whereas the failure of one of the distributed machines would not become very serious for several weeks.

So, the analyst must carry out a comprehensive risk analysis in this area, using outside expertise in support if necessary, and must keep the resultant constraints in mind at all times.

5.4 IT FOR COMPETITIVE ADVANTAGE

In the early days of information systems, their justification seemed straightforward enough. For the most part, the systems were 'number-crunchers' which could carry out routine repetitive tasks, like the calculation of payrolls, much more quickly and cheaply than an army of clerks. The payoff was thus clearly in staff savings plus perhaps some additions in the form of better or more timely information for management.

There are few, if any, of these first time applications available in the 1990s and many administrative systems are now into their third or fourth incarnations. Justification for the new developments has generally been that:

- The old ones are incompatible with newer technological platforms resulting perhaps from a switch to 'open' architectures.
- They have become impossible to maintain because of the poor documentation or configuration management of many early systems.
- They need skills and resources to maintain them which are no longer available or are prohibitively expensive.

None of these however are business reasons in the sense that they support some key business objective; rather, they are technical reasons justified only in terms of the inherent nature of IT itself.

As expenditure on IT has risen, so managements have become increasingly keen to ensure that the money spent contributes in some tangible way to the achievement of business objectives. At the same time, some more enlightened boards, and some IT directors with a wider interest than in the technology itself, have become interested in the idea of 'IT for competitive advantage'. The concept is simple enough. If IT can give your company some unique offering, or contribute to providing some unique offering, then it will give you a competitive edge over your competitors and hence contribute directly to increased sales and profits. Two examples may serve to illustrate the idea.

Example 1
One of the continuing headaches for retailers is the level of inventory, or stock, they keep in their stores. If it is too high, excess funds are tied up in it, profit margins are depressed by it and it occupies floorspace which could be more usefully employed to display and sell goods. If inventory is too low, they risk 'stock outs' and customers cannot buy what is not on the shelves. In high-volume operations like supermarkets, if customers cannot buy what they want immediately, they will go elsewhere and that particular sale is gone for ever.

How can information systems help? A system is implemented which constantly monitors sales at the checkouts and signals a distribution centre when stocks of items fall below predetermined levels. Replacement stock is loaded onto trucks and, provided the operation has been set up correctly, arrives at the store just as the last item is sold. The stock is unloaded straight onto the display shelves and, without having excessive local storage space, the store is able to avoid customers going away empty-handed. Additionally, the cost savings can be passed on to customers as lower prices, thus attracting more business and improving the market share and profitability of the chain.

Example 2
In the late 1970s, many airlines were interested in seeing how they could tie travel agents and customers to their services and, by the same token, exclude their competitors. Several of them formed consortia with the idea of developing powerful booking systems which they could provide to travel agents. These would make it very much quicker and simpler for the agents to deal with the participating airlines than with their competitors. It is interesting to note that, in the major shake-up of the international airline business which has occurred in the 1980s and '90s, it is the airlines which have invested most in this technology which have moved to dominant positions in the marketplace.

In both of these examples, then, information systems have been used to gain competitive advantage over the competition. However, there are some other important points that need to be made here. Firstly, since information systems are now literally at the heart of many businesses (one has only to consider the two examples already quoted, or the clearing banks) it follows that these organisations are totally dependent on them. So huge ongoing investments are needed to keep the systems reliable and secure. Secondly, although the company which develops these systems first will enjoy initial advantages, it will also most likely suffer the setbacks and problems encountered by all pioneers. The competitors will be able to study these difficulties and avoid them. Thirdly, if one starts a competition with information systems, as with anything else, one can seldom land a knock-out blow with them. The competition will respond, then the original firm has to fight back and so on, and a spiral of increasing system sophistication and cost is created. Finally, whoever introduces their system first will also reach system obsolescence first and, while the competition is enjoying the benefits of any later technological gains, the pioneering firm will have to be investing large sums of money to catch up and get ahead again.

None of this should be taken to suggest that 'IT for competitive advantage' is an illusion. Far from it. The point is that one cannot invest once and then forget about new systems development for years. It is necessary to revisit the situation regularly, to watch the competition and to be constantly looking for ways to get ahead again.

This would seem to suggest, then, that if IT is to be used to gain competitive edge, the gains from any particular investment ought to be sufficiently significant to give the business breathing space to enjoy the advantage before the opposition catches up.

Finally, let us go back to the discussion of the skills and background of the analyst. We said there that, ideally, we need analysts with previous experience of a business but, if we are seeking to use IT for competitive advantage, this poses two problems. Analysts with long experience in a particular area of business will inevitably tend to become somewhat hidebound by the norms for that area and this will tend to militate against the emergence of the radical solutions which are generally needed if major competitive breakthroughs are to be achieved. Also companies will wish to prevent analysts who have worked on 'competitive edge' systems from moving to competitive organisations and taking their knowledge with them.

5.5 SUMMARY

If they are to be effective and provide value for money, information systems must only be developed if they can be linked in some way to the achievement of the organisation's business objectives. The need for a strong business-orientation for information systems means that systems analysts must possess a mixture of all-round business acumen, specific understanding of the area(s) to be studied and a broad-based technical knowledge. To ensure that the information systems are acceptable to the organisation, and are used to maximum effect, the analysts must take into account a variety of constraints: the structure of the business; its work practices; financial controls; security and privacy considerations; legal requirements; audit requirements; and the need for fallback and recovery arrangements. Increasingly, businesses are looking to information systems not just to reduce costs and administration but also to give them some competitive edge over other companies in the marketplace. This requires radical thinking and continuing effort and investment to gain and retain that edge.

CHAPTER 5 CASE STUDY AND EXERCISES

Q1 Bearing in mind

- what you know about System Telecom and the reasons for setting it up
- its style of operation and
- the information in this chapter about business

what do you think would be a reasonable set of business objectives for Systems Telecom for the next three years? What might be the critical success factors for these objectives?

Q2 Having identified some critical success factors, suggest ways in which new computer-based systems in System Telecom might help to achieve them.

CHAPTER 6

Project Management

6.1 INTRODUCTION

In this chapter, we look at project management for the analysis and design of information systems. Because of the increasing need for formality and structure, we have used examples from the UK Government's PRINCE (PRojects IN Controlled Environments) method. Its general approach is a codified form of the procedures that successful project managers have always used and it therefore has widespread applicability.

Analysis and design are part of a process of development which leads towards an operational information system: before analysis starts, there may have been strategic and feasibility studies; after design finishes will come the coding, testing and installation of computer programs.

Unfortunately, in systems development – as in many other fields – different people have adopted different names for the same things; or the same names for different things. The terms here are those widely used in the IT industry but if you are used to different terms, don't worry – the meaning will be clear from the context.

6.2 STAGES OF SYSTEM DEVELOPMENT

6.2.1 Before analysis and design

Ideally, the need for information systems will grow from the development of business strategy, as explained in chapter 5. This strategy may have been developed internally, by the business's own senior management, or may be the result of the work of management consultants. In either event, it forms the starting point for IS projects. Once the business knows where it is going, it is possible to sketch an IS strategy which will support it on its journey. Management consultants may have been involved here, too, or the business's IT director may have developed a strategy for the approval of the board. An IT strategy will typically cover the overall scope of information systems in the business. It may also include a general hardware policy, for example the organisation's preference for mainframe/minis, proprietary or open systems and so on, and some commitment to specific development methods and tools. New systems projects usually begin with a feasibility study. Its purpose is to examine the proposed development at a high level and to make preliminary business decisions on whether to commit funds to it.

In carrying out a feasibility study, the major analysis techniques will be employed but not, this time, to produce a detailed specification of the requirement. Instead, the analysts will be trying to define:

- The overall scope of the proposed system
- An idea of the system's probable size

- The general development approach to be used, for example bespoke versus packaged, in-house versus bought-in
- The costs and benefits which will flow from the development
- The impact which the new system will have on the business, particularly on areas outside of the system's scope
- The resources and timescales which will be required for the full development

6.2.2 Analysis and design

Analysis and design begins with requirements analysis. If there is already an existing computer system, then analysis can start there. In particular, the analysts will be looking for problems with the existing system or additional requirements which the replacement must have. Some effort must be made to categorise these into, for example, vital, useful and 'nice to have' so that decisions can later be made on which features to include in, or leave out of, the new system. A particular challenge for the analysts in this stage is to think about 'what' the existing systems are doing, rather than how they do it – in other words to look at the business needs rather than the technical implementation.

It is unusual nowadays to find a requirement for a computer system where there has been none previously, but it can happen. The need for systems to manage the late Community Charge is a good example. Here, the analysts must really use their business acumen to work out, with the system users, what will be required.

At some stage during requirements analysis, decisions must be taken on precisely what is to be included in the new system. Some of the requirements will turn out to be vital, others less so, and there will always be a trade-off between functionality and cost. It is crucial that these decisions are made by the system's users or by their senior management and that the implications of these decisions are carefully evaluated.

Once the requirements for the new system have been properly documented, it must be specified in sufficient detail to form a basis for development. This specification must include:

- a specification of business requirements, and
- a specification of a technical platform and development path.

In general, it is desirable to keep these two aspects as separate as possible, otherwise the business requirement gets confused with the technical means used to meet it. This can have adverse consequences later during the maintenance of the system or if it is desired to change it to some other technical environment.

Business needs must be documented in such a way that they can be reviewed by, commented on and accepted by the system's users. It has been generally found that the diagrammatic approach favoured in structured methods has definite advantages here over the more traditional narrative approach. The implications of the choice of technical platform must also be spelled out and brought to the attention of the users. Whatever means of specification is employed, it is very important that the users formally agree the requirement before system design begins. This is not to prevent future changes – pretending that there won't be changes in requirements is to defy reality – but so that the developers have a baseline against which to measure and control change. Finally, looking forward to delivery of the final system, the requirements specification should provide the criteria against which the users will ultimately test and accept the system. Ideally, these acceptance criteria should be in the requirements specification itself. Making sure of the acceptance criteria at this point will save a lot of time, trouble and argument for both developers and users later on in the project.

Armed with a comprehensive requirements specification, the developers can now set about designing a system which will meet the users' needs. The design will include:

- Database or files required to support the new system
- Update facilities, as online screens or batch programs, to be available to the users
- Range of reports and enquiries to be provided
- Detailed processes to be invoked when invalid or incompatible data are encountered
- Procedures for fallback and recovery after a system failure

Design is a more purely technical process than analysis and the users will be less involved in the detail. However, the users must see and approve such things as screen or report layouts – after all, they will have to work with them. Their consent may also have to be gained if it appears from the technical design that the defined performance criteria cannot be met – if, for example, an online response time looks like being longer than expected.

6.2.3 After analysis and design

The design is now transformed into an actual computer system through the development and testing of programs. Program testing, or unit testing as it is also called, is carried out by the programmers, who check that each program meets its own design specification. Once it has been established that each individual program works on its own, it is necessary to integrate the programs and to check that they work together as a system. Generally, integration testing will be incremental, that is programs (a) and (b) will be fitted together and tested, and then program (c) will be added. However, in a very large or complex system, it will be necessary to explore a large number of testing threads to ensure that the programs all work correctly in their different combinations.

When all the integration tests are complete, the developers now carry out their own system test to ensure that, as far as they are concerned, the system works together as a whole and meets its design objectives. The system test criteria will be partly technical – derived from the design documentation – and will also relate to the functional and non-functional needs contained in the requirements specification. Once the developers are satisfied that the system operates properly, the users are invited to carry out their acceptance tests. Acceptance criteria should have been derived from the requirements specification. However, the important point is that users should only be accepting against the *expressed and documented requirement*, not against what they might now think they really want. If the users' requirements have indeed changed then the changes can be discussed and, if agreed, can be implemented. This process must not however be allowed to prevent acceptance of the current system if it does really meet its documented objectives.

The accepted system is now installed on the computer on which it will operate and it is commissioned. At this point, the developers will hand the system and all its associated documentation and test regimes over to those who will operate it. Commissioning may be a progressive affair, with functions being added incrementally. There may also be a need for file creation or data take-on and for progressive cutover from a previous system.

All of these activities will require careful planning and management.

Finally, the system starts to operate live and, we hope, to deliver the business benefits for which it has been designed. It is very likely that some sort of support arrangement will be agreed with the users so that problems which arise during live running can be dealt with and also so that the system can be enhanced to meet the users' changing business requirements.

6.3   PROJECT PLANNING

Two important things should be understood about planning: firstly, that it is essential for a project's success and secondly that it should be undertaken as early as possible. There is often a reluctance to plan, perhaps stemming from a feeling that this will unduly constrain the development and perhaps also from a fear that the developers will be committed too early to a course of action which later proves untenable. Whilst this is understandable, it does not really make sense. A plan should not be seen as a straitjacket but as a map setting out the route to be followed. The thing to remember is that the plan is not the project – it is only a model of the project. It is created so that the project manager can use it to check progress and adjust the work to changing circumstances.

So what will a good plan look like and what will it contain? Firstly, a good plan must be a flexible, revisable document. We have already said that any plan can only be regarded as a model and will have to be modified and revised as the project progresses. So it makes sense to devise a structure which will allow for this revision rather than constrain it.

From this it follows that we will not want to treat the whole project as one gigantic task. Rather, we will want to break it down into more manageable sub-tasks that we can modify more easily. There is another reason for this breakdown too: when we come on to estimating later in this chapter, it will be clear that much more accurate estimates can be produced for small tasks than for large ones. Alternative methods of achieving this project breakdown are considered in the next section.

Apart from a breakdown of the work involved, a plan will also contain:

- A description of the organisation of the project, showing who the personnel are and their roles and responsibilities
- Descriptions of the products to be produced, with their completion and quality criteria
- Descriptions of the individual work packages for team members
- An analysis of the interdependency of the various tasks, expressed perhaps as a network diagram
- An analysis of the risks involved in the project, with the possible counter-measures for each risk.

There is some debate in project management circles as to whether the quality plan should be part of the project plan or a document in its own right. We have adopted the second approach here but in practice it makes little difference. The important thing is that the quality issues are thought through and the project's approach to them properly documented; this topic is covered later in this chapter.

6.3.1 Stages in planning

Planning requires a methodical approach. One which has been found to be successful over many projects is described here.

(i) Break the project down   There are two slightly different ways of breaking down the project into smaller components – by work breakdown structure (WBS) or by product breakdown structure.

With the conventional *work breakdown structure*, we start by considering the overall project and progressively break it down into its component hierarchy of stages, steps and tasks. For systems development projects, an obvious set of stages is:

- Business strategy
- Information systems strategy
- Feasibility study
- Requirements analysis
- Requirements specification
- System design
- Program development and testing
- Integration testing
- Acceptance testing
- Installation and commissioning
- Live operation/support.

These stages are, though, still too big to control properly, so we need to break them down into steps. If we take requirements analysis as an example, we might break it down into these steps:

- Interview users
- Examine document flows
- Study rules and regulations
- Build a data model
- Develop dataflow diagrams
- Review results with users.

Finally, we need to decide the individual tasks which make up each of these steps. For 'build a data model', these might be:

- Identify data entities
- Produce entity descriptions
- Identify relationships between entities
- Carry out normalisation
- Validate data model against processing requirements
- Complete documentation
- Enter information into data dictionary.

Of course, this is not the only way in which we could have approached this breakdown. Suppose, for example, we were going to analyse the requirements for a new system to support a business. At the top level (stages), we could have broken our project down into functional areas like marketing, production and accounts. Then we might break down each area again into steps so that 'accounts', for example, could become accounts payable, accounts receivable and banking. And finally, we could have the same set of tasks (interviews, data modelling and so on) as the bottom-level within each step. The actual method of decomposition will be decided by the project manager, taking into account the nature of the work to be undertaken. The important thing, though, is to create a set of low-level tasks against which we can make our estimates and control our project.

The *product breakdown structure*, which is a feature of the PRINCE project management method, approaches this decomposition from a slightly different angle – that of the products that will result from the project. Considering our development project again, we would find some top-level products like 'delivered system' and 'requirements specification'. We could then break these down so that the requirements specification might be found to consist of:

- Function definitions
- Dataflow models
- Logical data model
- User descriptions.

Each of these would have component products: the data model, for example, will consist of a diagram plus entity and relationship descriptions.

These are all what PRINCE calls technical products – that is, they are the things which the project is explicitly set up to develop. But PRINCE also recognises other types of product: management products, like the plans and reports which the project will generate; and quality products, like quality definitions and review sign-offs. The point to grasp about PRINCE's use of products is that this approach forces the developers to focus their attention on the *deliverables* from the project – what is going to result from it. Each PRINCE product will have a description and a set of completion criteria which will be used to determine if it has been properly developed and tested.

With the product breakdown structure approach, we ultimately consider what work has to be done to develop each product – and thus, by a different route, we get back to a list of the fundamental tasks which need to be undertaken to complete the project.

(ii) Estimate durations With the tasks clearly defined, it is now possible to estimate the duration of each task. The first thing to remember is that there is a difference between the amount of work needed to complete a task – the 'effort' – and the actual time it will take – the 'elapsed time'. This is because, however hard people work, they can never spend 100% of their time on project-related activities. People have holidays, they become sick, they go on training courses and they have to spend time on non-project work like attending management meetings.

When planning a project, therefore, it is necessary to make allowances for all these time-stealers and a common way of doing this is to assume that only four days in each week are available for productive project work. In other words, if the effort to complete a task is estimated as four days, then allow an elapsed time of five days and remember to record both calculations.

(iii) Calculate dependences In planning project work, it is important to know which tasks are dependent on other tasks. In many cases, dependences will be fairly obvious – one cannot test a program until it has been written for example. Let us assume that we have a task called 'collate and print requirements specification'. Clearly, we can only carry out this task when all the component materials of the specification have been completed. So, this task might have three dependences:

- on the completion of the function descriptions
- on the completion of the data model
- on the completion of the dataflow models.

If these three tasks have different durations, then the 'collate and print' task can only begin when the *longest* of the three has been completed.

For the moment, we need only to consider all of our tasks carefully and record which ones are dependent on which; we will consider what we do with this information in the next stage.

(iv) Produce network diagram The network diagram is one of the most valuable tools for the project manager. In the example here, the diagram consists of a series of boxes representing tasks connected by arrows which show the dependences between the tasks. Each box contains the data as shown in figure 6.1. Figure 6.2 illustrates our description of the process of creating a network diagram.

We begin our time analysis with a forward pass through the model to calculate, by addition from the start date, the earliest start and finish dates for each task. We then complete a backwards pass to calculate, by subtraction from the end date, the latest start and finish dates for each task. The difference between the earliest and latest finish dates for each task represents its *float* – a useful planning resource – and those tasks which possess no float are said to be on the *critical path*.

We said earlier that three tasks were prerequisites for the task 'Collate/print requirements specification'. If we look at figure 6.2, we learn these things about the three tasks:

- The earliest finish date for 'function description' is day 10 but the latest finish date is day 11; so there is one day's float in that task.
- The earliest finish date for 'dataflow models' is day 9 and the latest finish date is day 11; so there is two days' float in that task.
- The earliest and latest finish dates for 'data model' are the same; so there is no float, the task is on the critical path and any slippage in this task will cause the whole project to slip too.

This is a very trivial example and you could probably see which was the critical task without drawing the diagram. But you can also see that, on a large project with many tasks, the diagram is essential to understanding the interdependency of the tasks.

The value to the project manager of working out which are the critical tasks is obvious. If, in our example, the project manager finds out that the task 'data model' is going to run late, then the manager knows that the project will be delayed. So it might make sense to switch resources from 'dataflow models', which has two days' float, to 'data model' to bring the project back on course.

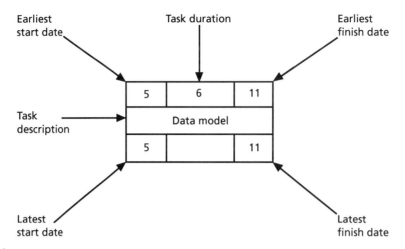

Fig. 6.1

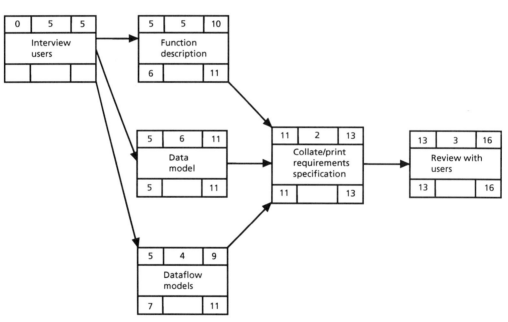

Fig. 6.2

(v) Barcharts Barcharts provide a highly visual way of showing when each project task will be tackled and its duration. A typical barchart lists the tasks vertically and the duration of each is indicated by the length of horizontal bar.

Figure 6.3 shows our same sequence of analysis activities and you will notice that the fact that 'function definitions' and 'dataflow models' are not on the critical path is clear because 'Collate/print' does not start immediately they finish.

When drawing up the barchart, the project manager will take the dependences into account and juggle tasks between team members so as to achieve the optimum staffing plan.

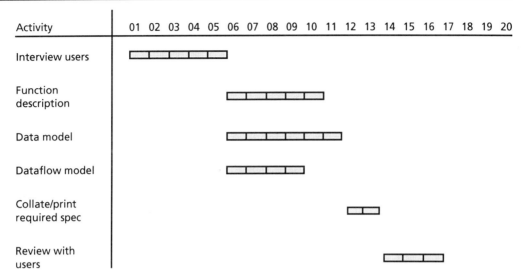

Fig. 6.3 Barchart

(vi) Individual work plans Each team member should be provided with an individual work plan showing the task/s they are to work on and how long they should take. A common way of doing this is to provide each person with a personal barchart containing a subset of the tasks shown on the overall plan. Individual work plans should also contain the following:

- Descriptions of the products to be produced
- Definitions of the completion criteria which will be applied and
- Information on the methods and standards to be used

(vii) Review/reappraise/revise Plans are not static documents. Things go wrong on the best-organised project, delays and problems are encountered and even, on some rare occasions, tasks take less time and effort than expected. The project manager must, therefore, constantly review the plans, reappraise where the project has got to and revise the plans as necessary. Whenever plans are revised, everyone concerned should be informed of the changes and this is particularly important where changed work plans need to be issued to individuals.

6.3.2 Planning for quality

If a project is to be completed to the correct level of quality – which we can define as meeting its stated requirement – this will not happen by itself. Quality must be planned for like all other activities, as explained in chapter 4, and a good way of doing this is to produce a quality plan.

The quality plan may be part of the project plan, or it may be a document in its own right. This does not matter too much but what is important is what is included. A quality plan should cover, at least, the following topics:

- A description of the technical methods to be used during the project
- Details of the standards to which the work will be performed – either by including the standards in the quality plan or, more usually, by cross-referencing
- An outline of the means, both how and when, by which quality will be checked during the project, for example by reviews or structured walkthroughs
- An analysis of the risks to project quality and how these can best be met.

As with the main project plan, it may be necessary to carry out quality planning in outline only at the start of the project and to fill in the details as they become available.

6.4 ESTIMATING

It has to be said at the outset that estimating for software projects does not have a very good track record. The reasons for this are many and various but they include:

- The difficulty of defining precisely enough the scope of software projects
- The fact that most projects involve a degree of innovation, so there is little history to go on
- A lack of experience in estimating in the people carrying it out
- The general lack of suitable metrics.

To these should be added something we might call 'political bias', where the estimator is trying to produce a result it is thought that the boss wants to hear, rather than an accurate one.

Of these difficulties, though, the most common are the lack of scope definition and the lack of metrics. If we compare software production with, say, a mature industry like civil engineering, we find that in the latter:

- Nearly all projects, no matter how apparently innovative, can be broken down to elementary tasks which have been done before – like laying a course of bricks.
- There are abundant metrics available, in the form of standard reference works with tables on, for example, the time to produce so many square metres of concrete flooring.
- Civil engineering estimates are produced once the design is known – once the architect has drawn the plans – not *before* design commences.

The morals for software engineers are therefore twofold:

- Only estimate firmly on the basis of a clear programme of work
- Collect and use metrics.

6.4.1 Estimating for analysis and design work

The main problem with estimating for the analysis phase of a project is that, by definition, the developers do not know the scope, size or complexity of the system until they start to do the analysis work. The first thing to do, then, is to get some overall idea of the scope of the system to be studied. A sensible approach is to carry out a high-level analysis to find out where the departmental boundaries are, how many people are involved and some broad measure of the volumes of data handled. For example, if examining a sales order processing system, the numbers of invoices produced per week would be a very useful metric.

This high-level information will enable some important basic questions to be answered, for example:

- How many people must be interviewed?
- How complex are the procedures we need to study and document?
- Are there outside parties, like auditors, whose view must be taken into account?
- Are there existing computer systems to be studied and, if so, is there business-level documentation available or only program listings to work from?

Any assumptions on which the estimates are based should be included with the estimates and stated in the contract for the analysis work. Then if, for example, it turns out that more interviews are needed than planned, the developer has a good case for asking the customer for more time and money to complete the analysis. If the work is being done by an internal IT department, an informal 'contract' with the users should still be created so that all parties agree the basis on which the analysis has been planned. Remember as well that even though there is high-level commitment to a project, to the users themselves the analysts will initially be a nuisance and interviews will have to be fitted in around their 'real' work. So, without building undue slack into the estimates, a realistic view must be taken of how many interviews can be fitted into each day, and how often the same user can be revisited.

To some extent, estimating for the design work is rather easier than for analysis, since at least the scope of the system should now be agreed and its dimensions and complexity will be understood. However, the designers must plan to take the users along with them as their design proceeds and adequate provision must be made for this. Design is essentially the process of taking the documented requirements and translating them into an implementable computer system.

The estimates for design can therefore be based upon the various components of the design, for example:

- Logical data design
- The design of physical files
- Process design
- The human/computer interface
- Security and control requirements.

As with analysis estimates, any assumptions should be fully documented. So, if you are planning to design for one of the so-called fourth generation languages, you may not intend to produce full program specifications; but if you switch to, say, COBOL, these may be required after all and you will have to revise your design estimate.

6.4.2 Advantages of the structured approach

We have seen already that one of the problems of estimating is trying to decide the tasks that need to be accomplished. In this area, the very obvious advantage is that the structured method is well-defined and therefore provides a detailed list of the tasks involved. However, care must be exercised even here as there may be implicit or unstated tasks, which a given method does not cover. In SSADM for example, the first three substantive steps in requirements analysis are:

- Step 120 - Investigate and Define Requirements
- Step 130 - Investigate Current Processing
- Step 140 - Investigate Current Data

It is likely that suitable metrics will be available from previous projects. Since the same approach is used on each, it is only necessary to make allowances for scope and complexity to be able to use data from one project to produce estimates for another. Of course, this is rather a simplification and does depend on the collection of the metrics in the first place but, even without these, a project manager with previous experience of the method can make more realistic estimates for the new work.

We have already said that systems development is bedevilled by a lack of metrics. To rectify this will take time and effort but it is important that project managers collect metrics to assist them and others on later projects. To be of value, the metrics must:

- be collected accurately and honestly, even if this shows up shortcomings in the estimates
- have qualifications attached, so that project peculiarities can be taken into account when re-using the statistics
- be collected on a consistent, like for like, basis.

6.5 PROJECT MONITORING AND CONTROL

The plan is only a model, an idealisation, of what you want to happen on your project. It is vital to check constantly that the planned things are happening, that your ideas are working out in practice. This requires the project manager to be rigorously honest in measuring progress and facing up squarely to the problems that will surely arise. Only on the basis of accurate information can proper decisions be made to keep the project on course. Work must be checked using some regular and systematic method. Typically, this will involve project staff completing activity logs or timesheets and the holding of regular progress meetings.

Team members should report regularly – usually weekly – on:

- the tasks they have been involved in during the week
- the effort spent on each task
- the effort they estimate will be required to complete each task and the likely completion date
- any problems encountered.

It is much better for estimates of 'time to complete' to be made as objectively as possible, even if the result is a shock to the project manager. Corrective action can then be taken to keep the project on track.

It has been observed that it is not the underestimated tasks which usually sink projects; rather it is tasks which were not suspected at all. We might also add that growth in the scope of work is another factor which leads projects into disaster. So, the project manager must be vigilant to spot additional tasks creeping in and the boundaries of the project expanding. However small these changes may seem at the time, they can have a considerable impact on the final outcome of the project. It should also be remembered that input – effort – is not the only thing which needs monitoring. Output – the quality of delivered work and products – must also be kept under constant review and, again, rigorous honesty is required if the project is to deliver acceptable products at the end. The monitoring and control process can be reduced to a convenient five-stage model which provides a standardised approach:

(i)   Measure – *what progress has been made*
(ii) Compare – *the measured work is compared with the work planned*
(iii) Evaluate – *are we on plan, ahead or behind? What corrective action can we take?*
(iv) Predict – *the result of each possible corrective action, or of no action*
(v)   Act – *tackle the problem now.*

6.5.1 The control of quality

The quality control procedures will have been documented in the quality plan. There are three elements to quality control:

- WHAT you are going to control
- WHEN you are going to apply quality control, and
- HOW you are going to do it – the methods you will use.

In general, all finished technical products – as defined in the project plan – should be the subject of quality control (QC) but you may also want to apply QC to interim products. You may also want to apply QC during production, to make sure that the team is on the right track. The project manager must devise a review regime which does keep the products under constant review but without actually proving a hindrance to doing the work.

As to QC methods, some of the most-used, and most effective, have proved to be:

- Management review, whereby the project manager or team leader examines the work of the team members and provides feedback and criticism; this does require that the manager shares the same discipline as the team members, which may not always be the case.
- Peer review, which is similar to management review except that analysts, say, review and criticise each others' work; this method is useful when the project manager does not have the right background or the time to examine all of the work personally.
- Structured walkthrough, whereby a piece of work is examined more or less line by line by a team of reviewers (see chapter 4).

The Fagan Inspection has proved very effective and is becoming more widespread (again, see chapter 4).

When carrying out reviews, remember that the purpose is to discover defects and inconsistencies – not to find solutions. The problem should be documented, perhaps with an assessment of its severity, and the rectifications remitted to its author or another nominated person.

6.5.2 Documentation control

Every project needs a proper system to control the documentation it will produce. The documentation standards to be used on a particular project may be imposed on a supplier by the customer, determined by existing installation standards, or be procedures devised by the project manager, or more likely will be a composite of all three. Whatever standards are adopted, it is crucial that they are clearly set out – probably in the quality plan – and are understood by everyone concerned.

In systems development work, and especially where the developers are working to a quality system like BS5750/ISO9001, the issue of traceability is very important. That is, it is necessary not only that things are done as planned but that this can be proved to be so. Thus, it is vital that important decisions are recorded and can be located and referred to later. Quite often, important points of detail in analysis are resolved in telephone conversations with users. Where this is the case, it is a good policy to reflect back the decision in a written note to the user or, at the least, to make a note of the content of the call with the date, time and participants. Finally, an idea which has been used with success on some projects is the concept of a 'project log'. Progress is recorded on a day-to-day basis and also notes can be made of any significant issues that crop up.

6.5.3 Change control

Change is an inevitable fact of project life. Since IT projects tend to take a long time, the users' requirements are almost bound to change and their business itself may undergo major restructuring. The old idea that one can 'freeze' a specification and then work to it is obviously useless; what would be the point of delivering to the users a system which reflected how they used to work?

If change is inevitable, methods must be evolved to manage it. It is the lack of these methods that usually leads to problems. The methods needed to control change will depend to some extent on whether the work is being done in-house or by an external contractor; and in the latter case on whether the contract is for a fixed-price or not. However, in all cases, some general rules should be followed:

- No change should ever be accepted without thorough investigation of its consequences.
- All requests for change should be logged and then examined to decide whether they are feasible, how much more effort will be involved, how much it will cost and what the consequences are.
- Both user and developer must agree and accept the change, in the full knowledge of what it will mean for each of them.

With an internal development, changes will have an impact on the project's budget and this must be approved by whoever controls the finances. For an external project, the developer may not be too concerned if the work is being done on a time-and-materials contract since the change will probably involve more work and hence more revenue. So the customer must be careful that the costs do not jump alarmingly. If the contract is for a fixed-price, the developer must protect the profit margin and hence will resist all changes unless the customer agrees to pay for them.

6.5.4 Configuration management

Configuration management is the process of controlling the development and issue of the products of a development. On all projects it is an important matter but on large projects configuration management can be a major and central task of project administration. Configuration management includes:

- The establishment of 'baselines' for each product, so that when it is changed the new version can be clearly identified from its predecessor.
- Ensuring that information is readily available on which versions of each product are compatible with which versions of other products.
- Ensuring that changes in a late-stage deliverable are reflected back properly into its prerequisites.

Configuration management and change control are often confused but the two, though related, are clearly different. Change control consists of managing the alteration of a product from its initiation until its implementation; whereas configuration management is concerned with documentation and control of the changed product.

6.6 SUMMARY

Planning is crucial to the success of systems development projects and should be begun as early as possible. The project plan will describe what is to be done, by whom, and when, and the quality plan will define the methods and standards to be used. Estimating for analysis will be, to some extent, provisional until the scope of the project has been pinned down but design can be estimated with more accuracy once this has been done. All plans must be treated as provisional and must be refined and remade in the light of more information and actual project experience. So the project manager must put in place suitable mechanisms to monitor progress and must act decisively on what is thereby discovered. Change is inevitable in projects and procedures must be used to ensure that it is controlled properly. Similarly, a proper configuration management system is required to ensure that the products of the project are properly documented and controlled.

CHAPTER 6 CASE STUDY AND EXERCISES

Q1 As part of its expansion programme, System Telecom has brought forward the plan to develop a new system and is setting up a temporary development team for it. The team will be disbanded once the system has been implemented and as a consequence everyone except the Project Manager is either a contractor or employed on a fixed term contract. What particular problems will this give the Project Manager?

Q2 What techniques could be used during the systems development lifecycle to ensure that the developed system meets its objectives?

Q3 It is said that user involvement during systems analysis and design is very important. Describe how users can be involved in analysis and design, and in the management of the project.

Q4 System Telecom supports a number of social benefit and charitable activities in the countries in which it operates. It has decided for the first time to support La Societé du Troisième Age (STA) in France. This charity helps old, single people who live in their own accommodation and issues them with a System Telecom pager/panic transmitter. Local branches of STA will raise funds for local people so that pagers can be purchased. System Telecom wants to help these local activities to be organised efficiently. It has therefore offered to help by providing guidelines about fund raising activities. You have been asked to prepare a work breakdown structure for local car boot sales. This will then be given to local organisers to help them in organising such fund raising events. Prepare this work breakdown structure.

Q5 If you were the chairman of a local STA Committee how would you monitor and control the progress of your committee's projects.

CHAPTER 7

Systems Analysis: *Concepts*

7.1 INTRODUCTION

In chapter 1 we introduced the model of systems development shown in figure 7.1 where analysis is represented as a discrete stage, which fits neatly between feasibility and design.

Fig. 7.1 Stages in system development

While the model indicates the relative position of the stages in the development process, systems analysis cannot always be so easily compartmentalised and there is frequently an overlap between analysis and feasibility and between analysis and design. Indeed, the time on a project at which analysis ends and design starts can often only be identified because it says so in the project plan! High-level analysis begins during feasibility, high-level design begins during analysis and analysis continues as part of the design process. On small projects, analysis and design may be carried out by the same team of people, who have the job title analyst/designer or simply system developer.

Whilst it's important to appreciate this overlap between the stages of system development, for the purposes of this book we are treating analysis and design as separate processes. In this chapter we will answer the question, 'what is systems analysis?', consider a structured approach to analysis and introduce the five steps of the PARIS model. Each of the steps of the PARIS model of systems analysis is described in detail in chapters 8-12.

7.2 WHAT IS SYSTEMS ANALYSIS?

The Oxford Dictionary defines *analysis* as follows:

> '*separation of a substance into parts for study and interpretation; detailed examination.*'

In the case of systems analysis, the 'substance' is the business system under investigation and the parts are the various sub-systems which work together to support the business. Before designing a computer system which will satisfy the information requirements of a company, it is important that the nature of the business and the way it currently operates are clearly understood. The detailed examination will then provide the design team with the specific data they require in order to ensure that all the client's requirements are fully met.

The investigation or study conducted during the analysis phase may build on the results of an initial feasibility study and will result in the production of a document which specifies the requirements for a new system. This document is usually called the requirements specification or functional specification, and Tom De Marco has described it as a 'target document' because it establishes goals for the rest of the project and says what the project will have to deliver in order to be considered a success. In this book, we are defining systems analysis as that part of the process of systems development which begins with the feasibility study and ends with the production of this target document.

A systems analyst will be required to perform a number of different tasks in carrying out the analysis phase of a development project. As a result of discussions with practising analysts, five areas have been identified into which these tasks can be grouped, and these are represented in figure 7.2.

Investigation   This group of tasks consists of all the fact-finding activities which an analyst may have to undertake. At the heart of these activities is the key skill of asking questions, orally or on paper, which will yield the required information. However observing others and searching through documents can also be important tasks in gathering information.

Communication with Customers   Many analysts regard this as the single most important factor in ensuring a successful outcome to the analysis and producing an accurate specification of the client's requirements. It will include all the tasks which involve communicating ideas in writing, over the phone or face-to-face. This communication can be formal – presentations, meetings, walkthroughs and reports – or informal, but it does need to be regular and as open as possible. It may include giving explanations, providing reassurance and dealing with concerns expressed as well as exchanging factual information. In addition this group of tasks will also include regular communication with others on the analysis team and their internal customers.

Documentation   The production of documentation, like communicating with the customer, is a broad heading which encompasses many tasks. The writing of meeting minutes and interview records, the drawing of data models, the compiling of lists or catalogues of requirements and the reviewing of documents produced by others would all be included

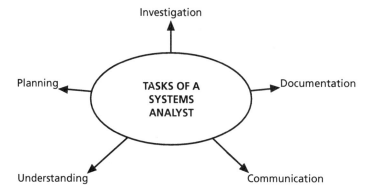

Fig. 7.2 The tasks of a systems analyst

in this group. To be useful to the author and to the rest of the analysis team, any documents produced must be complete, accurate and easily accessible to those who need them. The involvement of the users in checking these documents is a useful way of ensuring accuracy and has the added advantage of contributing to the building of a good working relationship.

Understanding This is a heading which really includes all the others, because at the heart of the analyst's job is the desire to understand the information collected, so that they can pass on this understanding to others on the project. The tasks in this group will include checking facts with the person who initially supplied them, cross-checking them, where possible, with others and recording them as precisely as possible. It also involves a number of interpersonal skills, especially listening, if *real* needs are to be documented and problems are to be understood from the *users'* point of view.

Preparation and Planning This group of tasks will include the planning of analysis activities, estimating how long these activities will take and scheduling them to fit in with the project plan. Also included are the management of time and other resources, detailed preparation for interviews, and the work involved in putting together presentations and walkthroughs. Analysts agree that these activities can be time consuming, but are essential if the analysis is to proceed smoothly.

We talked in chapter 1 about the role of the analyst. In thinking about the tasks the analyst has to perform, we can add the following guidelines which have been identified by practising analysts:

- Check and agree the terms of reference before beginning your work.
- Involve the client as much as possible, both formally and informally, in developing your understanding of the system.
- Don't take information at face value.
- Be prepared for some resistance. The analyst is concerned with change, and this is uncomfortable for many people.
- Be aware of political issues in the client's organisation, but don't get involved.
- Remember that ownership of the system must always stay with the users.

One other point about systems analysis. The model in figure 7.1 shows analysis occurring just once in the life of a project, and then the next phases of the project follow on from this. As we suggested, this is not really true, and another view is shown in figure 7.3. This diagram, the b-model of system development – devised by Birrell and Ould – shows the whole lifecycle of a system. Development is represented as a vertical straight line – similar to the horizontal path in our original model – and this leads into a maintenance cycle at the bottom. Each stage of the model is important and no stage is independent of the others. Analysts need to be aware of all the other stages in the life cycle and not just their part of it.

The life-cycle begins with *inception*, the identification of the need for a new computer system. This leads to the *analysis* stage, the objectives of which are to define the problem, to create a detailed specification of what the system has to do, and to agree with the customer the level of service and performance required. This stage on the b-model also includes the feasibility study – an initial investigation to enable a properly informed

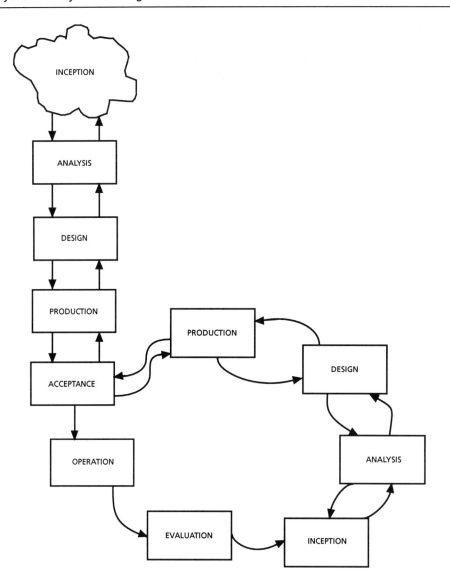

Fig. 7.3 The b-model of system development

decision to be made about whether to commit resources to the project. The next phase is *design*. The objectives of design are to define the structure and content of the system and specify how the system will be implemented. Within this phase, interfaces, dialogues, inputs and outputs are designed, and program and file or database specifications are produced as deliverables. Once the design is complete, the *production* of the system can begin. During this phase program code is created and tested, supporting manuals and documentation are produced, and work proceeds according to the agreed development schedule. In parallel with this activity, data may be converted into a form which can be used by the new system, and training courses designed and implemented in preparation for handover. *Acceptance* marks the point at which the system is installed, handed over and paid for by the client. Any testing at this stage is usually conducted by the client to make sure the system does what the client requested. Acceptance of the system by the client is a contractual issue and a project milestone.

Once development is complete, the system 'goes live' and is used by the client to meet the needs of the business. This is the *operation* phase. During the operation phase, there will be an *evaluation* of the system by the users which may lead to the *inception* of ideas for changes and improvements, and the beginning of the maintenance cycle. During the maintenance cycle, the system may be modified a number of times. For each modification, there will be another analysis phase where the problems associated with the current system and the requirements for changes would need to be investigated and understood. While this might be a much smaller piece of work than the initial analysis phase of the development project, the same principles will apply and the same types of task will need to be completed. Maintenance may account for the bulk of the total work done on the system and more than one change may be moving through the cycle at the same time. While a major change is moving slowly round the maintenance cycle, several smaller changes may move round it quickly.

You will notice in figure 7.3 that there are two-way arrows between most of the boxes; it is sometimes necessary to go back a step if there is a change in the requirement or if an error introduced earlier in the development only shows up in a later phase.

7.3 A STRUCTURED APPROACH

Analysis can be considered to be a four-stage process as illustrated in figure 7.4. This process begins with the analyst investigating and understanding the *current physical system*. This will involve fact-finding activities and the recording of information about how the current system operates. As part of this process, the analyst will also be constructing models to show the data and processing within the system, as well as documenting problems and requirements described by users of the system. The next stage requires the analyst to move away from the constraints which determine how the current system is physically implemented, and to put together a clear picture of the logical functions carried out by the system. In other words, to state exactly what the system is doing rather than how it is doing it. This view is described as the *current logical system*. To move to the *required logical system*, the customer's requirements for a new information system must be mapped on to the current logical system. This will state what the new system will do. By discussing the requirements with the users who specified them, priorities can be assigned and a number of alternative versions of the required logical system can be developed. These alternative versions can be presented to the client as part of the system proposal. Finally, when the client has given the go-ahead to the system proposal, the *required physical system* can be developed. This involves specifying in detail exactly how the new system will work and begins during analysis, with the high-level design included in the functional specification, and continues during the design phase of the project.

In traditional approaches to system development, there was a tendency to move from a description of the current physical system to a specification of the required physical

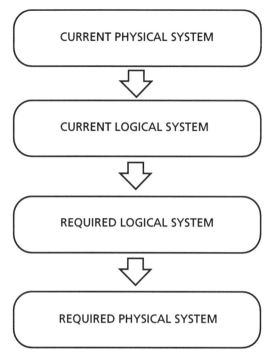

Fig. 7.4 A structured approach to systems analysis

system without considering the underlying logical view. Structured techniques such as data flow diagrams and data models support the four-stage model described above and ensure continuity between analysis and design, by developing logical views of the system.

7.3.1 Structured systems analysis

Structured systems analysis, which is based on the four-stage model described above, also has associated with it three general principles:

- Modelling
- Partitioning
- Iteration

Modelling refers to the use of graphic models which are employed wherever possible, in place of narrative text, to provide clear and unambiguous information about the system. They are produced to represent both the current system and data structure, and the required system and data structure. They enable detailed investigation to be made of the requirements and the design before money is spent in actually producing the system.

Partitioning describes a method of breaking the system down into a number of smaller parts so that it can be more easily understood, so that work can be allocated to the members of the project team. The system is first considered as a whole to establish the system boundaries. Once these have been agreed with the users, the system is partitioned, on a top-down basis.

Iteration: Since it is unlikely that the first representation of the current system and the requirements for the new system will be completely accurate first time, structured systems analysis provides opportunities for revisiting and amending the models of the system. If this iteration of the process of analysis is carried out in close consultation with the users, it will ensure that our understanding of the existing system is correct and agreed with the client before development of the new system begins.

7.4 THE PARIS MODEL

We have divided the process of analysis into five stages, each of which will be described in detail in subsequent chapters. The first letters of each step form the five-letter word, PARIS, which is a useful mnemonic to help you remember the steps. The five steps are:

1 *Planning the approach*
This is the vital first stage in the PARIS model and the success of the systems analysis phase of a project will depend on the thoroughness and care with which planning is carried out. During planning, objectives are set, constraints identified, terms of reference agreed and preparations made for fact finding. Planning is described in chapter 8, which also includes a section on the feasibility study.

2 *Asking questions and collecting data*
This includes all the fact-finding activities carried out as part of the analysis. The key technique here is interviewing, which applies many of the principles introduced in chapter 3 on communication. Interviewing is described in detail in chapter 9 under three headings – planning, conducting, and recording the interview – and there is also a section on difficult interviews. Other fact-finding methods described in chapter 9 include observation, designing and sending out questionnaires, document analysis and record searching.

3 *Recording the information*
The third stage in the model is about recording information. The fact-finding methods, used during stage two, yield many facts and details about the current and required systems. This information must then be recorded in a clear and unambiguous way. In structured systems analysis, a series of diagrams – or models – are drawn to represent the system, and these can be interpreted and built on by the analysis team, and may also be reviewed by the user to check that the information gathered by the analyst is complete and correct. Chapter 10 introduces two important models which are used to document the current system: the data flow diagram and the data model. Both are part of all structured methods, but we will be concentrating on the way they are implemented in SSADM to explain how they are constructed and interpreted.

4 *Interpreting the information collected*
Having documented the current physical system, we need to understand the underlying logical system, and then consider how the client's requirements can be built in. Again, diagrams can be used to help analysts through this stage of the PARIS model. Chapter 11 describes a number of techniques, illustrated with reference to SSADM, which build on the data and process models described in chapter 10.

5 *Specifying the requirement*

The final stage in the model, *specifying the requirement*, is described in detail in chapter 12. This involves the analyst in preparing a number of options, based on the models constructed earlier, for the development of the new system. These options are discussed with the client, costed, and then presented in a way which emphasises the benefits they will bring to the client's business. The analyst, during this stage, will usually be involved in writing a report, and preparing and delivering a presentation. Once a decision has been made by the client on the way forward, a detailed functional specification will be prepared so that the designers will know exactly what the system has to do to meet the requirements.

7.5 SUMMARY

This chapter has introduced the process of systems analysis, illustrated this with models, and explained where analysis fits into the development and maintenance life cycle. The PARIS model, which divides the job of analysis into five stages, provides the structure for the next five chapters, and each stage of the model will now be described in detail. *Planning the approach* is described in chapter 8; *Asking questions and collecting data* in chapter 9; *Recording the information* in chapter 10; *Interpreting the information collected* in chapter 11 and *Specifying the requirement* in chapter 12.

CHAPTER 8

Systems Analysis:
Planning the Approach

8.1 INTRODUCTION

One of the main causes of project failure is inadequate understanding of the requirements, and one of the main causes of inadequate understanding of the requirements is poor planning of system analysis. In this chapter, we examine the first of the five stages of systems analysis described in chapter 7 – Planning the Approach – and highlight the contribution that careful planning makes to the successful outcome of a project.

There are a number of starting points for the systems analyst. Feasibility studies or technical design studies may have been carried out or a high-level analysis may have been completed. Whatever the starting point, the first step taken by the systems analyst should be to plan the approach carefully, bearing in mind the old adage that 'failure to prepare is preparation for failure!'

Before beginning the detailed work of collecting, recording and interpreting data, the analyst will need to stand back, recall the objectives of the project and consider the following three points:

- What type of information is required?
- What are the constraints on the investigation?
- What are the potential problems which may make the task more difficult?

Having taken time to do this, the analyst is then in a position to plan the actions needed to take him successfully through the process. Let's consider an example to illustrate the importance of doing this. Imagine that you work for a computer services company based in the UK. Your company has just won the following contract with *The Instant Image Corporation* (TIIC). The contract covers:

- The analysis of the TIIC's current warehousing, stock control and manufacturing systems and the integration of these systems.
- An investigation of TIIC's current problems and of future strategic plans in this area.
- The production of a report outlining your company's proposals for meeting TIIC's future systems requirements in warehousing, stock control and manufacturing systems.

The Instant Image Corporation is a multinational company, based in the United States of America, with a presence in most countries throughout the world. It is world famous for the manufacture of photographic products especially film and paper. They used to manufacture small, cheap cameras but these products are less popular now and the Japanese make most of the new popular compact 35mm models. The company is split into a number of country-based organisations with the largest in the United States. *Instant Image UK* manufacture most of the products required by the UK, but import additional products from sites in France, Germany and Spain. Increasingly the smaller countries' organisations cannot compete with the US organisation in terms of productivity and profitability. The parent company in the United States has decided to split the worldwide operation into 3 divisions – America, Europe and Asia/Pacific.

Each division will act as an integrated whole, servicing all the product needs of its marketplace. This will mean closing down or modifying some production sites and radically changing the warehousing function in each country. There is no intention to integrate computer systems across all the European companies at this point.

Instant Image UK has four manufacturing sites. Cameras are manufactured in Luton, photographic paper in St Albans, 35mm film in Birmingham and Special Products are manufactured in Croydon. There are warehouses at each manufacturing site; the head office and distribution centre are both located in Croydon.

Against this background think about how you would plan this investigation. You may find it helpful to think about the following:

- Critical information you require before the investigation starts.
- How you will get this information.
- The fact-finding techniques which will be appropriate.
- The danger areas for the project and for your company

In producing your plan, you will have made a list of questions which, as an analyst on the project, you would need to ask before beginning your work. Questions covering the scope of the investigation – the resources available, the budget, the timescale, the key TIIC people to speak to and any restrictions on carrying out the analysis; questions about the business and the company's organisation ; questions, as well, about the objectives and expectations of the client. Standing back from the task and identifying key questions is the first stage of the planning process and, as we shall see, is critical in giving the best chance of success later on.

As part of the planning process, analysts must ensure that:

- they understand the objectives and terms of reference agreed with the client;
- they are aware of constraints which impact the analysis process;
- they plan the research, initial contact and other tasks to be completed during the investigation and manage time appropriately.

In this chapter, we will be discussing each of these areas. We will also turn our attention to the Feasibility Study, describing what constitutes such a study and considering its value as a piece of analysis in its own right.

8.2 OBJECTIVES AND TERMS OF REFERENCE

To understand more about the client's expectations, you need to ask a number of key questions at the beginning of the analysis phase of the project:

- Who initiated the project?
- What is their role in the organisation?

- What are their objectives for the project?
- What are the company objectives?

Once you know the answers to these questions, you can begin to understand the context in which the analysis is to be carried out. A project will usually originate to meet the needs of one or more parts of an organisation. For example:

- Senior management may need earlier and more accurate information to improve their control over the business and enable strategic planning to be carried out.
- Line managers may need a new system, or enhancements to an existing system, to better support the activities of the company.
- The IT department may have identified a more cost effective or efficient solution to a problem as a result of new technologies or methods becoming available.

Whatever the source of the initiative, the senior management of the organisation will expect to see measurable benefits resulting from their investment in the project. Benefits such as:

- Increased profitability
- Improved cash flow
- More effective utilisation of resources, including people
- Improved customer service (resulting in higher levels of customer satisfaction)
- Faster access to management information
- Better management control.

Understanding which of these objectives are the most important to the person, or people, who initiated the project means that you are in a better position to address these areas when planning the analysis and also when presenting system proposals – described in chapter 12 – at the end of the analysis phase.

The stated objectives of the client will usually be recorded in the Terms of Reference. These will have been agreed with the client before the start of analysis and should define the scope of the investigation about to be undertaken. Indeed the word SCOPE is a useful mnemonic which we can use to summarise the main areas included in the terms of reference.

System boundary This will define the area of the organisation under investigation and may also specify the limit of any new system implemented as a result of the project.

Constraints Factors, including budget, timescale and technology which may restrict the study, or the solution, in some way. These constraints will be considered in more detail later in this chapter.

Objectives An unambiguous statement of the expectations of those in the client's organisation who have initiated the project. These may be broken down by function or department. Well defined objectives are clear and measurable.

Permission This will indicate who in the client's organisation is responsible for the supervision of the project and, if permission needs to be granted – for example to extend the scope of the analysis – who has the authority to do so. Points of contact and the appropriate reporting structure may also be defined.

End products   A description of the deliverable or end products of the investigation. This will usually take the form of a written report and a supporting presentation to managers of the client organisation.

As we pointed out earlier, it is important that the terms of reference, once they have been agreed with the client, are clearly understood by everyone in the analysis team. They are useful, not only in the planning stage of systems analysis, but also later on during recording and interpretation to answer disputes which might arise about, for example, the system boundary. If no written terms of reference exist, the analysis team would be well advised to prepare a draft based on their understanding and present it to the client for agreement.

For example, the terms of reference for an analysis team carrying out an investigation in an organisation called *Computer Training Services (CTS)*, has been agreed with Andrew Smythe, the company's Training Manager, and reads as follows:

'The analysis team will:

1. *Investigate the existing booking and invoicing operations within CTS.*
2. *Document the characteristics, deficiencies and problems in these operations.*
3. *Determine where there is duplication of effort in the current system.*
4. *Obtain and/or produce documentation of the current manual procedures.*
5. *Identify those parts of the booking and invoicing operations suitable for automation.*
6. *Produce a written report of the findings of the review which will be delivered to Andrew Smythe on Thursday 31st March 1994.'*

In these terms of reference, the system boundary is defined (booking and invoicing operations), a constraint (the timescale) is specified and six clear objectives are listed. The issue of permission is also covered: it is Andrew Smythe who initiated the study, and to whom the report – the end product – is to be delivered.

8.3   CONSTRAINTS

Terms of reference have been agreed with the client. The objectives of the analysis phase are clear. Now it is essential that all constraints on the analysis are understood early in order to help with planning and to avoid problems arising during detailed analysis. Some constraints may have been set by the customer, which will limit the options that can be presented as part of the system proposals. They are illustrated in figure 8.1.

These constraints will probably be included in the terms of reference and cover:

Technology   The customer may be committed to a particular hardware or software solution. Indeed, there may be a corporate strategy in this area. For example, they may require any new system to run on their present mainframe computer, or to be developed using a specified programming language – such as COBOL – in order that it can be integrated with existing software.

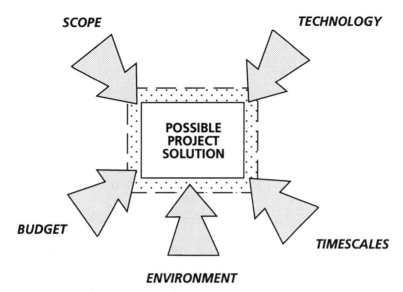

Fig. 8.1 Constraints on the possible solution

Environment   The system may have to work in exceptional conditions, for example in a missile, in a submarine, or on a factory floor. The system may be used by skilled or unskilled operators, perhaps by operators wearing protective clothing.

Timescales   Project delivery times based on the customer's immovable time scales may also be specified. These could in turn be determined by the introduction of new government legislation or a key date in the business cycle of the client's organisation.

Budget   If the work is to be done by a systems house, the contract may well be a fixed price one. This poses real problems for the analysts on the development team. How much resource is committed to the analysis phase? What other budget constraints exist? For example, how much money is available for the purchase of new hardware or software and are there any limitations on the annual operation budget?

Scope   What is the area under investigation in this project? What is the boundary of the system? What parts of the organisation are off limits?

All of these factors will constrain any solution presented by the analysis team. But there is also another set of constraints which need to be identified by the analyst early on, because these will limit the way in which the investigation is conducted. These will include the areas shown in figure 8.2.

These constraints, which are listed below, will enable the analysis team to determine which fact-finding methods are most appropriate, as well as helping them to put together a detailed plan for the investigation.

- The project resources available during the analysis
  How much support will be available to the analysis team? For example, will CASE tools be available for producing documentation, will there be administrative support

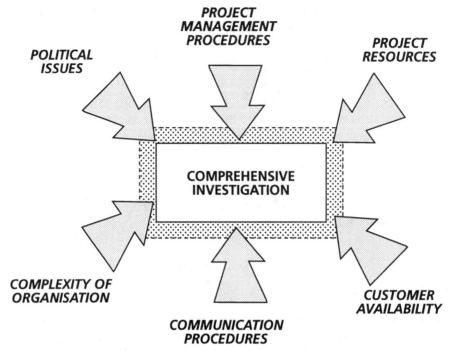

Fig. 8.2 Constraints on the investigation

from the project office, are there colleagues who have worked with the client before who would be available to give advice to the team?

- The availability of customer contacts
  It is important to know if appropriate staff will be available for discussion and to review analysis products. For example, an organisation might require that all interviewees must be given 3 days notice, or that review meetings involving client staff cannot be held on any Monday or Friday.

- The political issues important in the customer's organisation
  There may be politically sensitive issues in the organisation which an analyst should be aware of before embarking on the investigation. For example, the introduction of a new computer system might mean that fewer employees are needed to do the work, or that one department will have to close down. In this case, the analyst should be clear about whether the staff to be interviewed are aware of the implications of the change, and what line should be taken if difficult questions are asked.

- The complexity and size of the organisation
  This will influence the choice of fact-finding techniques and could mean that sampling methods have to be used. If, as part of an investigation you are undertaking, you need to collect information from all parts of a large organisation, a number of options are available, including:
  – interviewing everyone
  – sending out questionnaires

– interviewing a representative sample.
  It is important that the implications, costs and benefits of the alternatives are understood and discussed with the client.

- The project management procedures used by the project team
  An analyst will need to be aware of the expectations of the client organisation with regard to the management of the analysis phase of the project. It might be, for example, that the client organisation uses a structured approach to project management such as PRINCE, or the client may require the team to give detailed information about the planning and monitoring techniques to be used during the analysis phase of the project.

- Communication procedures
  It is important that the analyst is aware of, and follows, the communication procedures set up within the project team and between the project team and the client. This may involve agreement on appropriate channels of communication and the identification of named individuals as 'contacts'.

8.4 PREPARING FOR DETAILED ANALYSIS

In order to be well prepared for the later stages of systems analysis and to increase their credibility in interviews with client staff, the time spent by analysts on research during the planning stage represents a good investment. This will mean that, in addition to reading and understanding the terms of reference, the contract should be studied, as well as any background documents, such as preliminary studies, which put the current work into perspective. Researching the client's organisation will involve reading the annual report of the company, obtaining organisation charts showing departments involved, personnel and job responsibilities and building up a list of any special terminology used together with definitions. By talking to people in your own organisation who have had dealings with the client in the past, you can gather valuable 'off the record' information about working relationships. In addition your organisation might have information about similar work which has been done for other clients, for example previous studies and their outcome together with user decisions.

Having completed the initial research described above, you can begin to identify the types of information you will need to collect during your investigation. To help in this process, a list of the areas to explore should be prepared. While this will clearly vary according to the business of the client and the nature of the project, the following are examples of topics which the analyst may wish to investigate:

Growth — What plans does the organisation have for future growth, and what would be the information requirements to support this growth?

Functionality — What are the major areas of the business – the functions – that will be investigated during the system and what are the client's requirements for the functionality of the system?

Procedures — What procedures, standards and guidelines govern the way in which the organisation conducts its business – and are they written

	down somewhere?
Volumes	What are the volumes of data which pass through the system? For example, how many orders are processed by the sales department in a week, how many amendments are made to customer records each month?
Fluctuations	The identification of bottlenecks or hold-ups in the system, as well as the 'peaks' and 'troughs' (busy and slack periods) in the operation. Where and when do these occur in the current system? What steps are taken to deal with these?
Information required	What information is currently required by the business in order to carry out its functions effectively, and what are the sources of this information? What information, if it were available, would bring significant benefits to the organisation?
Environment	In what type of environment is the business conducted, and how does this affect the way in which information is exchanged?
Problems	In the view of users, what are the main problems with the system, what are the implications of these problems, and how might they be overcome?

In planning the approach to analysis, an important area to consider is the first face-to-face contact with the client. This initial contact should be formal and at the highest level possible. In addition to any fact-finding objectives, a key aim of this contact will be to build a good relationship with the client and to establish the analyst's credibility. In this meeting, and indeed in all subsequent meetings with users, the following guidelines represent good practice.

- Focus on confidentiality, integrity, respect and confidence building.
- Recognise expertise in the users and welcome their input.
- A key objective is building the client's confidence.
- Keep everybody informed. This includes client contacts and project staff.
- Be discreet and diplomatic.
- Double check any information gathered.

There are many tasks to complete during systems analysis, time is limited and often different stages of the analysis process will be taking place at the same time. One other area, therefore, for the analyst to consider, when putting together the plan for detailed analysis, is the management of time. Time is a resource to be budgeted, managed and used. To help you manage your time as effectively as possible, here are some guidelines:

- List objectives and set priorities.
  Decide what you are really trying to achieve during the analysis, list your objectives and, having done this, decide on the priorities of each. The temptation is to start with the small, trivial jobs to get the desk clear, so that the difficult, involved tasks can be dealt with later. But interruptions, telephone calls, visitors and distractions can mean that later never comes!
- Make a daily 'to do' list.
  Daily, as a routine, at the same time, whenever best suits you, make a list of the

things you actually want to do today to move a step - even a short one - towards your objectives. Some people find it helpful to prepare their 'to do' list in advance, at the end of the previous day.
- Handle paper only once.
  Once you have picked up a piece of paper, don't put it down until you have taken some action on it. 'Pending' should mean 'awaiting the completion of some action I've already initiated' - not 'too difficult today'.
- Set and keep deadlines.
  With our own jobs, and with those we delegate to others, the work expands to fill the time available. So setting deadlines at the outset is a good discipline. A long, complicated job, like the analysis of a system, can be split up into steps or stages, with a planned deadline for each stage. And don't forget to build in some contingency when budgeting time to allow for the unexpected.
- Ask yourself frequently 'What's the best use of my time right now?'
  You must train yourself to take frequent breaks, to come out of the trees to take a look at the wood.
- Always carry a notebook.
  To collect information or ideas as they occur, a useful idea is to carry some type of notebook rather than collecting numerous little scraps of paper which can get lost.
- Do it now.
  In other words, avoid procrastination!

8.5 THE FEASIBILITY STUDY

A feasibility study is really a small-scale systems analysis. It differs from a full analysis only in its level of detail. The study involves analysts in most of the tasks of a full systems analysis but with a narrower focus and more limited time. The results of the study help the user decide whether to proceed, amend, postpone or cancel the project – particularly important when the project is large, complex and costly. However, a feasibility study is no substitute for a full, detailed and thorough analysis of the client's system. Different people can provide different parts of the answers in a feasibility study: those who initiated the study, the technical experts, and those who will have to use the new system. The job of the analyst is to pull all this information together and present it to the client in the form of a coherent report. Detailed investigation of operational and procedural activities during a feasibility study is very limited. Analysts should concentrate on providing the answers to four key questions:

How much ?	the cost of the new system
What ?	the objectives of the new system
When ?	the delivery timescale
How ?	the means and procedures used to produce the new system.

During the feasibility study, a number of structured techniques can be used to record the findings in an effective way, and later to present data in a graphical form. These techniques are described in more detail in chapters 10 and 11. At the end of the study a report is prepared for the client.

The feasibility study report has to address three levels of feasibility:

Technical feasibility   Is it going to work?
Business feasibility   Are cost and timescales right for the business and will potential returns justify the initial outlay?
Functional feasibility Will the solution satisfy the end users?

If, for example, you were building a house, it wouldn't be enough just for the house to stand up (technical feasibility) and for the price to be right (business feasibility). You'd have to want to live there as well (functional feasibility).

Figure 8.3 illustrates the main sections of a feasibility report. The contents of these sections could be as follows:

Background

- Terms of reference
- Reasons for the study

This section will outline the background to the project and the way it relates to the stated objectives of the organisation.

The current situation

- Overview of current situation
- Problems and requirements identified

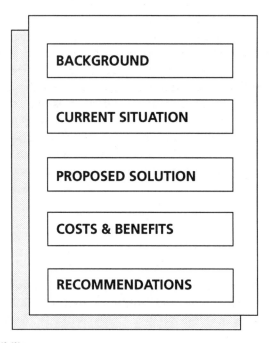

Fig. 8.3 The feasibility report

Proposed solution

A description of the requirements of a new system along with a number of options explaining how this solution might be implemented. Each option will address:

- Technical implications – how it meets the requirements, the hardware and software needed
- Operational implications – the impact the solution will have on the business in terms of human, organisational and political aspects
- Cost implications – both initial (capital) and continuing (operational). There are a number of methods of assessing the costs of solutions. In the feasibility report, the analyst should use the cost assessment method specified by the client.

Cost/benefits analysis

- A comparison of costs and benefits prepared using whatever evaluation technique is favoured by the organisation.

Recommendations

- Summary of the previous sections of the report
- Recommendations as to how the client should proceed

Three distinct types of recommendation can be made in a feasibility report:

1. Advising the client to progress with the full detailed analysis. If this is the case, a plan would also be included for this phase of the project.
2. Advising the client to review the terms of reference or the scope of the study before proceeding further or making any judgement on feasibility.
3. Advising the client to scrap the project as it is not feasible, that the resources could be better spent elsewhere.

Once the feasibility report has been delivered, and assuming that the recommendation made by the author(s) is to proceed, the detailed systems analysis phase can begin. This is described in the following four chapters, but it is worth noting here that any preliminary work done as part of the feasibility study will be valuable in planning and carrying out the later stages.

8.6 SUMMARY

In this chapter we have looked at planning the approach to analysis. In the past this stage has often been rushed or omitted and this has led to costly problems later on, but by careful planning many of these problems can be avoided.

We have discussed the importance of understanding the client's objectives and agreeing terms of reference which reflect the SCOPE (System boundary, Constraints, Objectives, Permission and End-products) of the investigation. We have considered constraints which need to be identified as early as possible by the analyst. These may limit the possible solutions to, or the investigation of, the client's problems. Stages in preparing for detailed analysis have also been described in this chapter beginning with research and moving on to planning the initial contact with the users. Time management, another skill needed by an effective analyst, has also been discussed.

We looked finally at the feasibility study and feasibility report. The existence of a feasibility study does not mean that detailed analysis is not required – a full and comprehensive analysis is the only way to provide a thorough understanding of the user's requirements. However a feasibility study does provide a useful starting point for a full analysis.

The overall message behind all the sections in this chapter has been this: that by taking time, standing back from the problem and carefully planning your approach to analysis, you give yourself the best chance of success.

CHAPTER 8   CASE STUDY AND EXERCISES

Q1 Imagine that you are the Personnel Director for Systems Telecom. Prepare the terms of reference for a project to develop a new personnel system.

Q2 Using the example of the Instant Image Corporation, how would you plan the approach to the systems analysis needed for the contract won by your computer services company.

Q3 This chapter talks about face-to-face contact with the client and the need to build a good credible relationship. If you were an analyst meeting the System Telecom Personnel Director for the first time to discuss the terms of reference mentioned in Q1, how would you begin to build this relationship? Be as specific as you can.

CHAPTER 9

Systems Analysis: *Asking Questions and Collecting Data*

9.1 INTRODUCTION

Fact-finding takes place from the start of the project – during the feasibility study – right through to the final implementation and review. Although progressively lower levels of detail are required by analysts during the Logical Design, Physical Design and Implementation phases, the main fact-finding activity is during Analysis. This fact-finding establishes what the existing system does, what the problems are, and leads to a definition of a set of options from which the users may choose their required system. In this chapter we consider the second stage of our Systems Analysis model – Asking Questions and Collecting Data. Fact-finding interviewing, an important technique used by analysts during their investigation, is described in detail, and a number of other methods used to collect data are also described.

In the last chapter, we discussed the importance of planning and of preparing thoroughly before embarking on detailed analysis. If this has been done, you will already have collected quite a lot of background information about the project and the client's organisation. You may also have been involved in carrying out a feasibility study. You will have some facts about the current system; these may include details of staff and equipment resources, manual and computer procedures and an outline of current problems. This background information should be checked carefully before beginning the detailed fact-finding task. You should now be ready to begin asking questions, and as a result of your planning, these questions will be relevant, focused and appropriate.

In carrying out your investigation, you will be collecting information about the current system, and, by recording the problems and requirements described by users of the current system, building up a picture of the required system. The facts gathered from each part of the client's organisation will be primarily concerned with the current system and how it operates, and will include some or all of the following: details of inputs to and outputs from the system; how information is stored; volumes and frequencies of data; any trends which can be identified; and specific problems, with examples if possible, which are experienced by users. In addition, you may also be able to gather useful information about

- departmental objectives
- decisions made and the facts upon which they are based
- what is done, to what purpose, who does it, where it is done and the reason why

- critical factors affecting the business and
- staff and equipment costs.

In order to collect this data and related information, a range of fact-finding methods can be used. Those most commonly used include interviewing, questionnaires, observation, searching records and document analysis. Each of these methods is described in this chapter, but it is important to remember that they are not mutually exclusive. More than one method may be used during a fact-finding session; for example, during a fact-finding interview, a questionnaire may be completed, records inspected and documents analysed.

9.2 FACT-FINDING INTERVIEWS

An interview can be defined as 'a conversation with a specific purpose'. This purpose could be selection in a recruitment interview, counselling in a performance appraisal interview or collecting information in a fact-finding interview. An interview is a form of two-way communication which requires a range of interpersonal skills to be used by the interviewer to ensure that the purpose is achieved. The interviewer will need to be a good listener, to be skilful in the use of questions so that the conversation flows smoothly, and to be able to control the interview while at the same time building and maintaining rapport with the interviewee. In a fact-finding interview, these skills are needed to collect and record information in order to build up a picture of the current system, and to catalogue the requirements for a new one. In describing this fact-finding technique, we will look at three stages: planning, conducting and recording the interview.

9.2.1 Planning the interview

When planning a fact-finding interview, you are trying to answer five questions:

- What do I wish to achieve as a result of these interviews?
- Who should be interviewed?
- How will these interviews be conducted?
- What topics will be covered during the interviews?
- Where will the interviews take place?

The answer to the first question helps you to identify a set of objectives; question two leads you to a list of interviewees and to a sequence in which they will be interviewed; the answer to the third gives you a format or structure for the interview; in answering question four, you are putting together an agenda; and there are implications for both interviewer and the interviewee depending on the answer to question five with regard to location. We will consider each of these issues in turn.

The first stage in planning the interview is to set clear, specific and measurable objectives. If such objectives are set, not only will you understand what you are trying to achieve, but there will also be some criteria for measuring the success of the interview when it has been completed. An objective is written in the form:

> *By the end of the interview, the analyst will have put together a list of the major problems encountered by Pat Clarke when using the current system.*

which contains a time (by the end of the interview), an action (putting together a list) and a deliverable (a list of the major problems encountered by Pat Clarke when using the current system). This can be used to evaluate the success of the interview because the question 'has the analyst achieved this objective?' can be answered with either a 'yes' or a 'no'.

An objective can be defined as either passive or active. Passive objectives are concerned with collecting information from the interviewee, while active objectives are about decisions or actions which you require the interviewee to take. An example of a passive objective is

By the end of the interview, the analyst will have identified the interfaces between bookings and sales and collected copies of any forms which pass between the two departments.

An example of an active objective, on the other hand, would be

By the end of the interview, the interviewee will have agreed to contact the managers who report to her to enable further fact-finding interviews to be set up.

In putting together an interviewing plan, you will identify who in the client organisation you wish to see and the order in which you wish to see them. Key people and decision makers will need to be interviewed first. Often they will not be able to provide the detailed information, being removed from the day-to-day operation of the business. However, they can provide you with a high-level view of the current system and a strategic view of the business, as well as information about their requirements for the new system. They will also be able to suggest the best people to fill in the detail. When planning a logical sequence for the individual interviews, it is helpful to tie this in with the structure of the system under investigation. This should enable you to gradually build up the whole picture, while also giving the opportunity to hear the views of a range of users of the system.

Most fact-finding interviews follow a similar structure, which is shown in diagrammatic form in figure 9.1. This consists of four stages – social chat, overview, questions & answers, and closing. The length of each stage will vary, depending on the individual being interviewed and the amount of detail required, but the percentages shown in figure 9.1 represent a typical breakdown of the total time available.

(a) Social chat An interview begins with a casual, friendly opening to create a relaxed atmosphere and put interviewees at their ease. This is an opportunity to reassure the interviewee, who might be feeling nervous, as well as calming any nerves of your own. During this stage you can give the interviewee some background information about the reasons for the investigation and answer any questions they have. The early minutes of

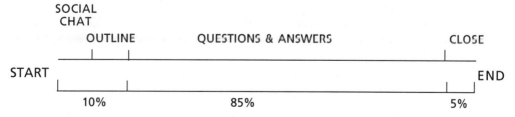

Fig. 9.1 The structure of a fact-finding interview

an interview are critical in building rapport with the interviewee and making it feel more like a conversation and less like an interrogation.

(b) Overview Having created a relaxed atmosphere in the first stage of the interview, you now move on to outline what will happen next. In effect you are 'signposting' the various parts of the interview and it is usually helpful at this point to have the agenda visible to both parties. During this overview you can also explain the objectives of the interview, the time you will need, and the main topics to be covered as well as asking for the interviewee's permission to take notes.

(c) Questions and answers This is the fact-finding part of the interview. In this stage you ask questions to find out as much as possible about the interviewee and their role in the organisation. By listening carefully to the answers, making notes and checking understanding of the information collected, a lot of useful information can be gathered. It is important to keep control and direct the interview during this stage to ensure your objectives – both active and passive – are met. It is also important to maintain the rapport, remembering that you are not there to evaluate the interviewee but to hear their views.

A useful model for structuring the questions and answers stage of the interview is shown in figure 9.2. As you can see, there are four steps in this questioning model, designed to guide the analyst through the fact-finding process and provide information about the user's problems and requirements.

In conducting an interview, the questions should initially be at a high level so information is gathered about the background and work environment of the interviewee (user) within the organisation. These are called CONTEXT QUESTIONS, and often contain the words 'tell', 'explain' or 'describe'. For example,

> 'Can you describe your main responsibilities as sales manager?', 'take me through a typical day in this office', 'will you explain this department's role in the organisation?', 'tell me about the main stages in the bookings cycle'.

The context questions are followed by a set of DETAILED QUESTIONS which enable you to obtain specific information about the areas explored in step one. This information will usually include facts and figures, which indicate volumes as well as operational peaks and troughs. For example,

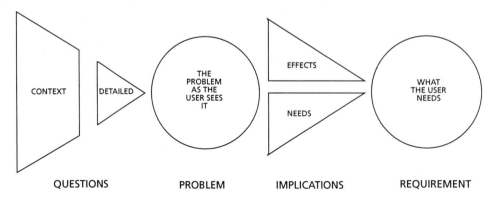

Fig. 9.2 The questioning model

'You mentioned a problem which occurred last February – can you say more about that?', 'on average, how may invoices would be processed each month?, 'you said that the current procedures cause difficulties for your staff – what sort of difficulties?', 'which months in the year are the busiest for you?', 'do you have examples of the documents you're describing?'

The context and detailed questions should help you to understand the current system and enable you to identify the nature and cause of specific PROBLEMS.

Once problems have been identified, you should ask further questions to find out what EFFECTS the interviewee believes each problem has on the organisation and, more specifically, on their area of responsibility.

'So what effect does this problem have on your work?', 'can you give me examples of the sort of comments which members of staff in your area have made about the current purchasing system?', '....and when the orders are held up in Keith Smith's section, what is the impact of that in your section?'

The fourth step, which can take place alongside step three, is to explore the user's views about the solution to a problem and to find out what the user NEEDS to correct the problem. It is important during fact-finding to discover what the user's *real* needs are, so that any solution developed will be efficient and effective.

'In your opinion, what steps could be taken to solve this problem?', 'if you had a free hand to make changes, how would you improve the system?', 'which of the requirements you've described is the most important for the business at the moment?', 'what is the single most important change which would make life easier in this department?'

An appreciation of effects and needs, from the point of view of the user, will enable you to identify the user's REQUIREMENTS.

(d) Closing As the model indicates, time should be left at the end of the interview to bring it to a formal close. In closing the interview, summarise the points discussed, checking key facts with the interviewee, and describe what will happen next as a result of the information obtained. In most cases it is appropriate to offer to send a copy of the formal record of the interview for the interviewee to check and confirm. Finally, arrangements can be made to re-contact the interviewee if there are any problems, thereby 'leaving the door open' for future discussion.

The content of the interview must also be carefully considered during the planning phase and an agenda produced. The starting point for this task is the list of objectives prepared earlier. Using the objectives – especially the passive ones – the information to be collected can be prioritised as follows:

1. Facts which you *must* find out in order to develop the new system. This relates to the key objective, or objectives.
2. Facts which you *should* find out to add flesh to the bare bones of the system.
3. Details which, given the time, you could find out to add the final polish to the system.

Prioritising your fact-finding according to these three headings provides an outline to guide you through the interview. This approach can be represented as a tree (figure 9.3).

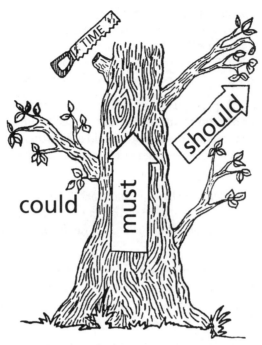

Fig. 9.3 Fact-finding – the tree structure

The key fact-finding objectives – 'must find out' – form the *trunk* of the tree. From these, an agenda can be prepared listing these areas as main headings. We can leave the trunk and go down the *branches* to get supporting information – 'should find out' – which will provide additional information about the key areas. The *leaves* provide the final level of questioning – 'could find out' – representing detailed questions about specific areas within the list of objectives. Having travelled out to the leaves to collect the detail, we can return to our main objectives on the trunk of the tree, and so avoid the danger of 'falling off' the tree by getting lost in details and losing sight of the main purpose of the interview. The *saw* in the diagram is a reminder that the agenda might have to be pruned if the available time is used up. Such pruning should enable the interview to be properly closed and a second interview to be arranged.

A final point to consider when planning the interview is where it will take place. There is usually a choice of location – either on the client's premises or at the analyst's office. Both choices have associated advantages and disadvantages which are summarised in the table in figure 9.4.

As can be seen from the table, the analyst's offices have many advantages for the interviewer, while on the other hand there are clearly good reasons – in terms of access to information – for travelling to the client premises. A midway point, which is often the most satisfactory, would be a dedicated office on the client's premises in which the analyst would conduct the interviews. This would be a private area, with a layout decided by the analyst, where client staff could be interviewed without such distractions as telephone calls. All the interviewee's information would be readily available and in addition other staff would be nearby.

CLIENT PREMISES	ANALYST'S OFFICE
Advantages Little inconvenience for client – no travel Client will be relaxed Client will have information to hand Other client staff are available	*Advantages* Interviewer in control Interviewer more relaxed Client away from day-to-day pressures Fewer interruptions & privacy guaranteed Interviewer has information to hand Interviewer lays out room
Disadvantages Possible interruptions, e.g. phone, people or intercom Privacy not always possible, e.g. open plan Layout not always acceptable Analyst has no access to own information Not as easy for interviewer to control	*Disadvantages* Other client staff are not readily available Client may not have all information to hand Client may not feel relaxed More inconvenient for the client

Fig. 9.4 A table comparing possible interview locations

9.2.2 Conducting the interview

Having emphasised the importance of careful planning to make sure that the time spent conducting a fact-finding interview is used effectively, we now turn our attention to the interview itself to consider how the work done during the planning phase may be put to good effect. In this section, we'll discuss two key skills – listening and questioning – as well as the important issue of control.

Listening As we pointed out in chapter 3, listening is an essential component of any effective face-to-face communication. It is at the heart of the process of interviewing and the quality of the information gathered by analysts depends to a large extent on the effectiveness of their listening. An effective listener will not only encourage the interviewee to provide them with clear and accurate information, but will also understand much more of what they hear. This said, listening is not a well developed skill for most people. Although we are taught in school to read and write, few of us have been formally taught to listen. Being an effective listener means: not being switched off just because the other person is not clear or concise; not reacting emotionally; not letting the other person's mannerisms distract you; keeping an open mind until the other person has completely finished; being patient with slow and ponderous speakers and trying not to interrupt.

To be effective listeners, analysts need to work on developing their skills in this area as well as on adopting an open, receptive attitude when engaged in listening. We will deal first with the skills, described in chapter 3 as *active listening*, and then go on to discuss attitude and the blocks to effective listening.

'Active' listening has been defined as a set of techniques through which one person can obtain information from another. It involves the listener communicating their interest and their understanding to the speaker, encouraging them to continue and giving them the opportunity to talk without constant interruption. Active listening also enables the interviewer to better control the flow and direction of the interview. There are both non-verbal and verbal signals which demonstrate that active listening is taking place.

Non-verbal signals You can communicate that you are actively listening by showing that you are paying full attention. This involves:

- Looking at the person
- Maintaining eye contact, but not staring
- Nodding your head from time to time
- Appropriate facial expressions, especially smiling
- Attentive body posture, which is often shown by a listener leaning forward.

Verbal signals There are a number of ways which will indicate clearly that you are not only listening but also interested in what the other person has to say.

- Repeating in your own words, e.g. 'So what you are saying is...'
- Summarising key points
- Encouraging the speaker to continue, e.g. 'That's interesting, tell me more'
- Asking questions to obtain further information or clarification
- Making encouraging noises.

Attitude and blocks Perhaps the chief requirement for effective listening is to have 'room' for others. It is not sufficient just to develop appropriate verbal and non-verbal listening skills. We have all seen politicians on television who have been coached to display the skills associated with active listening but who demonstrate from their subsequent remarks that an open attitude was not present behind their behaviour. If we are pre-occupied with our own thoughts, ideas and views we are not mentally 'available' to listen effectively. When listening it is helpful to really try to understand the other person's view without superimposing one's own views or judgements prematurely – a major block to effective listening.

Other blocks to effective listening include:

- Listening for an opportunity to interrupt to talk about your own experience
- Listening for agreement rather than understanding
- Assuming you know, and understand, what the person is going to say
- Rehearsing your response to something the speaker has said
- Preparing in your mind the next question you are going to ask.

With all of these blocks, the listener is moving concentration away from the speaker and onto personal issues. Once this happens, the listener is in danger of missing not only the words being spoken, but also, and perhaps more importantly, the subtle non-verbal messages being communicated in the tone of voice and body language of the interviewee.

Questioning The other key interpersonal skill which needs to be put into practice by the analyst during fact-finding interviewing is questioning. Asking the appropriate question to obtain the information required is a technique which has many other applications and which is central to fact-finding interviewing. As we shall see later, it is also an important technique to use when designing questionnaires, another means of collecting data during systems analysis.

Different types of question elicit different types of response and are, therefore, used for different purposes. A selection of question types are described below:

- *Closed questions*
  Closed questions close down the conversation. They produce a definite 'yes' or 'no', or sometimes 'perhaps', and can help to control the talkative interviewee. They are also useful for checking and clarifying facts. For example, 'can you…?', 'will you…?', 'have you…?', 'is it…?'.
- *Open questions*
  Open questions, on the other hand, open up the conversation. They are used to obtain information, provide insight into a client's feelings and motivations, and encourage the quiet interviewee to relax and be more forthcoming. For example, 'tell me ….', 'please explain …', 'could you describe …?', 'how is your purchasing organised?', 'what is your opinion of …?'
- *Probing questions*
  As the name suggests, these questions are used to obtain more detail to probe further when a previous question has not yielded sufficient information. For example, 'why?', 'please could you tell me more about…?'
- *Probing techniques*
  These include planned pauses, periods of silence, or an unfinished sentence. These are more subtle, and often more powerful, ways of probing which can be used instead of, or in conjunction with, probing questions.
- *Reflective questions*
  This involves repeating a key word or phrase to encourage the speaker to say more about it. For example, 'difficulties?', 'a significant number?' Used as described, reflective questions act as probes to elicit more information. They can also be used to demonstrate active listening when they reflect not a key word but an unspoken message or feeling. For example, 'it sounds like the new system is really causing you a problem', 'you seem very pleased about that…'.
- *Limited choice questions*
  The objective here is to direct the other person's attention to a range of options, while leaving them with the final choice of answer. For example, 'which of these three courses of action is most suitable …?', 'is your busiest time in April or in September?'
- *Leading questions*
  Leading questions imply the correct answer. Examples: 'would you agree that…?', 'don't you feel that…?, 'so all your problems result from the introduction of these new procedures?' They can be useful in confirming information, but the danger is that the interviewee will agree because this is the easiest option. They should be used sparingly, if at all, and are usually better framed as a different question. Another danger, if they are used too often, is that the interviewer can appear pushy and irritating.
- *Link questions*
  For example, 'in view of what you have just said about the importance of credit terms, how important are delivery times?', 'the point you've just made about

customers leads me to another question; how do new customers appear on the system?' Link questions like these are excellent for steering the discussion from one topic to another while allowing the other person to do most of the talking.

Putting questions together to take the interviewee through a particular area is another part of interviewing which the analyst should address. The sequence of questions asked during an interview can be critical in the successful achievement of objectives. One useful model in building a sequence of questions is the funnel which is illustrated in figure 9.5. In this model, the interview begins with open questions at the wide end of the funnel, followed by probing questions to focus on specific points raised in response to the open questions. These are followed by closed questions to obtain factual data and lastly by summarising to check that the facts collected are correct.

Control It is important to remember that however much you open up the interview during questioning, you, as the interviewer, must always be in control of the process. Guiding conversations by using the right questions is one way of maintaining this control. Other controlling techniques include:

- *Signposting* Giving the interviewee an idea of the path you are planning to take and therefore the reason for your questions. An agenda is very useful, and can be used to bring the interview back on course if control begins to slip away.
- *Confirming* Re-stating or repeating the interviewee's statement to ensure that you have clearly and correctly received their message.
- *Summarising* Consolidating what has been covered in the discussion; clarifying and checking for understanding.
- *Note-taking* Memory is fallible; recorded notes are less so. It is a compliment to the client to demonstrate a desire to be accurate, and note-taking can also be useful in controlling the interview by giving the analyst some thinking time during the interview.
- *Listening* As well as being a useful skill in its own right, listening is one of the best methods of maintaining control of the interview. By showing that you are involved and interested in the discussion, through the correct use of eyes, body, head and voice, you are confirming to the interviewee that he or she is the most important person to you at that moment. If in addition to this, you are concentrating hard on what is being said, it is much easier to find a point at which to enter the discussion and then move it on to the next 'branch' of the 'tree' – the next topic on your agenda – without antagonising the interviewee.
- *Pausing* A well placed pause or silence is a good way of keeping control and allowing you to decide the next step. Silence will also often encourage a quiet person to provide more information.

Recording the interview Having looked at planning and conducting a fact-finding interview, we now turn to the subject of how to record it. This is an area which is particularly important when asking questions, as the record of the answers given will the form the basis for constructing models of the system. No matter how skilled the interviewer, how good their questioning and listening, the interview is likely to be less than successful if the recording techniques are incomplete or inappropriate. Often

Systems Analysis: Asking Questions and Collecting Data   137

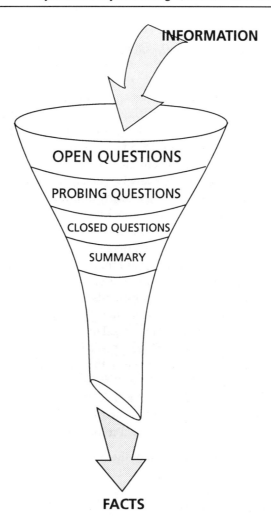

Fig. 9.5   The fact-finding funnel

interviewers comment on how difficult it is to develop a recording technique which is easy to use, unobtrusive during the interview, non-threatening to the interviewee and understandable after the interview.

Some general guidelines for note-taking include:

- Always ask for permission to take notes.
- Ensure that you use an 'open' note-taking style and don't hide what you are writing from the interviewee.
- Pause during the interview while recording important information, don't try to listen and take notes at the same time.
- Check with the interviewee that the information recorded is correct and make sure that any actions on either party are accurately recorded.

Many fact-finding interviews result in the interviewee describing a lot of complex procedural details and it can be difficult to capture this information unambiguously. A graphical technique can simplify the capture of this sort of information, but to be effective it must be simple and quick to use, and be easily understood by the interviewee so it can be checked for accuracy. For example, a mind map can be used to record information. This consists of a central idea, or topic, enclosed in a circle with related topics radiating out from the circle. The data flow diagram, which provides a pictorial view of how data moves around an information system, is another recording technique which has been used effectively by many analysts. It is described in the next chapter.

In order to take reliable information away from a fact-finding interview, the time immediately following the interview is critical. It is essential to read through your notes and ensure they make sense. Complete any unfinished sentences, tidy up any parts which are difficult to read, and fill in any gaps while the information is still fresh in your memory. If this task is left too long – and overnight is often too long – you will not be able to remember all the important points, or what you intended by the hieroglyphics on the page! At the same time as carrying out this check you can add any additional comments to your notes, such as informal comments or feelings which might be useful in subsequent interviews or which may help other members of the team.

The other recording task after the interview, although this one is not so urgent, is to prepare a formal record of the interview. This serves at least two purposes: it is the official record of the interview which can be accessed by others on the project, and it is a document which can be checked by the interviewee to make sure it is accurate. The NCC Interview Report provides the headings required for a formal record of this kind. Figure 9.6 is a formal interview record based on the NCC form. This is a comprehensive form, and the amount of detail may not be appropriate in every case. However a formal record should contain the following information as a minimum:

- the date, location and duration of the interview
- the names of the attendees
- the agenda or objectives
- the main points discussed
- any conclusions
- any actions
- the date of next meeting if appropriate.

Difficult interviews The interviewing techniques described are based on good practice and have been successfully used in numerous fact-finding interviews. However, there will be times when you find yourself in a situation where, even before you ask the first question, you realise that because of the attitude of the interviewee, the interview is going to be difficult. It might be that the interviewee is very talkative, and begins to explain at high speed, and not very well, the procedures in their job. In this case, the control techniques outlined earlier would be appropriate. However an interviewee who appears quiet and defensive is often more challenging, and an aggressive interviewee is, for many analysts, the most difficult of all. In both of the latter situations, it would first be necessary to attempt to build rapport with the interviewee if the interview is to be effective. This is because real communication takes place at an emotional level as well as at logical verbal level. Building rapport means finding connections between yourself and the interviewee and developing these 'bridges' by:

Title Interview record	System BBS	Document 2.1	Name CTS/IR 14	Sheet 1 of 2
Participants Pat Clarke, CTS Bookings Manager			**Date** 5th January 1994	
Objectives/Agenda To understand Pat Clarke's role, including details of procedures; to determine the problems encountered when using the current system; and to establish her requirements for a new system			**Location** CTS offices	
			Duration 10.00am – 11.15am	

Results	Cross-reference
1. BACKGROUND Pat Clarke has worked for Computer Training Services (CTS) since it was set up by the parent company, Industrial Services Ltd, in 1986. She is responsible for maintaining the booking system in CTS. Customers book courses, and PC keeps a record of these (on the bookings board), sends acknowledgements and joining instructions, deals with enquiries and cancellations, supervises resourcing of courses and prepares a monthly report for the Training Director. PC reports to the Training Director and has an assistant, Sandy Southgate, who is responsible for providing resources for scheduled courses. PC's role is central to CTS's booking and billing system and as the company has grown, so have Pat's responsibilities. 2. DETAILED PROCEDURES. 2.1 Bookings Customers may book courses on the official CTS booking form and this is treated as a confirmed booking. Any other booking is described as provisional. These may be telephoned bookings, memos or letters from customers, or memos from the sales department. PC phones customers who have booked provisional places, five weeks before the course to gain confirmation. Customers may confirm in writing at any time. If a provisional booking is confirmed, it is marked with a tick on the bookings board to indicate a confirmed booking. 2.2 Cancellations If a customer cancels up to five weeks before the course, they will not be charged. If they cancel less than five weeks before the course, they will be charged the full amount. 2.3 Under subscribed courses If there are still places on a course five weeks before it runs, PC notifies Chris Hislop in Sales who will try to fill the remaining places. If the course does not reach the minimum number of delegates, CH and the Training Director will decide whether this course should run, and PC is informed (in writing) of their decision.	 Doc. IR 14/1 Organisation Chart Doc. IR 14/2 Booking form

Fig. 9.6   A formal interview record

- involving the interviewee in discussion
- listening to and understanding their point of view
- gaining their trust.

It takes time and energy to develop empathy by trying to see the situation from the interviewee's point of view, but if the interviewer is able to do this, the interviewee will be less likely to defend or justify their position. They may even look at their own point of view more objectively.

The analyst contributes to the building of rapport by:

- *Demonstrating personal warmth*
  By smiling and showing interest .
- *Removing barriers*
  Communication consists of one person, the sender, sending a message and another person, the receiver, receiving and understanding it and then responding in some way. In effective communication there will also be some degree of rapport between the sender and the receiver. Barriers can block this communication, as shown in figure 9.7. These barriers may be *physical*, involving furniture arrangements, noise, or interruptions. They may be *semantic*, concerning the choice of words and how they are said – for example being too technical, using jargon, over-simplifying or engaging in sarcasm. The barriers might also be to do with *body language*: either the inappropriate use of body language by the interviewer or a failure to 'read' the body language of the interviewee. On this last point, the interviewer should be on the look-out for any significant change in the body language of the interviewer which suggests the interview is moving into a difficult area – for example turning away, avoiding eye contact, leaning back in the chair. Another signal to look out for is body language which doesn't match what is being said

We have already discussed the key role of active listening in the fact-finding interview. In addition to listening, two other approaches to building rapport are:

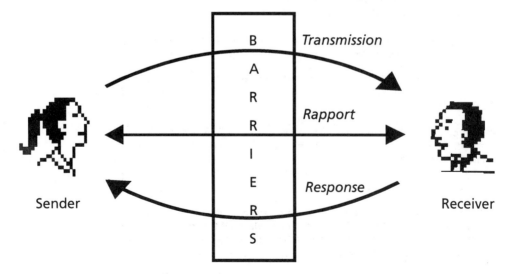

Fig. 9.7 Barriers to effective communication

- *Involving the interviewee*

  We can involve the interviewee by asking for information or investigating their opinion. This can often *draw out* the intention, goals and position of others. For example, ' I'd be interested to hear your suggestions about how to deal with these problems'.

  The interviewer can also *encourage* the interviewee by giving credit to their contribution. For example, ' That's a very interesting point you have raised. Could you explain how that might work'.

  By being *responsive* to the interviewee's questions and concerns, the interviewer is building an atmosphere of trust. For example, ' I can understand your concern about your job, but at present this investigation only covers the current situation'.

- *Openness*

  By being open about their motives and intentions, interviewers help to create that same behaviour in the interviewee. Disclosing information is one way to achieve openness. For example, ' If you are concerned about the recommendations, it may be possible to give you a copy of the report before it is published.'

  By accepting criticism without being defensive, we can increase our credibility on major points where it really matters with the interviewee. For example, ' You're right. I haven't taken all the points into consideration'.

  Asking for help lets the interviewee know that you need their input to find the solutions to their problems. For example, ' You know, I can't do this without your help.'

  Remember that your objective is always to build trust. As the confidence of the interviewee grows so will their openness and honesty. In situations where the interviewer is met with an aggressive attitude or some other strong emotion, it is important to bear in mind that this feeling is usually directed at the company or the system represented by the analyst rather than at the analyst personally.

9.3 QUESTIONNAIRES

The use of a questionnaire might seem an attractive idea as data can be collected from a lot of people without having to visit them all. While this is true, the method should be used with care since it is difficult to design a questionnaire which is both simple and comprehensive. Also, unless the questions are kept short and clear, they may be misunderstood by those questioned, making the data collected unreliable.

A questionnaire may be the most effective method of fact-finding to collect a small amount of data from a lot of people. For example, where staff are located over a widely spread geographical area; when data must be collected from a large number of staff; when time is short; and when a hundred per cent coverage is not essential. A questionnaire can also be used as a means of verifying data collected using other methods or as the basis for the question and answer section of a fact-finding interview. Another effective use of a questionnaire is to send it out in advance of an interview. This will enable the respondent to assemble the required information before the meeting which means that the interview can be more productive.

When designing a questionnaire, there are three sections to consider:

A heading section which describes the purpose of the questionnaire and contains the main references – name, staff identification number, date, etc.

A classification section for collecting information which can later be used for analysing and summarising the total data, such as age, sex, grade, job title, location, etc.

A data section made up of questions designed to elicit the specific information being sought by the analyst.

These three sections can clearly be observed in the questionnaire in figure 9.8: the heading section is at the top (name and date), the next line down is for recording classification information (job title, department and section), and the data section is designed to gather

XYZ CO. LTD – DUTIES LIST			
SURNAME AND INITIALS		DATE FORM COMPLETED	
YOUR JOB TITLE	DEPARTMENT	SECTION	
Enter each main duty you perform, and indicate how many hours per week it requires–			
No.	DESCRIPTION OF DUTY		Approx. hours per week
	Other activities (lunch, tea-breaks, etc.)		
	TOTAL HOURS WORKED PER WEEK		
	WHEN COMPLETED HAND THIS FORM TO YOUR SUPERVISOR		

Fig. 9.8 Sample questionnaire

information about the duties associated with a particular job.

The questionnaire designer aims to formulate questions in the data section which are not open to misinterpretation, which are unbiased, and which are certain to obtain the exact data required. This is not easy, but the principles of questioning described earlier in this chapter provide a useful guide. A mixture of open and closed questions can be used, although it is useful to restrict the scope of open questions so that the questionnaire focuses on the area under investigation. To ensure that your questions are unambiguous, you will need to test them on a sample of other people, and if the questionnaire is going to be the major fact-finding instrument, the whole document will have to be field tested and validated before its widespread use.

One way of avoiding misunderstanding and gaining co-operation is to write a covering letter, explaining the purpose of the questionnaire and emphasising the date by which the questionnaire should be returned. Even with this explanation, an expectation that all the questionnaires will be completed and returned is not realistic. Some people object to filling in forms, some will forget, while others delay completing them until they are eventually lost!

To conclude then, a questionnaire can be a useful fact-finding instrument, but care must be taken with the design. A poor design can mean that the form is difficult to complete which will result in poor quality information being returned to the analyst.

9.4 OBSERVATION

The systems analyst is constantly observing, and observations often provide clues about why the current system is not functioning properly. Observations of the local environment during visits to the client site, as well as very small and unexpected incidents, can all be significant in later analysis, and it is important that they are recorded at the time.

The analyst may also be involved in undertaking planned or conscious observations when it is decided to use this technique for part of the study. This will involve watching an operation for a period to see exactly what happens. Clearly, formal observation should only be used if agreement is given and users are prepared to co-operate. Observation should be done openly, as covert observation can undermine any trust and goodwill the analyst manages to build up. The technique is particularly good for tracing bottlenecks and checking facts that have already been noted.

A checklist, highlighting useful areas for observation, is shown as figure 9.9.

A related method is called *systematic activity sampling*. This involves making observations of a particular operation at predetermined times. The times are chosen initially by some random device, so that the staff carrying out the operation do not know in advance when they will next be under observation.

9.5 RECORD SEARCHING

Time constraints can prevent systems analysts from making as thorough an investigation of the current system as they might wish. Another approach, which enables conclusions to be drawn from a sample of past and present results of the current system, is called

☐ **Working conditions**
    light
    heat
    noise
    interruptions

☐ **Layout**
    ease of access
    movement possible
    proximity to colleagues, filing systems and telephones

☐ **Ergonomics**
    workstation arrangements for microcomputing, use of terminals and printers
    furniture layout
    adequacy of furnishings

☐ **Supervision**
    management style
    availability when needed

☐ **Workload**
    light, heavy, variable, bottlenecks

☐ **Pace and method of working**
    peaks and troughs of activity
    procedures and standards

Fig. 9.9 Observation checklist

record searching. This involves looking through written records to obtain quantitative information, and to confirm or quantify information already supplied by user staff or management. Information can be collected about:

- the volume of file data and transactions, frequencies, and trends
- the frequency with which files are updated
- the accuracy of the data held in the system
- unused forms
- exceptions and omissions.

Using this information, an assessment of the volatility of the information can be made and the usefulness of existing information can be questioned if it appears that some file data is merely updated, often inaccurate or little used. All of the information collected by record searching can be used to cross-check information given by users of the system. Whilst this doesn't imply that user opinion will be inaccurate, discrepancies can be evaluated and the reasons for them discussed.

Where there is a large number of documents, statistical sampling can be used. This will involve sampling randomly or systematically to provide the required quantitative and qualitative information. This can be perfectly satisfactory if the analyst understands the concepts behind statistical sampling, but is a very hazardous exercise for the untrained. One particularly common fallacy is to draw conclusions from a non-representative sample. Extra care should be taken in estimating volumes and data field sizes from the evidence of a sample. For instance, a small sample of cash receipts inspected during a mid-month slack period might indicate an average of 40 per day, with a maximum value of £1500 for any one item. Such a sample used indiscriminately for system design might be disastrous if the real-life situation was that the number of receipts ranged from 20 to 2000 per day depending upon the time in the month, and that, exceptionally, cheques for over £100,000 were received. It is therefore recommended that more than one sample is taken and the results compared to ensure consistency.

9.6 DOCUMENT ANALYSIS

When investigating the data flowing through a system another useful technique is to collect documents which show how this information is organised. Such documents might include reports, forms, organisation charts or formal lists. In order to fully understand the purpose of a document and its importance to the business, the analyst must ask questions about how, where, why and when it is used. This is called document analysis, and is another fact-finding technique available. It is particularly powerful when used in combination with one or more of the other techniques described in this chapter.

The NCC have prepared a form, the Clerical Document Specification, to assist with the process of document analysis, a copy of which is shown as figure 9.10. Even if the form cannot be fully completed at the first attempt, it will point to those topics which need further investigation.

9.7 SUMMARY

As we have seen, there is a variety of methods which can be used by the systems analyst to gather information. These methods involve asking questions and collecting data.

Asking questions is a key activity which can be achieved either by interviewing or through the use of a questionnaire. Interviewing is the most widely used fact-finding technique, and is at the heart of systems analysis. To be an effective interviewer, the analyst must work on their interpersonal skills, such as listening and building rapport, as well as developing effective recording techniques. Questionnaires should be used with care, and are only really appropriate when a relatively small amount of information is required from a large number of people or from remote locations.

Data can also be collected by other methods which do not involve asking questions directly, although their use may lead the analyst to further investigation. These methods include observation, record searching and document analysis. Observation can help in understanding where the current system is working well and where problems are being experienced, and can also provide clues as to why this is happening. Searching records enables the analyst to confirm or quantify information which has been provided by the client, while document analysis is a systematic approach to asking questions about documents collected during the investigation.

These techniques can be regarded as a tool kit which can be used flexibly by systems analysts during their investigation. They can be used in isolation, or in combination, to gather information about the current system and to understand the requirements for any new system. The way in which the data collected is recorded by the analyst is the subject of the next chapter.

Clerical Document Specification	Document description		System	Document	Name	Sheet
	Purchase Order		POS	3	PUORD	1
NCC	Stationery ref. DS 46	Size A4		Number of parts 4	Method of preparation Typed	
	Filing sequence by order number		Medium loose leaf binder		Prepared/maintained by HO Admin	
	Frequency of preparation as required		Retention period 3 months after payment		Location HO Admin supervisor	
	Monthly VOLUME 5	Minimum 20	Maximum 300	Av/Abs 120	Growth rate/fluctuations no growth likely	

Users/recipients	Purpose	Frequency of use
– HO Admin – Purchase accounts – Originator of order request	Raise order To check against supplier invoice To check against delivery & authorise payment	daily monthly weekly

Ref.	Item	Picture	Occurence	Value range	Source of data
1	Supplier name		1 per order		POR
2	Item to be ordered	9 (6)	5 per order	000001–999999	POR
3	Quantity of item	9 (6)	as ref 2	000001–999999	POR
4	Est. cost of item	£999999.99	as ref 2	000001–999999	POR
5	Total order value	£999999.99	1 per order	000001–999999	POR
6	Delivery address		1 per order		POR
7	Authorised signature		1 per order		HO Admin
8	Date of order	99AAA99	1 per order	rec'd date	HO Admin
9	Order number	9 (5)	1 per order	00000–99999	pre printed

Notes

S 41

Author DY

Issue 3

Date 12.8.93

Fig. 9.10 NCC clerical document specification

CHAPTER 9 CASE STUDY AND EXERCISES

Q1 In this chapter, five different methods for the collection of information have been described. They can be used separately or in combination during fact-finding. Choose different circumstances from System Telecom to illustrate the usefulness of these methods.

Q2 Your boss is thinking about writing a book about data processing management. To help her come to a conclusion she has asked you to prepare a report identifying what's involved in writing such a book and what the benefits might be. You will have to interview her and one or two authors in addition to gather information. How will you go about

- preparing for these interviews
- conducting the interviews
- analysing the information

What conclusion will you come to, and how will you tell your boss!

Q3 During the fact-finding stage of a big project, different activities are carried out by different members of the project team; for example some people may investigate processing requirements while others analyse the data. This may mean that there are many different views about how the existing system works, and the nature of the new requirements. How can you ensure that all of the systems investigation work is consolidated into a coherent whole?

CHAPTER 10

Systems Analysis: *Recording the Information*

10.1 INTRODUCTION

Having looked at planning in chapter 8, and at the process of asking questions and collecting data in chapter 9, we now look at the third stage of the PARIS model of systems analysis – Recording the Information. Typically, the analyst collects a considerable amount of information during the investigation phase which may include interview reports, observation records, sample documents, completed questionnaires and lists of problems and requirements. Some of this relates to the current system, and some to the new system required by the client.

It is important for the systems analyst to record this information in an unambiguous, concise manner which will be clear and accessible to others, and which can be used by other analysts and designers involved in developing the new system. Structured techniques were developed to help system developers to record information in this way, using diagrams and a limited amount of text, and in this chapter we introduce some of the key techniques.

As we saw in chapter 2, modern methods of structured analysis and design view information systems principally in two ways, firstly as data, the information that the system records, and secondly as processing, what the system does with this data. The most common techniques used for describing these two aspects of systems are data flow diagrams for processing, and entity models – also described as logical data structures or entity relationship diagrams – for data. Neither of these techniques however describes the precise order in which the system processes data. A third major view of the system, which describes the order in which external and internal events affect the system and the data, and brings together a view of the processing to complete the description of the system, is described in the next chapter.

Structured techniques are used to model the existing manual or computer system using information gathered from users. These models are then modified and extended with new user requirements in order to produce models of the new, required, system. The model of the required system is based on the current system because there is usually a core set of processing which will still be required unless there has been a drastic change in the way the enterprise operates. The data held by the existing system is unlikely to change although in most cases new data will be required.

This chapter introduces the techniques of data flow diagramming and entity modelling and describes how they are used, together with a data dictionary, to record information about the current system, and explains how a requirements catalogue is used to record new requirements. These activities are closely related and are conducted in parallel during the early stages of requirements analysis.

Before introducing these recording techniques, we will briefly describe data dictionaries and CASE tools which help the analyst to create and edit diagrams, and to store data in a structured and meaningful way. Without data dictionaries and CASE tools, the large-scale application of structured methods would be difficult, if not impossible, to achieve.

10.2 DATA DICTIONARIES AND CASE TOOLS

An essential concept underlying the use of structured methods is that of the *data dictionary*. A data dictionary is used to record all those pieces of information about a system (textual or numeric) that cannot be recorded on diagrams. It is the underlying structure which links the different views of the system presented by different types of diagrams. In SSADM, for example, the importance of the data dictionary is recognised in the concepts of data flow models (DFMs) and logical data models (LDMs) in which diagrams and their associated background documentation are considered as a whole. In effect it is a database which supports analysts and designers in their tasks of analysing, modelling and designing computer systems.

It is possible to use a paper-based data dictionary but purpose-designed dictionaries based on commercial database management systems (DBMS) are now generally used. This is because they show the relationships between all the different components of the system's models and designs that are constructed in a project using structured methods. This enables the impact of changes in one part of a model to be traced through to other components making it much easier to keep the models consistent and accurate.

CASE stands for Computer Aided Systems Engineering and is a term used to describe any software tool designed to make the development of computer systems easier. A data dictionary system can therefore be considered a CASE tool although the data dictionary concept predates that of CASE. The term is also used to describe drawing tools specifically designed for the production of the types of diagrams used in structured methods. These tools make it much easier and quicker to create and change diagrams than using paper and pencil, or even general-purpose computer-based drawing tools. They also produce more professional and readable results than manual methods. In addition the more advanced CASE tools incorporate drawing tools and a data dictionary making it possible to check sets of diagrams for consistency and to see the results of changes. These tools can be used to generate components of the eventual physical system, such as the database design, from the system models produced during analysis, and can partially automate many trivial analysis and design tasks.

10.3 DATA FLOW DIAGRAMS

Data flow diagrams are the most commonly used way of documenting the processing of current and required systems. As their name suggests, they are a pictorial way of showing the flow of data into, around and out of a system. They can be understood by users and are less prone to misinterpretation than textual description. A complete set of DFDs provides a compact top-down representation of a system which makes it easier for users and analysts to envisage the system as a whole.

10.3.1 DFD components

DFDs are constructed using four major components: external entities, data stores, processes and data flows. Figure 10.1 shows how a DFD would be constructed using SSADM conventions, in which a process is represented by a rectangle, a data store by an open rectangle, a data flow by an arrow and an external entity by an ellipse.

External entities represent the sources of data that enter the system or the recipients of data that leave the system. Examples are clerks who enter data into the system or customers who receive letters produced by the system. In SSADM they are shown as ellipses with a name and a unique identifier consisting of a single lower case letter.

Data stores represent stores of data within the system. Examples are computer files or databases or, in a manual system, paper files held in filing cabinets. They are drawn as open-ended rectangles with a unique identifier, a box at the closed end and the name of the store in the open section. Manual data stores are identified by the letter M followed by a number and identifiers for computerised stores are prefixed by a D. Data stores cannot be directly linked by data flows either to each other or to external entities without an intervening process to transform the data.

Processes represent activities in which data is manipulated by being stored or retrieved or transformed in some way. They are shown as larger rectangles with a numeric identifier in a box at the top left corner. The location where the process takes place or the job role performing it is recorded in a box in the top right corner and is only used in diagrams of the current physical system. The name of the process is recorded in the remaining area at the bottom. Process names should be unambiguous and should convey as much meaning as possible without being too long. In general, names should take the form of an imperative

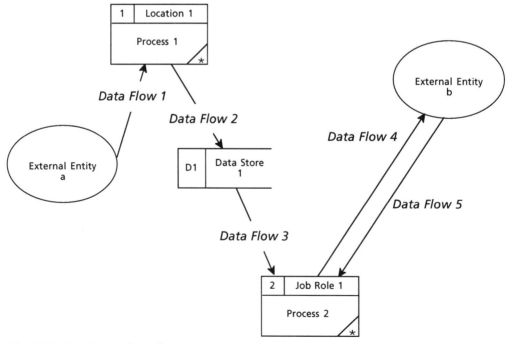

Fig. 10.1 Basic data flow diagram components

and its object as in Open Account. Generalised verbs such as Process or Update are not usually helpful.

Data flows represent the movement of data between other components, for example a report produced by a process and sent to an external entity. They are shown as named arrows connecting the other components of the diagram. Data flows are generally shown as one-way only. Data flows between external entities are shown as dotted lines.

It may be necessary to draw the same external entity or data store more than once on the same DFD in order to make the diagram clear and to avoid too many crossing lines. An external entity which appears more than once in the same diagram is shown with a diagonal line at the top left. Duplicate data stores are given an additional vertical line on the left side of their reference boxes.

10.3.2 DFD hierarchies

A system is rarely simple enough to be shown on a single DFD and so a hierarchical set is produced. This consists of a top-level DFD in which the processes are major system functions described in more detail by one or more associated lower-level DFDs. The process of breaking a higher-level (parent) DFD into its constituent lower-level (child) DFDs is known as *levelling*. There are no particular rules to levelling, the aim being simply to make the diagrams less cluttered and therefore easier to read and understand. As a rule of thumb however processes on the lowest level, called elementary processes, should correspond to single events or actions affecting the system, for example cashing a cheque in a banking system.

Figure 10.2 shows a level 1 (top level) DFD and figure 10.3 shows a DFD from the next level down. The diagrams illustrate a number of additional conventions which have not yet been described.

The first thing to notice is that the child DFD – level 2 – has a box round it. This corresponds to a process box on the level 1 DFD and contains the same identifier. Processes on the child are given identifiers which show that they are sub-processes of the one on the parent by adding a number to the identifier of the parent, separated by a point, for example process 1 on the level 2 DFD for process 4 has the identifier 4.1. The values of these numbers have no significance since DFDs do not show the sequence of processing, only the flows of data between them. In this they differ from flowcharts which are specifically intended to show a sequence of actions and decisions. Some of the process boxes on the diagrams are shown with an asterisk enclosed by a diagonal at bottom right. This means that the process concerned is not broken down any further and is therefore an elementary process. These occur in figures 10.1, 10.2, and 10.3.

External entities, data flows and data stores can also be broken down into more detail on lower-level DFDs. External entities which are components of a parent external entity are distinguished by suffixing numbers to the parent identifier. In a similar way child data stores are referenced by their parent's identifier followed by a lower case letter. Data store 1 within process 4 for example has the identifier D4/1. Data flows are not given any reference other than their name but a data flow shown as a single line on a parent DFD may represent a number of flows on its child. On level 1 DFDs double-headed arrows may be used as a short-hand way of representing two or more flows in opposite directions in lower-level DFDs.

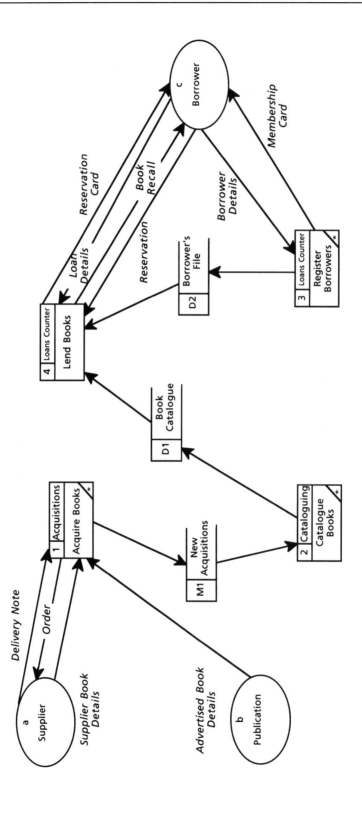

Fig. 10.2 Level 1 DFD for a library system

Systems Analysis: Recording the Information 153

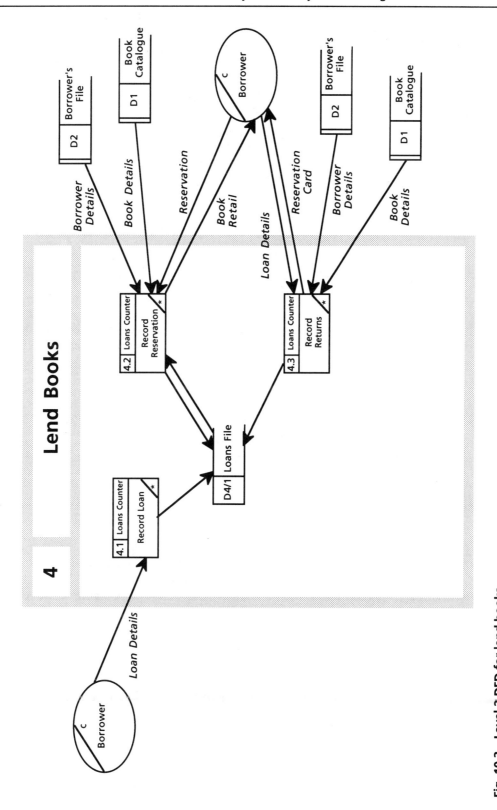

Fig. 10.3 Level 2 DFD for lend books

Diagrams alone are not sufficient to convey an understanding of the system being modelled. It is important that DFD components are described in more detail than can be conveyed by the short names they are given in the diagram. For example data flows consist of a list of data items which must be documented in a data dictionary so that the data items can be related to attributes in the data model. The term data flow model (DFM) is used in SSADM to refer to a complete set of DFDs describing a system, plus the necessary data dictionary documentation of objects shown on the diagrams. The extent to which higher-level DFDs are documented may vary but it is essential that objects at the bottom levels are clearly described in order that the diagrams can be unambiguously understood and to ensure that information about the system is carried through to later stages of the project.

Generally speaking only external data flows, data flows into and out of the system, need to be described in detail. These *Input/Output* (I/O) descriptions contain:

- Identifier of the object from which the data flows (e.g. the process number)
- Identifier of the object to which the data flows
- Data flow name
- Description of the data flow.

In a similar way only the bottom level – the elementary processes – need to be described since this is where the detailed processing occurs. Elementary process descriptions consist of:

- Process number
- Process name
- Description of the process – this may be a simple textual description or may use more rigorous techniques such as structured English or decision tables.

External entity descriptions contain:

- External entity identifier
- External entity name
- Description of the external entity.

Data stores are initially documented in a similar way, by a text description associated with the data store identifier and name. However, once the data flow model and the data model have become reasonably stable, data stores are documented in terms of the entities which they contain, as this provides a good way of cross-referencing the two models.

10.4 MODELLING CURRENT PHYSICAL PROCESSING

So far we have described the structure and content of DFDs and DFMs but, given a clean sheet of paper at the beginning of a project, how do you build up your model of the existing system? The basic answer to this question is that you talk to users of the system and, using the techniques already described, gradually work down from a high-level view showing major functional areas by inserting more detailed processing into DFD levels which you create below these as you become more familiar with the system. It is important that this is done with constant feedback from the users to make sure that you have got things right. The identification of functional areas and lower-level processes is subjective but provided the diagrams are used as a means of communication between analysts and users a realistic consensus view of the system can be agreed.

Two areas may cause difficulty when starting out. One is defining the boundary of the system. What are the parts of the business to be included in the system? The other is, once the system boundary has been defined, what are the boundaries of the top-level functional areas included in the system? Two subsidiary techniques help to answer these questions: context diagrams and document flow diagrams.

A *context diagram* is similar to a top-level DFD but with the whole system shown as a 'black box'. In other words external entities and data flows into and out of the system are drawn but no processes or data stores are shown. An example is shown in figure 10.4. Context diagrams are used early in a project as a means of describing and agreeing the scope or boundary of the system to be developed.

Document flow diagrams may be used as a preliminary to producing DFDs for the current system in the early stages of a project. As their names suggest they are used to show how documents move round in a manual system. The example in figure 10.5 shows how

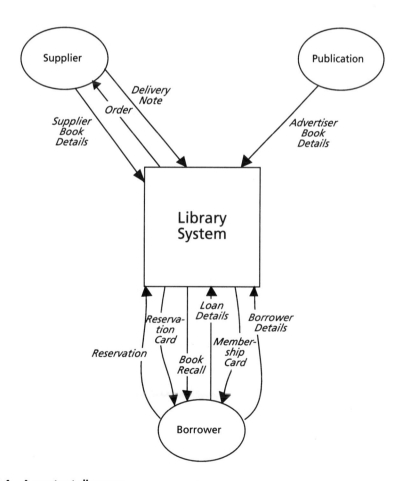

Fig. 10.4 A context diagram

documents and the people or departments who handle them are represented. By drawing boundaries around parts of the diagram, different functional areas of the system can be distinguished. Areas of the diagram outside the boundaries are external to the system and will appear as external entities and external data flows on the current system DFDs. The bounded areas will appear as processes on the level 1 DFD.

Top-level processes are usually fairly easy to identify as they often correspond to departments in the organisation. In our library example Acquisitions, Cataloguing and Loans are likely to be separate sections in a large library, run by different librarians. They indicate the presence of processes called Acquire Books, Catalogue Books and Lend Books. However these activities may not be organised into sections, and the staff in one section may perform a number of processes. The presence of major data stores such as the book catalogue and the loans file can also indicate different areas of activity even if the work is done by the same section.

When the top-level processes have been identified they should be looked at in more detail to see if they need to be broken down further. Lend Books, in our example, has been broken down into Issue Books, Reserve Books and Recall Books. These may in their turn be broken down still further. Issue Books may be split into Record Book Loan and Record Book Return. The aim is to break the processing down until the bottom-level processes each handle a single event such as the return of a book.

10.5 ENTITY MODELS

An entity model represents the network of relationships between classes of things which need to have data recorded about them in the system. The term entity type or entity is used to describe a 'class of things'. Having drawn an entity model, it is possible to show how the system can use these relationships by following them as paths for obtaining related pieces of data either for update or for reporting and enquiry purposes. For example many different transactions may be made on a bank account during a month. In order to print a monthly statement for the account, all the transactions must be found and a relationship between the *entities*' 'bank account' and 'transaction' must be present. The general concept of a bank account corresponds to an entity while the bank account for a particular customer is known as an *occurrence* or *instance* of the entity. As this example shows, entities represent not only physical objects but also conceptual objects, such as 'bank account', or events, such as 'transaction'.

10.5.1 The logical data structuring technique

Since we are using SSADM as our example structured method, we'll look at the Logical Data Structure (LDS) which is the name given to the entity model in SSADM. LDSs are simpler than DFDs in that they have only two major components, entities and the relationships between them.

Entities, as already described, are classes of things about which data is recorded in a system. An entity is usually represented as a rectangle containing its name, written as a singular noun. It is important that entities are given meaningful names, preferably ones

Systems Analysis: Recording the Information   157

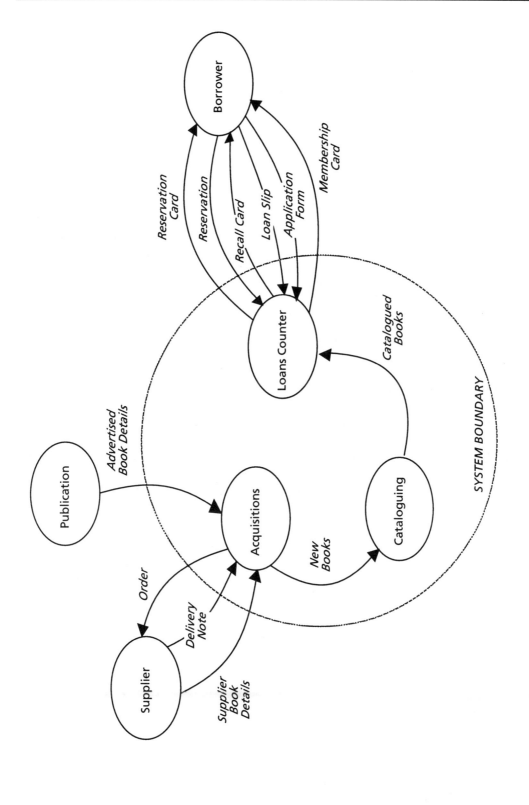

Fig. 10.5 Document flow diagram

which are used by or will be understood by the users. In SSADM the entity rectangles have rounded corners.

Relationships are shown as lines linking entities. Relationships can be traversed in both directions and so each end of a relationship is named in order to describe it from the point of view of the entity at that end. Most relationships are between one master entity and many detail entities. This is known as a *one-to-many relationship* and is shown by giving the line a 'crow's foot' at the many or detail end as shown in figure 10.6 which represents a relationship between one 'bank account' and many 'transactions'. Other methods may use an arrow rather than a crow's foot. Relationships may also be many-to-many or one-to-one. This classification of relationships is known as the *degree of the relationship*. One-to-one relationships are uncommon as it is usually found that two entities which are linked in this way can be combined to give a single entity. Many-to-many relationships are more common but it is usual to resolve them by introducing a new 'link' entity which is a detail of the two original entities, as shown in figure 10.7. In this case the many-to-many relationship between 'book' and 'borrower' is resolved by introducing the entity 'loan' which has a relationship with both 'borrower' and 'book'.

Relationships are also classified in terms of their optionality. *Optionality* is where the analyst considers whether an entity occurrence at one end of a relationship can ever be present in the system without the presence of a corresponding occurrence of the entity at the other end of the relationship. Figure 10.8 shows the different types of optionality that may occur.

Relationships can be described as *exclusive*. One type of exclusivity occurs if a detail entity has two (or more) masters and an occurrence of the detail may only be linked to one of the masters but not both. The other is the converse situation where a master may be linked to only one of two or more sets of details. Exclusive relationships are shown by drawing an exclusion arc to connect them as shown in figure 10.9.

The first diagram shows that an appointment may only be made with one doctor or one nurse or one midwife. An appointment may not be made, for example, with a doctor and a nurse or with two doctors. The second diagram states that a doctor may have responsibility either for one or more research programmes or for one or more patients

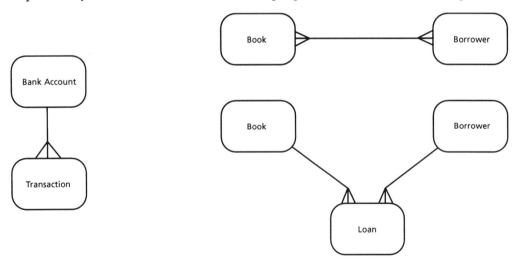

Fig. 10.6 A one-to-many relationship   **Fig. 10.7** Resolving a many-to-many relationship

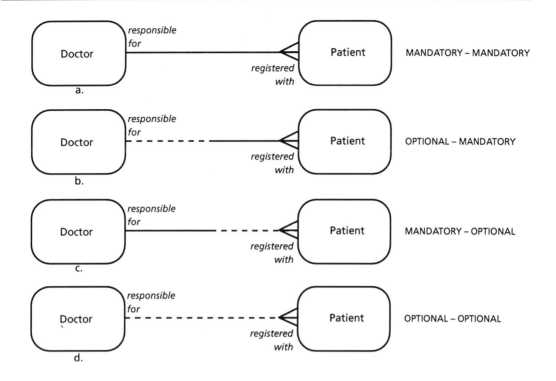

a. a doctor <u>must</u> be responsible for one or more patients and a patient <u>must</u> be registered with one and only one doctor.

b. a doctor <u>may</u> be responsible for one or more patients and a patient <u>must</u> be registered with one and only one doctor (for instance if some doctors only do research).

c. a doctor <u>must</u> be responsible for one or more patients and a patient <u>may</u> be registered with one and only one doctor (for instance if temporarily registered patients are not allocated to particular doctors).

d. a doctor <u>may</u> be responsible for one or more patients and a patient <u>may</u> be registered with one and only one doctor.

Fig. 10.8 Optionality in relationships

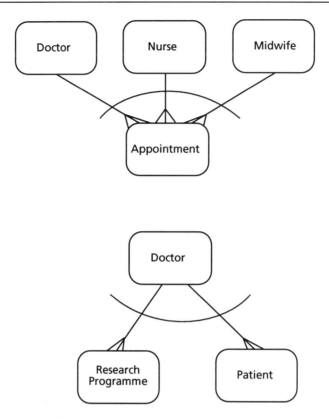

Fig. 10.9 Exclusivity in relationships

but cannot have responsibility for both patients and research programmes.

It is possible for entities to be related to themselves in what are called *recursive relationships*. In other words individual occurrences of entities can be related to other occurrences of that entity. There are two ways in which this can happen.

The first is where there is a one-to-many or hierarchical relationship between entity occurrences. An example of this is the relationship between managers in a company. The senior manager has a number of middle managers working for him, each of whom has a number of lower managers working for them. This can be shown either by identifying three entities called senior manager, middle manager and lower manager or by a single entity called manager which has a recursive relationship with itself. Figure 10.10 shows these alternative approaches. Notice that the recursive relationship is optional at both ends. This is because there is one manager, the senior manager, who does not report to another manager and because junior managers do not supervise other managers.

The second way in which an entity can be related to itself is where there is a many-to-many relationship between occurrences, indicating a network structure. The classic example of this sort of relationship is known as the bill of materials processing or BOMP structure. This is where, for example, a piece of machinery is made up of many different parts, some of which are sub-assemblies which themselves contain a range of different parts. This structure is not a hierarchy as some parts may be listed as components at a number of different levels of assembly. One way in which the structure can be shown is

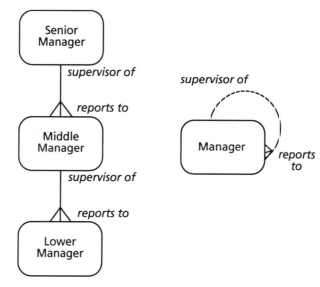

Fig. 10.10 Recursive relationships

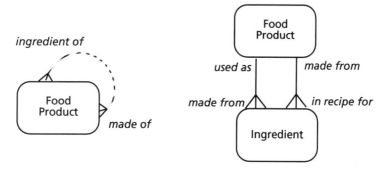

Fig. 10.11 More recursive relationships

as a single entity linked to itself by a many-to-many relationship. However a more helpful representation is gained by breaking the many-to-many relationship into two one-to-many relationships and creating a new entity which acts as a link between different occurrences of the original entity. These two representations are shown in figure 10.11 where food products are used as an example. For instance jam is a food product made up of other food products such as fruit and sugar. However jam may appear as an ingredient of a cake which also has fruit or sugar specified as ingredients independently of their use in the jam.

10.5.2 The logical data model (LDM)

As with DFDs, LDS diagrams are not able to carry all the information that analysts need to record about a system. A logical data model (LDM) consists of an LDS diagram and a set of entity descriptions and relationship descriptions which give more detail about the diagram components.

Relationship are usually documented in both directions. In other words they are described from the point of view of each of the two entities which make up the relationship. For each relationship 'half', the following information should be recorded:

- First entity name (the entity at this end of the relationship)
- Second entity name (the entity at the other end)
- Relationship name or phrase shown on the LDS
- Description of the relationship (in business terms)
- Degree of the relationship (one to many, many to one, etc.)
- Cardinality of the relationship (the number of second entities expected to be linked to each first entity – this may be an average or, better, a distribution)
- Optionality of the relationship
- List of users and their access rights to the relationship (update, read etc.).

Entity descriptions should contain at least the following information:

- Entity name
- Alternative names (synonyms)
- Description of the entity
- The owner. This is the user to whom the data in the entity belongs
- List of users and their access rights to the entity (update, read etc.)
- Expected number of occurrences of the entity and growth rates
- Rules for archiving and deleting entity occurrences.

One of the things we also need to record about an entity is the list of attributes it contains. An *attribute*, or data item, is a piece of information which the system needs to record about an entity. Attributes may be held by an entity purely as information or they may play a role in relationships between entities in which case they are known as *key attributes* or keys. Keys are principally of two types, prime keys and foreign keys.

Prime keys are used to identify different occurrences of the same entity. The entity 'bank account' is a generalisation of many different occurrences of individual bank accounts. To extend our previous example, in order to produce a statement for a particular account on demand it must be possible to pick that account out from all the other accounts in the system. The way to do this is to identify, or invent, a piece of information about the account that is unique and distinguishes it from all other accounts. The obvious candidate for this in our example is the account number.

Foreign keys are attributes which are also present as prime keys on other entities. They are another means of indicating that there is a relationship between the two entities. The entity which has the attribute as its prime key is the master of the entity in which it is a foreign key.

It is not always possible to use single attributes as keys. Sometimes a combination of two or more attributes is needed to uniquely identify an entity and act as its prime key. Keys consisting of more than one attribute are known as compound keys while keys which contain only one attribute are called simple keys.

10.6 MODELLING CURRENT DATA

Modelling the data in a system is in some ways more difficult than modelling the processes. Most people tend to think of what they do and how they do it. They think about the processing aspect of their job, more than the data or information they handle.

The starting point in developing a logical data structure (LDS) is to identify the entities which must be included. An initial set of the major entities is often fairly easy to develop. When reading documents which are used in or which describe the current system, or when talking to users, certain nouns will crop up over and over again. Frequent repetition of nouns such as 'customer' and 'account' will suggest the need for entities to represent them in a banking system, for example. A bank statement will reinforce the importance of 'account' as an entity by having an item called 'account number' on it. This indicates both an entity and its possible key. The example of a bank statement is also presented as a warning however. It might appear at first sight that an entity 'statement' should also be present on the LDS. This would only be so, however, if information about the bank statement itself needed to be held on the system. Virtually all the information on a statement belongs to some other entity such as 'customer', 'account' or 'transaction'. Probably the only information about the statement itself is the date it was produced and so the only reason for having a 'statement' entity would be if a chronological record of when statements were sent to customers was required.

Once an initial list of entities has been identified, a first attempt at an LDS can be made by considering the relationships between them. Some relationships may seem fairly obvious. For example a customer may have many accounts and, at first sight one, might think that an account may only be held by one customer. This would be represented by a one-to-many relationship between 'customer' and 'account' on the LDS. However all possible relationships between entities should be examined and decisions made on whether they are required or not. Drawing a matrix which cross-references entities to each other may help here. The exact rules for each relationship must then be established by questioning the users. It may turn out, in our example, that joint accounts are allowed and that the correct relationship is many to many. In early versions of the LDS produced by the analyst, many-to-many relationships can be left on the diagram, and they will normally be resolved into two, one-to-many relationships with a new 'link' entity at a future stage. In our example this might be called 'account holder'. This example shows how some, less obvious, entities may be identified in the course of examining the relationships between the major entities.

An example of a simple LDS for a library system is shown in figure 10.12. This should be compared with figure 10.7 to show how additional entities and more complex relationships may be discovered by investigating data requirements in more detail.

An LDS should be clearly laid out, with as few crossing lines as possible. The convention is normally to have masters positioned above details so that entities with the least number of occurrences tend to be at the top of the page and those with most at the bottom. This often means that the major entities appear in the top half of the diagram and lesser entities in the bottom half. An LDS is likely to need amending and/or extending many times before it reaches its final state and so use of a purpose-designed CASE tool is recommended to reduce the amount of time this takes and to give the most readable results.

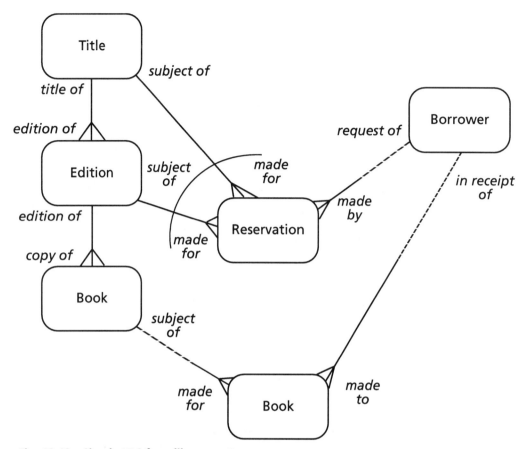

Fig. 10.12 Simple LDS for a library system

10.7 THE DATA CATALOGUE

As its name suggests the data catalogue is a list of all the data items or attributes which have been identified as being required in the system. Attributes are the individual items of data which are used to describe entities in the logical data model and which travel along data flows in DFMs where they are listed on the I/O descriptions. The data catalogue is in fact a subset of the data dictionary and is concerned with individual data items and the values they may take.

The information which should be recorded about attributes includes:

- Attribute name
- Alternative names (synonyms)
- Description of the attribute
- Attribute location (entity or data flow)
- Relationships to other attributes
- Format (including units and length)
- Values (or ranges of values) the attribute is allowed to have
- Rules for deriving the value of attribute occurrences

- Optionality of the attribute
- The owner, i.e. the user to whom the data in the attribute belongs
- List of users and their access rights to the attribute (update, read etc.).

10.8 RECORDING THE REQUIREMENTS

Requirements for the new system are recorded in the requirements catalogue together with the source of each requirement, the user or user area to which it belongs, how important it is and how it is to be satisfied by the new system. Requirements should also be cross-referenced to any processes in the current system DFM which carry out related processing. This makes it easier to map the requirements onto the current system DFM in order to develop the required system DFM later in the project.

The requirements catalogue is begun early in a project, as soon as any information has been gathered, but entries are developed and refined throughout the analysis and design stages as more is learned of the system and as the views of users and developers crystallise.

A requirement is commonly documented as a functional requirement – what the system must do in business terms – and a one or more associated non-functional requirements which specify measures of how the system is to deliver the functionality. For example a functional requirement might be to print library membership cards. An associated non-functional requirement may be that the cards should be printed within 30 seconds of confirming membership details.

The information which should be recorded for each requirement consists of:

- Requirement identifier
- Source of the requirement (the person who raised it)
- The owner of the requirement (the person or department to which it applies)
- Priority or importance of the requirement
- Description of the functional requirement
- List of the associated non-functional requirements and their values or measures
- Benefits of satisfying the requirement
- Possible solutions to the requirement
- Cross-references to related documents and requirements.

10.9 SUMMARY

This chapter has introduced logical data modelling for recording data and data flow modelling for processing. These techniques are central to virtually all structured analysis methods. The techniques use diagrams to make it easier to visualise the system as a whole and how the components fit together. They also record background information which cannot be shown on diagrams. This detail is recorded in a data dictionary or CASE tool which enables it to be structured in a way which ties in with the diagrams.

In the early stages of a project the techniques are used to describe the workings of the system which is to be replaced. This system is usually a mixture of manual and computer processing. The reason for modelling the current system is that it is rare for the underlying data and core processing to need to change radically. The aim is usually to improve and extend the system model and to take advantage of more modern hardware and software to provide better performance and facilities. The current system model therefore provides a starting point for developing the new system model.

At the same time that the current system model is being developed requirements for the new system are captured and recorded in a requirements catalogue. The new data and processing requirements are then applied to the current system model to produce a data flow model and a logical data model for the new system. The way in which this is done is described in the next chapter.

CHAPTER 10 CASE STUDY AND EXERCISES

Q1 The case study information you have shows a level 1 DFD and two level 2 DFDs (see pages 18, 19, 20). Using the level 2 DFDs as models, and the other case study information, draw a level 2 DFD for the process 'Issue Bills'.

Q2 Having drawn the level 2 DFD for 'Issue Bills' write the Elementary Process Description for one of the process boxes.

Q3 Using the level 1 DFD for System Telecom and other case study information, what are the major entities of the Customer Services System. There are more entities than this, so how have you chosen to identify the major ones? A good place to start is to list the criteria for identifying entities.

Q4 Having identified the major entities from the Customer Services System choose one of them and identify the data items which make up that entity.

CHAPTER 11

Systems Analysis: *Interpreting the Information Collected*

11.1 INTRODUCTION

At this point in the process, the analyst has collected information about the current system by asking questions, gathering documents and recording details about the current system using diagrams and text. It is now important to make sense of the data, to draw out the underlying logic of the system and to map on the requirements for the new system. This involves checking details with the user, amending existing models, and constructing new ones. We have called this stage – the fourth in the PARIS model – Interpreting the Information Collected.

The previous chapter introduced two structured techniques, data flow modelling and entity modelling, and showed how they are used to describe the existing manual or computer system. This chapter explains how the models created using these techniques are extended to incorporate the new requirements and it introduces new techniques which are used to model the required system in more detail and which form the basis of logical design. The most important of these new techniques adds the dimension of time to the picture of the system we have developed so far. It does this by modelling the sequence in which internal and external events trigger processes which change the data. The models introduced in this chapter replace data flow diagrams as the primary process models for the required system.

Throughout the chapter we will refer to SSADM, although similar techniques are used in other structured methodologies.

11.2 CREATING A LOGICAL MODEL OF CURRENT PROCESSING

Three different data flow models (DFMs) are produced during a project: the current physical DFM, which represents the current system 'warts and all' and was described in the previous chapter; the logical DFM which is produced by removing any duplicated or redundant processing or data from the current physical DFM; and the required system DFM, which shows how the new processing and data required by the users are incorporated into the logical model. Existing systems are often badly structured because they were not designed from first principles but were simply developed over time. As a result they can include inefficiencies such as the same data being held in more than one file or the same processing being performed more than once by different programs. In addition to this being a waste of effort it also increases the likelihood of error. Structured methods aim to increase efficiency and reduce error by creating systems in which common data and processing are shared wherever possible. One of the ways this is done is by rationalising or 'logicalising' the current system DFM before incorporating the requirements for the new system.

What does logicalisation mean in this context? Put simply, it is a tidying up process. We create a DFM which shows the existing system with all its inefficiencies and duplications removed. This results in a well organised and clear picture of the system as it should be rather than the way it actually is. By adding the new requirements to a logical DFM we do not perpetuate all the things which were wrong with the old system.

The major tasks in turning a physical DFD into a logical DFD include rationalising data stores and processes. Let's see how this is done.

An existing system may store customer information in more than one place, for instance paper records in one department and a PC database in another. These are combined into a single logical data store called Customer which includes the data from both physical data stores. In the library system, introduced in the previous chapter, the New Acquisitions and Book Catalogue data stores would be merged into a Books data store. Sometimes we also need to split physical data stores if they contain two completely separate types of data such as Customer information and Product information.

In addition to the main logical data stores there may be a need for transient or temporary data stores. These are common in physical systems and many of them are not logically necessary and will be removed in the logical model. However there may, for example, be a need for a batch of data to be entered to the system by one process and then checked and authorised by a manager using another process before the main data store is updated. Data in transient stores is always deleted by the process which reads it since the data is either accepted for inclusion in the main data stores or rejected.

Bottom-level processes are merged either because they are duplicates of each other or because they represent a sequence of small tasks which can be combined into one larger task. Duplicate processes often occur because two or more files contain the same information and the information is entered into each file by a separate process. We can combine these processes so that the data only has to be entered once.

Processes are removed if they do not update the information in the system but only reorganise it – for instance by sorting it – or output it in reports or enquiries. In the latter case the processes are not removed altogether but are simply moved to the requirements catalogue in order to make it easier to see the essential update processes on the DFDs.

When we are satisfied with the new sets of logical processes and data stores they need to be put together to form the new set of logical DFDs. The elementary processes are grouped to form higher-level processes or functions if they use closely-related data or if they are used by the same external entities. Processes may also be grouped if they have similar timing requirements such as year-end processes.

It is important that the logical DFM is checked against the current physical DFM to make sure that nothing essential has been lost and nothing added during logicalisation. It is also important that the model fits together and makes sense and that it ties in with the logical data model. This latter activity is aided by producing a logical data store/entity cross-reference table which shows which entities are contained in each logical data store. In the logical model an entity is only allowed to appear in one data store and cannot be split across different data stores.

11.3 MODELLING THE REQUIRED SYSTEM

Having created logical models of the current system, the requirements for the new system, documented during the analyst's investigation, can be added to these models.

Existing processes which are to be automated in the new system are carried over from the logical DFM to the required system DFM making sure that changes resulting from the new requirements are included. Existing processes which are to remain manual tasks are removed, along with any associated data stores, data flows and external entities which are not used by the remaining processes. New processes are modelled based on the information contained in the requirements catalogue, and added to the DFDs at the appropriate points with any new data flows, data stores and external entities that are needed. For instance a Record Fines process could be added to the set of processes included in the library system.

In a similar way entities and their relationships are carried over, added or removed to create the required entity model, and the data dictionary is updated to reflect the changes to the models.

Once the new models are fully documented in the data dictionary they are reviewed and checked against each other and against the requirements catalogue. Every requirement in the catalogue must have a cross-reference to the process or processes which are intended to satisfy it. To make sure that data requirements are satisfied, the contents of the data stores in the process model are checked against the contents of entities in the entity model using the updated logical data store/entity cross-reference.

There is also an additional technique which can be used to validate the entity model. Entity models are produced by a top-down approach, identifying the 'things' about which a system must hold data and then defining the relationships between them. The attributes of the entities – the items of data they contain – are identified and described in the data catalogue. An alternative, bottom-up, method is to identify the individual data items that the system must hold independently, and then to build data models by identifying the relationships between the items and grouping them according to these relationships. This technique is known as relational data analysis and is described fully in chapter 17. It can be used to build a model of the data which can then be compared with the required system entity model. If there is a difference between the models then the necessary corrections can be made.

11.4 ADDING THE TIME DIMENSION

We now come to the third major view of the system, one which will enable us to define the order in which processes update the data. This view models the events which affect the system and is important in structured methods, particularly in SSADM where it forms the basis of detailed process modelling and analysis. A number of different techniques have been devised to model this sort of information, such as state transition diagrams and petri nets, but we will continue to use SSADM to illustrate the structured approach, by describing two techniques: *entity life histories*, which provide a data-oriented view, and *effect correspondence diagrams* which provide a process-oriented view.

An event is something which happens, an occurrence. It is useful for the analyst to distinguish between 'real world' events, business events and system events. Real world events cause business events which in turn cause system events. In the earlier stages of analysis the distinction between the last two may be difficult to make as it will not yet be clear which parts of the business are to be computerised. An example of a common real world event is when a person moves house. This has an impact on businesses for whom that person is a customer – a business event – and will require the details held on a database by that business to be updated – a system event.

Analysts and users often speak in terms of the business event when they mean the corresponding system event, and vice versa. For an analyst, it is important that the effects of a system event are clearly defined to help with the process of logical design specification. A system event acts as a trigger for a process or set of processes to update a defined set of data.

An initial set of system events can therefore be identified by picking out all the data flows entering data stores in bottom-level DFDs in the required system DFM and tracing them back to an initial trigger either inside or outside the system. In the simple example used above we can trace a data flow into the 'Customer' data store back to the 'Maintain customer' process and thence back to the 'Customer address change' data flow which crosses the system boundary from the 'Records clerk' external entity. This data flow suggests one event, 'Customer change of address recorded', while the other data flow 'New customer details' suggests another event 'New customer recorded'. Not all events will be obvious from the DFDs. In this example other events can be envisaged such as 'Customer change of name recorded' even though there is no data flow corresponding to it. So it is important to carry this task out critically rather than treating it as a purely mechanical exercise. The process of reviewing the DFM to identify events may uncover omissions from the model, particularly if the user's help is enlisted, as it should be.

The type of event we have been discussing so far is described as an *externally triggered event*, as it requires data to enter the system from an external entity. The data flow represents the trigger for the processing associated with the event. There are also two other types

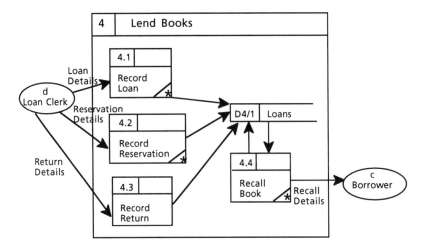

Fig. 11.1 Event recognition

of event both of which originate from within the system rather than outside it. These are the *time-based event* and the *system-recognised event*. They are both likely to be handled by batch processes since they do not have input from an external entity. The presence of all three types of event may be deduced by considering figure 11.1. 'Loan recorded', 'reservation recorded' and 'return recorded' are all externally triggered events. 'Recall issued' may be either a time-based event or a system-recognised event.

Time-based events are triggered by the arrival of a specified date or time or by the passage of a specified period of time following another event. This type of event is quite common, an example being 'Year end' for financial and accounting systems. In figure 11.1 'recall details' may be triggered by a loan becoming overdue a specified number of days after it was made.

System-recognised events are caused by the system recognising a change in its state caused by a piece of data it holds being changed to a specified value. The system monitors itself for this change and triggers a process when the change is detected. For example in figure 11.1 'recall details' could also be triggered when the system detects that a reservation has been made for a book title.

11.5 MODELLING THE EFFECTS OF SYSTEM EVENTS

The activities described in this section define the effects that system events may have on entities and the order in which these effects are allowed to occur, and provide the basic framework around which detailed models of individual update processes can be built. In SSADM, these techniques are known as entity-event modelling.

Two types of model are produced in entity-event modelling

- Entity life histories (ELHs) are diagrams that show which events affect particular entities and in what order.
- Effect correspondence diagrams (ECDs) show all the entities which may be affected by a single event plus any entities which may need to be read for navigation or reference purposes.

ELHs in particular are an important analysis tool since their production requires a detailed understanding of the business rules which control updates to the database. Close user involvement is needed in their production and this frequently highlights anomalies and omissions in the DFM or the LDM.

The first step in entity-event modelling is to create a matrix which cross-refers entities to the events which affect them. The list of entities which forms one axis of the matrix is taken from the required system entity model. The list of events for the other axis will come from the definitions of system functions. For every entity in the matrix, determine which events create it, which modify it and which delete it, and letter them C, M or D as appropriate (see figure 11.2).

When the entity/event matrix is thought to be complete it should be checked to make sure there are no empty rows or columns as this would imply redundant entities or events. The matrix should also be checked to make sure that every entity is created by an event and, preferably, is also modified and deleted by one or more events. Once we are satisfied that the matrix is as complete as it can be, work on ELHs and ECDs can begin. The starting

Entity Event	Title	Edition	Book	Loan	Reservation	Borrower
Book Acquired	C	C	C			
Book Catalogued			M			
Book Borrowed				C	D	M
Book Reserved				M	C	
Book Returned				D	M	M
Borrower Registered						C
Membership Terminated			M	D	D	D

Fig. 11.2 Part of an entity/event matrix

point for creating an ELH is the column for the relevant entity, while an ECD is constructed by taking all the entities which appear in its row in the matrix and putting them in the correct order as shown in section 11.9.

11.6 ENTITY LIFE HISTORIES (ELHs)

An entity life history (ELH) is a diagrammatic way of representing the different types of events that may affect an entity, the order in which they may occur and the effects that they may have. ELHs effectively summarise all the different life paths that occurrences of an entity may take between their creation on the system and their deletion.

The ELH technique is based on concepts developed by Michael Jackson for structured program design. These concepts are described in more detail in Chapter 21 but the essential idea is that all data processing can be described in terms of sequence (order), selection (choice) and iteration (repetition) of processing components, which are derived from the data structures. In an ELH these ideas are used by analogy to model sequences, selections and iterations of events affecting an entity.

An example of a simple ELH is shown in figure 11.3. This ELH looks very much like a Jackson structure diagram and contains examples of sequence, selection and iteration. It shows, in a simple library example, that a borrower entity is created when a borrower first registers with a library and that no more changes may be made to the entity after the borrower closes membership. In between these 'birth' and 'death' events there may be a number of 'life' events. Jackson rules are observed in that the diagram shows that it is possible for there to be no changes between creation and end of life for a particular instance of borrower since an iteration may occur zero, one or many times.

Figure 11.4 shows some of the other conventions which are used, in a completed ELH for borrower. The additional features are parallel lives, quits and resumes, operations and state indicators.

Parallel lives are used when there are two (or more) independent sets of events which can affect an entity. Since events from the two sets are not dependent on each other, but only on events from their own set, they cannot be ordered together in a predictable way. In figure 11.4 for instance, change of address or of name are events which can happen to people quite independently of the way they conduct their borrowing activities and

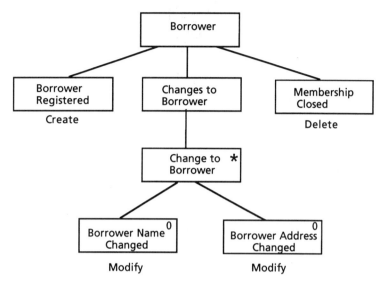

Fig. 11.3 Simple entity life history

therefore they have no direct relationship to events which are specific to their roles as borrowers. The diagram illustrates the convention whereby events which *are* specific to the system under consideration are treated as part of a primary life and are shown on the first branch of the parallel bar. The less application-specific events are shown as an alternative or secondary life on a second branch.

Quits and *resumes* are a means of jumping from one part of the diagram to another in order to cope with exceptional or unusual events. If used indiscriminately they can undermine the apparent structure of the diagram and make it more difficult to understand. Analysts should therefore only use a quit and resume when they are sure that there is no sensible way in which they can use normal Jackson structures to show what they want. Another reason for avoiding them if possible is that there is some controversy about exactly how they should be used. For example should a quit be mandatory or should it only come into effect if certain conditions are true. It is safer to make them mandatory and to annotate them to this effect but other analysts may not make it clear how they are using them so interpret with caution! Figure 11.4 shows a quit (Q1) from the Suspension Made Permanent event to a resume (R1) at the Membership Terminated event. This means that if a borrowing suspension is not lifted because the borrower does not return overdue books or pay for lost books then the system will allow librarians to terminate membership.

Operations are attached to the events to describe what update effects they have on the entity. An operation appears as a small box in which a number provides the key to a corresponding entry on a separate list of operations. SSADM provides a standard set of operations such as 'store', 'replace' and 'read' which are qualified by relevant attribute or entity names or by expressions such as 'replace balance using balance + transaction amount'.

State indicators are added to each event in order to provide a label for the state in which an occurrence of that event will leave the entity. These states may correspond to values of existing attributes or may require creation of a new 'entity status' attribute. The value of the state indicator on an entity occurrence shows the point it has reached in its life

174 Systems Analysis and Design

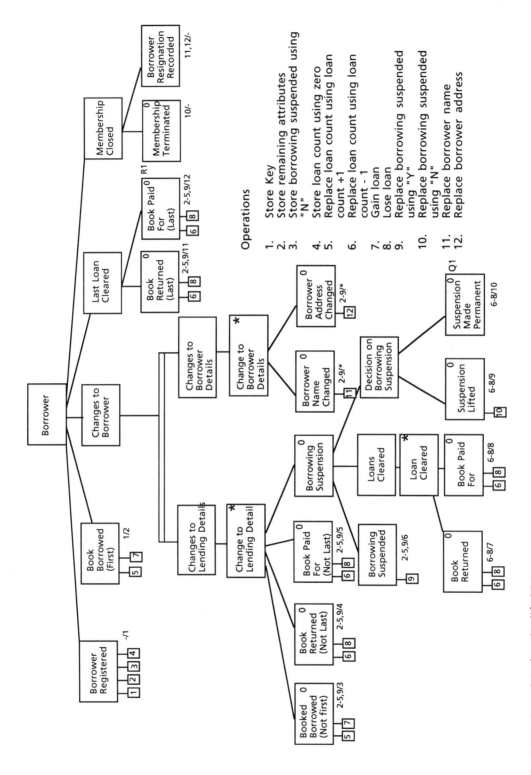

Fig. 11.4 Completed entity life history

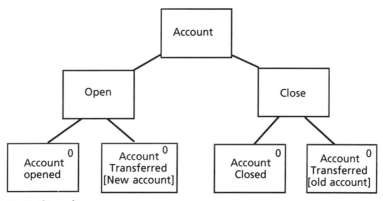

Fig 11.5 Entity roles

cycle and therefore what events are allowed to affect it next. State indicators are attached to each event on an ELH and valid before and after states for that event are shown, separated by an oblique. For example 2-5,9/3 means that the event to which this is attached puts the entity into state 3 and that the entity must be in states 2, 3, 4, 5 or 9 before this event is allowed to take effect. A state indicator consisting of a hyphen means no state, i.e. the entity does not exist (e.g.-/1 or 11,12/-). The purpose of state indicators is to indicate error processing which will be needed in the required system. Errors which result from allowing an event process to update data which is in the incorrect state are called *data integrity errors*. Where there is a parallel life a separate set of state indicators is needed to label the states in the second life. This is because there are no dependencies between events in the two lives and therefore no connection between the entity states that result from their effects.

A final point about ELHs is that there is not necessarily a 1:1 correspondence between events and their effect on entities. It is not unknown for a single occurrence of an event to affect two different occurrences of an entity in different ways. An example would be an account transfer event which involves the closure of one account and the opening of a new one. The account entity acts in two different roles in this situation and so the ELH shows entity roles as different effect boxes for the same event, with the role that the entity takes in each case shown in square brackets after the event name (see figure 11.5). It is also possible for separate occurrences of an event to affect the same entity occurrence in different ways. An example is an account closure request event. This event may have different effects depending on whether the account is in credit or whether it is in debit. In other words the effect of the event is qualified, depending on the state of the entity. Qualified effects are shown with the effect qualifier in round brackets after the event name. An example in figure 11.4 is the Book Borrowed event which has different effects depending on whether it is the first loan or a subsequent loan.

11.7 PRODUCING ENTITY LIFE HISTORIES

The power of the ELH technique is in its precision. In trying to describe *exactly* what should happen as a result of a given event, questions may be raised not only about the accuracy of the DFM and the LDM but also about the way the business is conducted in the real world.

A set of ELHs is usually created by making two passes through the LDS. The first pass works from the bottom of the LDS with those entities which do not act as masters in any relationship but only as details, and gradually moves up the diagram until all entities have been modelled. The reason for working from the bottom of the LDS upwards is that events which affect a detail entity may also have effects on its master and this is more easily seen by starting at the detail end of relationships. An example of this is where a balance attribute is held on an account entity. Whenever a new transaction entity is created for an account the balance on the account entity is modified by the transaction amount. The aim in this pass is to produce first-cut ELHs showing the 'typical' life of each entity. In the second pass the ELHs are reviewed starting with entities at the top of the LDS and working down to those at the bottom. This time all the less common or less obvious effects are searched out and the ELH structures modified to include them.

When developing an initial set of ELHs the entity/event matrix is used as a guide. For each entity, pick out its creation event from the entity/event matrix and draw it on the left of the diagram. If there are more than one represent them as a selection. Next look at the events which modify or update the entity and work out in what sequence they are allowed. Add them to the diagram after the creation events using a Jackson-like structure with sequence, selection and iteration as appropriate. This is usually the most complex part of the diagram and may need a lot of thought and consultation with the users in order to determine precisely what the rules are. Consideration of the detailed effects, in terms of operations, attributes updated and so on, should make the order clearer. In some cases the introduction of quits and resumes or parallel structures may be necessary, as described in the previous section. Lastly, deletion events must be added to the right of the diagram. The structure is not usually very complex but there may be selection between different events, as with creation.

When first-cut ELHs have been produced for all entities they are reviewed, starting with entities which have no masters, in order to incorporate less common or exceptional events. Situations which may need to be considered are :

- Interactions between entities
- Reversions
- Random events.

Interactions between entities often centre round death events. An example of how an event affecting a detail entity may also affect its master is where a bank account cannot be closed until all loans related to it have been paid off. Conversely, detail entities may be affected by events belonging to their masters. For instance an application for a bank account which is outstanding when a customer dies will not go through its normal life but will be terminated early as a result of the death of the customer. This is also an example of a *random event* the position of which on an ELH cannot be predicted. Random or unpredictable events are represented either by using the parallel life construct or by a special variant of quit and resume in which the quit is from anywhere on the ELH to a single resume event. Figure 11.6 shows a random event which could be added to the ELH in figure 11.4. This example shows an additional operation for recording the death of the borrower and would need a resume (R3) on the Membership Terminated event. This approach is normally used for exceptional events which lead to the death of the entity while parallel lives are used when the life continues.

Systems Analysis: Interpreting the Information Collected   177

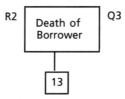

Fig. 11.6 Random event

Reversions occur when an event results in a return to an earlier point in an entity's life. The quit and resume notation is used, just as in figure 11.4, but the resume event is earlier in the life history than the quit event.

Once analysts are satisfied that the structures of the ELHs are correct the processing operations which produce the effects are added, though as already noted it is helpful to consider them in the earlier stages. If this has not been done it is quite likely that their addition will highlight errors in the ELHs.

The final task before going on to produce ECDs is that of adding state indicators to the ELHs. This is a mechanistic task which is done automatically by some CASE tools since the sequence of values for state indicators should follow the structure of the diagram.

11.8   EFFECT CORRESPONDENCE DIAGRAMS (ECDs)

Effect correspondence diagrams (ECDs) represent an alternative, event-focused, view of some of the information shown in ELHs. An ECD is produced for every event that can affect the system and shows all the entities which may be updated or read by the process corresponding to the event.

ECDs are named after the event they represent and show 1:1 correspondences between the entities which are updated by the event. If an event updates one occurrence of entity A and one of entity B then a rounded box is drawn for each entity and a double-headed arrow is drawn between them to show the correspondence. A Jackson-like construct is used to show if more than one occurrence of an entity is updated in the same way. In this case a box representing a set of the entity occurrences is drawn. This is connected by a plain line to a box below it which represents a single occurrence of the entity. This box has an asterisk in the top right corner to show that the effects of the event are repeated (iterated) for each occurrence of the entity.

Effect qualifier and entity roles, described in the previous section, are also shown and are called optional effects and simultaneous effects respectively. If the effect an event has may be qualified by the state of the entity this is shown using a Jackson-like selection or option structure. The entity is shown as a rounded box with two (or more) boxes below it, each connected to the entity box by a plain line and each containing the name of one of the possible effects and a circle in the top right corner to show that it is one of a selection. Entity roles mean that the event has simultaneous effects on different occurrences of the entity. This is shown by drawing a box for each occurrence affected, with its role in square brackets, and an outer box representing the entity type.

Two effect correspondence diagrams showing these conventions are given in figures 11.7 and 11.8.

11.9 PRODUCING EFFECT CORRESPONDENCE DIAGRAMS

A starting point for producing an ECD is gained by reading along the corresponding event row of the event/entity matrix in order to get a list of the entities which it affects. Then draw a box for each entity, drawing the boxes in roughly the same position to each other as they appear on the LDS. Add simultaneous, optional and iterated effects as described above. Next draw in all the 1:1 correspondences. The input data for the event must then be added. This comprises a list of attributes which usually includes the primary key of an entity which acts as the entry point of the process triggered by the event. A single headed arrow is drawn from the attribute list to the entity to indicate that it is the entry point. Lastly entities which are not updated but which provide reference data for the process or which must be read in order that other entities may be read (i.e. for navigation) are added to the structure.

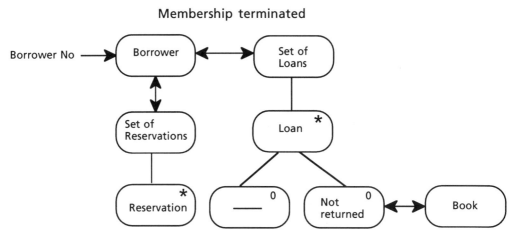

Fig. 11.7 Effect correspondence diagram

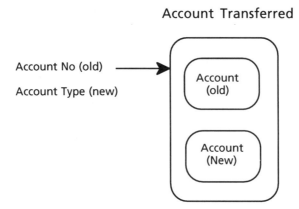

Fig. 11.8 Effect correspondence diagram – simultaneous effects

The completed diagram shows all the entities which must be accessed by processing triggered by the event.

11.10 MODELLING ENQUIRIES

Not all processing results in data being updated. The principal purpose of most information systems is to provide a source of data for screen enquiries and printed outputs such as reports and letters. Requests for information from the system are in effect non-update events and are called enquiry triggers in SSADM. It is usual to show only the major outputs of a system on the required system DFM in order not to clutter it up with too much detail. Details of other enquiries are documented in the requirements catalogue.

An enquiry access path (EAP) is the equivalent of an ECD where processing does not involve the update of entities. EAPs show the access path or navigation route through the LDM for enquiries and reports. For each enquiry or report an EAP with the same name is produced. EAPs are used as input to both data design and process design. They can be used to derive an output structure for use in logical design of the equivalent enquiry process. They also show data access points which must be built into the data design.

In appearance EAPs are very similar to ECDs but instead of double-headed arrows being used to show 1:1 correspondences between entities single-headed arrows are used. These access correspondences indicate the order in which entities need to be read in order to perform the enquiry. As with ECDs selections and iterations are shown using Jackson's conventions. An example of an EAP is shown in figure 11.9.

The basic approach to developing EAPs is to draw a view of the LDS showing the required entities. The entities can be identified by looking up the attributes in the enquiry function I/O structures in the data catalogue. Master to detail accesses are drawn vertically and detail master accesses horizontally. Relationship (crow's foot) lines are then replaced by arrowhead lines showing the direction of access. Where more than one occurrence of an entity may be accessed it is replaced by two boxes, one representing the set and below it, connected by a plain line, an asterisked box representing a single occurrence, as in ECDs. Selections are added where required, also in the same way as in ECDs. Lastly the entry point for the enquiry is shown by listing the input attributes required to trigger the enquiry and connecting the list to the first entity to be accessed. As with ECDs it may be necessary to add entities purely for navigation purposes. For instance if the attribute triggering the enquiry is the prime key of an entity which does not have data output from the enquiry but which is the master of detail entities which do provide output data then it may need to be added to provide the entry point for the enquiry.

Each EAP must be validated by checking that the accesses shown are supported by the LDM. It must be possible to read entities either directly by reading a master from its detail or by reading the next detail for a master. If not then either the LDM must be changed to support the access or a processing solution (such as a sort) must be found. The EAP must also be consistent with access rights granted to users of the function, if these have been defined.

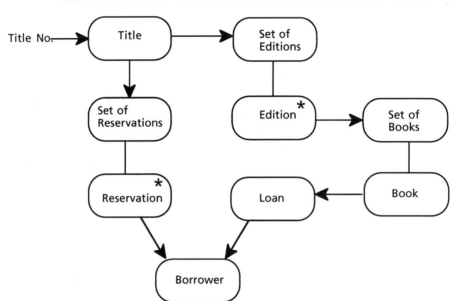

Fig. 11.9   Enquiry access path

11.11   DEFINING THE USER VIEW OF PROCESSING

The techniques we have been describing up to now model the *internal* processing and data needed for the required system. However we also need to look at how the system is organised from an *external* viewpoint, how access to data and processes is organised to suit the users.

Detailed models of processing are at the event level in SSADM because events correspond to the smallest unit of update processing that can be performed whilst preserving the integrity of the data. However different events or enquiries may be related to each other functionally or may need to be processed together for business reasons and so SSADM also has the concept of the function. *Functions* are made up of one or more enquiry or event level update processes which users need to run at the same time or in close succession. In a menu-driven system online functions correspond to the processing available from the lowest levels in the menu hierarchies. An example of an offline function would be an end-of-month batch run which may serve many different time-based events. A function can provide the processing for one or more events and, conversely, an event may be served by more than one function.

SSADM provides a model of a function in terms of separate input and output (external) and database (internal) processing components. See figure 11.10.

A typical example of the difference between a function and an event level process, which also shows the difference between logical and physical events, is that of the credit card bill. The payment slips attached to credit card statements often have change of address forms on the back so that a payment and an address change can be returned on a single piece of paper. Receipt of one of these forms is a single physical event, but it may

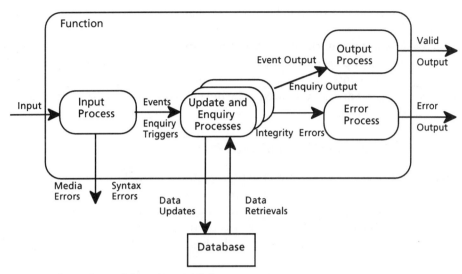

Fig 11.10 The universal function model

contain payment data, address data or both. In other words it is able to carry data for two different types of logical event which do not *have* to occur together. Users may nevertheless wish to have a single function which allows the data associated with both events to be entered on the same screen, particularly if changes of address are rarely notified in any other way. This would not prevent them having another function just for change of address if they wanted.

The important point to note about functions is that they package processing in the way that *users* want, in order to fit in with the way they wish to work. Generally speaking if users often wish to enter data for two or more events at the same time then a single function serving those events will be required, possibly in addition to functions serving some of the individual events. Different types of user may have different requirements for what processing they need packaged together. It is therefore necessary to identify the different user roles needed to run the system and the functions that must be provided for them. SSADM includes a user role/function matrix as one of its products and uses this as a guide or framework for defining menus and the navigation paths allowed between functions. The complete online system can be represented as a menu hierarchy which provides access to functions at its bottom level.

A function can be classified in three ways: as an update or an enquiry, as online or offline and as system- or user-initiated. These classifications overlap to some extent, for instance an update function may include some enquiry processing. It may also include both online and offline processing.

The first step in identifying potential update functions has already been described by recognising system-initiated and user-initiated processing on the DFD. Whether functions are online or offline may be a user choice or may need to be finally decided in physical design. The major source of potential enquiry functions is the enquiry requirements document in the requirements catalogue, though some may appear on the DFDs. The final decisions on the grouping of processes into functions should be made in consultation with the users and in the context of how use of the system is intended to fit into the user's overall job.

In SSADM important functions are chosen for prototyping so that the users are given a more concrete means of visualising what they have requested and can modify their detailed requirements if necessary. These prototypes are *only* intended for checking user requirements and are not intended to be developed as part of the final system.

11.12 MODELLING INPUT AND OUTPUT DATA

The method used to model processing structures for data input to and output from functions is based on Jackson-type structures as already described for ELHs. The same idea that processing components should be matched to the data structures they act upon applies here. The organisation of input and output data into sequences, selections and iterations in data flows can be used as the basis for modelling the corresponding I/O processes.

Figure 11.11 shows the SSADM I/O structure for an online update function and demonstrates the diagrammatic representation of these ideas. Each leaf on the diagram is known as an I/O structure element and the data items making up these elements are documented on the I/O structure description which is produced for each I/O structure. The data items in the I/O structure descriptions for update functions are taken from the corresponding data flow I/O descriptions in the required system DFM. For enquiry functions the data items have to be determined by discussion with the users. In both cases the structuring of the data needs to be derived from the views of the users. The I/O structures are used to help decide the grouping of data items when developing screen and report layouts during the design phase.

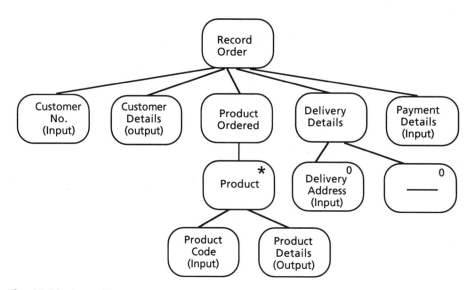

Fig. 11.11 Input/Output structure

11.13 SIZE AND FREQUENCY STATISTICS

One final point. It is not sufficient to collect information on what the system must do and then model the processing and data structures and relationships in it. It is also important to collect business data which may affect *how* it should do things. The basic information required concerns the frequencies of events and the numbers of entities. These, together with an indication of their growth or decline over the life of the system, can be used to provide estimates of performance and capacity demands that are likely to be made on the system. Event frequencies are recorded on function definitions in SSADM and provide the basis for estimates of the frequency with which the functions are likely to be used. Entity volumes are added to the LDM both by recording figures on relationship and entity descriptions and by adding them to the LDS itself. These figures are important in the design phase where they are used to estimate the amount of disc space required for the system, the traffic on communications lines and the likely response times and run times of online and offline processes.

11.14 SUMMARY

In chapter 10 we have described how details of the current system are recorded by producing a current physical data flow model and a current logical data model. We have also described how to record requirements for the new system in a requirements catalogue.

We have explained how this information is interpreted, by creating a logical version of the DFM, and applying the new requirements to it and to the LDM. We then described techniques for entity-event modelling. These are used to analyse and model the internal workings of the new system in more detail by adding a timing, or sequencing, element for update processing.

We have also shown how structured methods can be used to analyse and document the external aspects of the system, the way in which the user wants to access the system. The SSADM function model (figure 11.10) provides a good summary of the relationship between models of the internal and external aspects of the system. It is at this stage of a software development project that size and frequency statistics must be collected in order to help determine the performance and size criteria that the physical system must meet.

These models described in this chapter help in specifying the detailed requirement for the new system and are used as the blueprint for its design.

CHAPTER 11   CASE STUDY AND EXERCISES

Q1 Using the entity description for 'Bill', which is shown below, the case study scenario and the level 1 DFD, identify the events which will affect the 'Bill' entity during the lifetime of an entity occurrence.

Entity description :
  Subscriber no
  Rental period no
  Bill date

Bill amount
Bill type
Payment date
Bill status
Reminder date
Disconnection notice date

Bill type has the values :
  Bill due
  Request

Bill status has the values :
  0 - Bill issued
  1 - Payment made following bill
  2 - Following reminder
  3 - Payment made following disconnection notice.

Q2 Draw a partial entity/event matrix showing the effect of these events on this entity.

Q3 Draw the Entity Life History (ELH) for 'Bill'.

CHAPTER 12

Systems Analysis: *Specifying the Requirements*

12.1 INTRODUCTION

We have seen that the purpose of systems analysis is to find out what the users of the proposed information system want prior to beginning its development. However thorough the research is, though, it is necessary to specify the requirements in some form which is:

- accessible and intelligible to the user
- unambiguous
- reasonably practical and amenable to execution.

Let's examine these three criteria for a moment, as they are critical to the later success of the system's development.

We have seen in chapter 2 that one of the major objectives – and the main claimed advantage – of structured methods is that they deliver a specification which the user can understand. It is simply no practical use giving the user a multi-volume narrative to read and approve. Even if the user does, apparently, agree to such a specification, there will be endless arguments later over understanding and interpretation. It is, of course, true that it is sometimes difficult – very difficult even – to get users to review and agree specifications. However, securing such agreement is one of the analyst's skills and responsibilities and without it the development is unlikely to be successful.

First the user has to be persuaded to read the specification. Next, it is important that the user understands it in precisely the same way that the analyst intended. Here again, structured methods offer advantages in that it is less easy to introduce ambiguity into a drawing than into several thousand words of text. Ambiguity can however still creep in and every effort must be made by the analyst to ensure that it is eradicated from the final specification. One way of achieving this is to use a formal walkthrough technique, such as a Fagan inspection (see chapter 4). The users are asked to paraphrase the specification in their own words – and the analyst can see if the users think what she or he meant them to think.

Finally, the specification has to offer a practical solution. Although, in specifying the requirement, the analyst ought to consider business needs first and technological solutions second, it is not very sensible to specify something which is totally impractical or prohibitively expensive. Here, the analyst has a special responsibility to the user; after all, it is the analyst who should possess the better knowledge of the possibilities of technology and who must use this knowledge to steer the user in directions where a successful information system can be built.

Having disposed of these general points, let us now consider the specification process itself.

12.2 AGREEING THE OPTIONS

For any given system requirement, there are always a number of solutions available:

- At the one end will be the very simple solutions, providing a minimum level of functionality but requiring little time and money to develop and implement. Such solutions will be attractive to the users where funds are scarce or where timescales are very tight and it is important to have something, however basic, available as soon as possible.
- At the other extreme are the systems with full wide functionality providing every desired function, complete system help facilities and probably a very attractive user interface. Such a system will cost a lot of money, and take a lot of time to develop. This may well be necessary if the users' business is very complex or an objective is to get others to use the system as in the airline booking systems mentioned in chapter 5, where the airlines wanted their travel agents to embrace the systems offered.
- Between these extremes, there are all sorts of possibilities, including systems which have a limited number of fully developed elements, often referred to as 'core functions', and more rudimentary ways of providing less-used facilities.

The factors that will influence the way the system will be developed include:

- the speed of implementation
- the funds available
- the technical environments available, especially where the new systems must operate alongside existing implementations
- the technical sophistication of the target users. Can they reasonably be expected to grapple with the complexities of a multi-menu system?

A development route which is sometimes followed, especially where funds or timescales are tight, is to identify a 'core system' which will be implemented initially, with other functions being added as resources permit. This approach has the obvious advantage of spreading the costs and effort over a longer timescale but the danger with this is that the business requirements change during the project, so that the features to be developed later do not fit well with the initial facilities. Also, of course, technology moves on and the users may require more advanced facilities later in the project which cannot easily be reconciled with the more primitive features provided earlier.

12.2.1 Identifying options

In the initial stages of identifying options, both users and analysts should give themselves complete freedom of thought. In principle, nothing should be ruled out, however 'off the wall' it appears, as the idea is to bring the maximum of creativity to bear at this point.

However, the analyst has an important duty towards the users in this stage. The analyst is the one who should understand the possibilities and limitations of technology and must make sure that the users:

- do not limit their thinking too much by what they have seen, or had, in previous systems; in particular, the user must be encouraged to think about what they really want their new system to do and not just try to reproduce their old one
- do not think that just because an idea is suggested in this initial option-identification exercise it can necessarily be implemented; users must realise that cost and time constraints – not to mention the limitations of technology – will constrain the implementation route finally selected.

An excellent way of identifying options is to use the technique of 'structured brainstorming'. Get everyone involved together in a room and have them suggest ideas, capturing them on a whiteboard or on 'Post-it' notes stuck to the wall; in this initial stage, do not evaluate or judge the ideas, simply note them for later analysis. Then, identify similarities between ideas and rationalise the possibilities, thereby gradually narrowing down the list of possible options. Finally, analyse each option and produce a list of strengths and weaknesses for each. In this way, analysts and users generate an initial 'longlist' of all the possible options.

Before a proper selection exercise can take place, the initial list must be reduced to a shortlist of not more than two or three possibilities which can be presented to the senior managers who will decide how to proceed. Typically, this shortlist includes one option at each of the extremes already mentioned – high-cost, complex, full-features, and cheap, quick and simple – and a solution which, the users and analysts agree, offers a good balance between cost, complexity and functionality.

There is no magic formula for producing this shortlist – and especially not for identifying the balance or compromise option. Instead, users and analysts must engage in rational discussion and rigorously review the advantages and disadvantages of each possible solution. It is important that the options to be proposed are presented in a way that is:

- clear and concise to those who will make the decisions
- consistent as between the options offered.

High-level data flow diagrams are useful for illustrating the boundaries and functionality of the proposed systems, and tables present the pros and cons in a comparative way to the decision makers.

12.2.2 Choosing between the options

If the analyst wants a successful outcome to the selection process, it is vitally important to decide who are the decision makers. Quite often we talk about 'the users' but who are they exactly? Are they the people who will sit down at the keyboard and use our new system? Are they these people's managers? Or are they the senior management of the organisation who want a system to support some overall business objective? Very often, senior management makes the final decision based on the facts of the case – the evaluated costs and benefits of the proposed solutions – but also advised by those who will use the system directly and who may be influenced by more subjective criteria such as appearance of proposed on-line screens.

Whoever turn out to be the 'users', it is essential that they make the decision on how to proceed and that this is not left to the analysts. If the analysts feel that the decision is being edged back towards them, they must use their diplomatic skills gently but firmly to push it back where it belongs. Apart from any contractual implications here, it is vital to the success of the development that the users accept ownership of the project and the system, or they will not give their full commitment to its implementation. Without this commitment, the project is almost bound to fail.

The option selected may be one of the two or three proposed, or it will be a composite, embracing some features from each of the options. The selected option must be fully documented, including:

- what decision was reached
- who made the decision, and when
- what constraints or limitations were attached to the decision.

This documentation provides the terms of reference for the detailed design work which can now begin.

12.2.3 The use of prototyping

A persistent problem of systems analysis and design is that it is very difficult for the users of the proposed system to envisage, before it is built, what it might look like. Paper-based representations of online screens, for example, do not have the look and feel of the real thing and it requires a lot of imagination to see how the various menus and screens will work together. In addition, a set of routines which look fine in theory may turn out to be cumbersome to use in practice.

This problem is also recognised in more traditional engineering disciplines, and systems analysis has now adopted one of the techniques of the engineer to overcome it – the building of prototypes to test out ideas. In analysis and design, there are two distinctly different approaches to prototyping and it is important that the analysts decide in advance which of these to use:

- *Throwaway prototyping*, where a representation of the proposed system, or of parts of it, are created, probably using some screen-painting tool, and used to test out ideas with the users.
- *Evolutionary prototyping*, where elements of the system are created in outline in the actual chosen technical environment; these parts are then developed and refined and form part of the finished system.

Whichever approach is adopted, the process needs careful management. With throwaway prototyping, the users must be told, often and loudly, that what they are seeing is a simulation – not the real thing in embryo. They also need to realise that this simulation is not in any way shortening the development process; what it *is* doing is ensuring that the finished system more closely matches their requirements.

With evolutionary prototyping, on the other hand, the users are looking at part of the finished system. However, it is still necessary to manage their expectations since the part they are seeing – usually the online menus and screens – probably represents the lesser part of the whole system. Still to build, and taking more time, are the complex processing, data manipulation and validation routines which make the system work. In addition, there is the need for unit, system, and volume testing and all the other tasks which make for a soundly-built and reliable computer system.

The lessons learned from the prototyping process require documenting like all other analysis research. In addition, with evolutionary prototyping, the agreed screens must be placed under configuration control as soon as they are agreed so that they can be carried forward properly into the remainder of the development process.

In summary, then, the use of prototyping can prove very valuable in ensuring that the system will be acceptable to its users. But the prototyping process needs careful management to avoid raising over-high expectations either of the product to be delivered or of the delivery timescale.

12.2.4 Quantification of options

We mentioned before that when we present our options to the decision makers, they will want detailed information on which to base their decisions. Providing this information involves the application of some form of cost-benefit analysis to each of our options. Cost-benefit analysis is a very large subject in its own right, certainly beyond the scope of this book. However, if we think first about the costs side of the equation we need to consider the following:

Development costs Systems development is a labour-intensive business, so the cost of producing the proposed system will be based largely on the effort involved. The analysts need to produce estimates of the work involved in high-level and detailed design, in program coding and testing and in system testing, integration testing and implementation.

This is somewhat easier than producing the estimates for analysis since we now know more about the scope and complexity of the proposed system and we should have some idea of the development environment proposed. In addition, most development organisations have some broad metrics which provide ratios of effort as between analysis and coding and unit testing.

However, the option proposed may involve some new technology or some application of the technology which has not been tried before. This is particularly likely where the system is to achieve some competitive advantage for the organisation. In this case, the developers need to be cautious and careful with estimates and avoid raising the users' expectations unduly or underestimating the difficulties ahead.

Development timescales These are related both to the effort involved and to the resources which can be made available. Systems development though is not like digging a hole in the ground, where the time spent can be reduced just by using more people. There is an optimum size for each development team which enables the task to be accomplished in a reasonable time but which does not incur excessive overheads in coordination and control. If very specialist skills are needed for some part of the development – for example, to produce an optimal database design – then these will act as a limitation on the possible speed of development. These, and other, constraints, should be carefully considered and taken into account before proposing a development timescale for the option.

Hardware costs Broad-brush hardware costs can usually be arrived at by considering the overall size of the proposed system and what sort of platform is needed to support it. Will it be a mainframe, a minicomputer or a PC? Will it involve a network of some sort? Will the system be proprietary – involving components from one manufacturer – or 'open' – in which case the benefits of competition become available?

At the stage of deciding options, the actual final platform may not be decided, so these approximate figures may be all that can be offered with the options. However, the platform may already have been decided on – perhaps it was a constraint at the project initiation – or the organisation may have a hardware strategy – for example the requirement for open systems – which can be used to narrow down the hardware options. In this case, a more detailed estimate can be provided.

However, one word of warning on estimating for hardware costs. Two common problems seem to beset development estimates here:

- The final system turns out to be more complex and hence to require more computing power than was originally envisaged.
- The hardware itself turns out to have less available power than the sales specifications suggest.

The result is that initial estimates of the size of machine required almost always turn out to be too low. If the machine proposed initially is somewhere low to middling in the range, this is not insurmountable though it will involve going back to the users for more money. But if the machine proposed is at the top end of the range, the analysts should look at their estimates very hard indeed and consider whether they shouldn't be proposing a more powerful environment in the first place.

Other costs Development of both software and hardware is not the only area where costs are incurred. Other things the analysts need to consider and include in their options proposals include:

- Training of users for the proposed system
- Additional equipment which may be required
- Need to redesign office layouts
- Costs of possible parallel running of the old and new systems
- Need to recruit new staff with special, or additional, skills to operate the new system
- Possible costs of redundancy payments, or retraining costs, for staff displaced by the new system.

Other impacts Apart from the items which cost money, there are other impacts which the new system might have and which may have to be considered in the decision-making process. We have already suggested that there may well be retraining costs involved because people's jobs will be changed by the new system. The organisation may also have to reconsider the type of people it recruits in future and the training they need. Consider, for example, where an organisation has relied hitherto on receiving orders by post from its customers and entering them – via data entry clerks – into its computer system. If it introduces a system which can handle telephone orders, then its staff will now need to be good on the telephone and able to interact directly with customers.

The management style may also have to change because of the new system. The data entry operation just described requires no more than someone to oversee the process and supervise the clerks. But an operation where people are actively trying to make sales requires a management which is sympathetic and encouraging and prepared to train people and develop their skills. Some of these impacts may be intangible; it may not be possible to put a cash figure on them. However, the analyst should bring them to the attention of the user management so that they can make their subjective assessment of them and take them into account when choosing their system option.

Lifetime costs   When presenting the costs of a proposed system, it is important that analysts focus on the lifetime costs of the system, not just on the initial development costs. This is particularly important where a structured development method is being used since it is acknowledged that such methods do involve more effort, and hence cost, during development. However, the use of structured methods produces systems which are easier and hence cheaper to maintain in the long run.

12.3   IDENTIFYING BENEFITS

When we were discussing the costs of a development, we suggested that there were 'hard', that is tangible costs, and 'soft', that is intangible costs. The same applies to the perceived benefits of a new system, which can be tangible or intangible depending on whether a monetary value can be placed on them. The analyst needs to remember that when identifying and presenting benefits, they will appear rather differently according to who is considering them; one person will place more value on a benefit, particularly an intangible one, than another.

Tangible benefits are usually the easiest to identify and quantify, though not always. They could include:

- Increased throughput – the ability to handle a greater volume of business
- Greater productivity – the capacity to handle more work with the same or fewer staff, the reduced salary bill providing the quantifiable benefit
- The ability to enter a new market – the value being the increased turnover or margin which results
- Reduced running costs – if the system is replacing one which is very old, operating on unreliable hardware and requiring specialist, and therefore expensive, skills to maintain it.

Getting facts and figures on some of these can be difficult but some things, like salary costs, should be readily available. You can use salary costs, too, to work out other things like the cost of entering a single order in the system. The costs of operating existing systems are sometimes hard to establish. The simplest situation is where the system is run by a bureau, perhaps as part of a facilities management deal, but this is still fairly uncommon. Instead, you will need to get some overall costs for running the computer, including staffing, floorspace, environment, maintenance and so on, and work out what proportion of it is accounted for by the system to be replaced. Computer managers are sometimes reluctant to provide this information, indeed some do not even know themselves, but the analyst's usual persistence should eventually produce some useable figures.

It is very important when assessing benefits not to overstate them. Nothing so undermines the case for a particular course of action than if its proponents are over-optimistic about its benefits. If you are reasonably careful and conservative, and if you document the assumptions on which you have based your calculations, you will be able to present a robust set of costs to support your options. Intangible benefits are much more difficult to deal with. They are rather like an elephant! You have trouble describing it but you'd know one when you see it. Some intangible benefits which you might want to consider include:

- Greater customer satisfaction resulting from the use of the new system
- Lower turnover of staff working with the new technology
- Better company image caused by the use of up-to-date equipment and methods
- Better management decisions based on more up-to-date information
- The ability to react more quickly to changing circumstances in the marketplace.

Analysts often get themselves into trouble by trying to place a monetary value on these intangible benefits, by asserting for example that greater customer satisfaction will produce so many new orders per year, worth so much money. If the users do not share this assumption, then they will not agree with the analysts' evaluation of them. It is much better for the analyst to identify the *potential* intangible benefits and invite the users to place their own value on it. A managing director currently making many decisions 'in the dark' may rate very highly a proposed system that offers accurate and up-to-date management information. Finally, it is extremely unwise to base the whole case for a particular option on intangible benefits unless the users themselves identified these benefits when initiating the project. It is much better to quantify the tangible benefits and then to list the intangibles as additional potential gains.

12.4 PRESENTING THE REQUIREMENT

However good the analyst's ideas, however cogent the arguments for a particular solution, these will not sell themselves. It is very important that the proposed solutions be presented in a way which is:

- clear
- intelligible
- pitched at the correct level for the intended audience.

In most cases, the selection of options involves some sort of presentation, supported by appropriate written material. Later, the specifications for the proposed system will need to be expressed in paper form. So, the analyst needs to master the skills of expression both written and oral and, in the latter case, in small groups and to larger gatherings. We have covered these areas in detail in chapter 3, but we'll consider how they apply to the presentation of system proposals.

The three keys to a good presentation are preparation, preparation and preparation. Successful presentations may look effortless but that is because they have been planned carefully, and rehearsed thoroughly, in advance. Of course, a presenter needs to be able to 'think on his feet', to field questions from the audience but the basic structure of the presentation, the order and method in which subjects are introduced, must have been worked out well beforehand.

If, as is usually the case, a team is presenting, then there should be a thorough team briefing and everyone must be clear about:

- who will say what, in what order
- who will respond to questions in the various subject areas
- how questions will be fielded.

Nothing looks worse in a presentation than everyone in the team trying to talk at once or, just as bad, embarrassing silences because no-one had picked up a question.

Visual aids should be considered carefully. Overhead or slide projection can have a major impact but not if they are merely tedious lists of bullet points covering what the speaker is saying. They should illustrate vividly and directly the point the speaker is trying to make – like a pie chart showing the breakdown of costs for a proposed system. You need to practice, too, the use of the visual aids so that they supplement what you are saying, rather than acting as a distraction from it.

At the outset of the presentation, it is necessary to establish your credibility. You must understand the level of interest and understanding of your audience and pitch your talk correctly. Computer jargon should be avoided, unless you are addressing a very technical audience. You may reasonably use your audience's own jargon – engineering terms for engineers for instance – but only if you yourself have sufficient expertise; otherwise, you may well find yourself out of your depth and lose your credibility.

And finally, if you are trying to make a case, do not overemphasise the *features* of your solution; stress the *benefits* of the solution as they would be understood by the audience. So, for example, don't say 'the proposed hardware has six gigabytes of memory and the processor runs at 25 megahertz'; point out instead that 'the proposed hardware can handle 100 orders an hour more than your current system with no increase in manpower'.

Written reports must also be pitched at the intended audience and use appropriate language. A board-level report, for example, concentrates on high-level issues and overall costs whereas a program specification contains very detailed and technical implementation information.

If you have some standards for the format and content of a document, use them. You may find the standards a little constraining but this will be outweighed by the benefits that arise from everyone knowing how and where to find things in the finished document. At all times, you must consider the needs of the reader. The language in a report should be as simple as possible, given the subject matter. Of course, some points are difficult to convey in short, simple sentences but it can be done and it presents an interesting challenge to the skilled writer. If you are writing a long and detailed report, produce a management summary which distils the major points into a few sentences or paragraphs; it is this summary which arrests the attention, and secures the interest of the senior managers for whom it is intended.

The use of drawings and diagrams can save many hundreds of words and also, of course, helps to break up large areas of text. Use adequate 'white space' as well, to make the document more accessible and so encourage people to read it. If you are preparing a technical specification, then the need to gain the audience's attention is less acute. However, layout and presentation are still important, this time so that important points are not overlooked or buried in acres of text.

Finally, make sure that your document, report, specification or whatever contains adequate cross-references to other relevant documents or publications.

12.5 WRITING THE FUNCTIONAL SPECIFICATION

Once the client has selected one of the system proposal options presented by the analyst, the *functional specification* is written. This is a document which specifies in detail the functions to be performed by the new system and describes any constraints which will apply. Figure 12.1 represents the suggested contents of this document, with examples of the sort of information contained in each section. This is only a skeleton list, and it is likely that other sections would be added.

In writing this document, the analyst must ensure that it is unambiguous, and that it passes the '5 C's test' by being:

Clear
Complete
Consistent
Correct
Concise

The functional specification is passed to the client, who reads it and agrees the content. The client then signs it off as an acceptable deliverable and work begins on the design

- **System Performance**
  response times, throughput, dealing with hardware & software failures

- **Inputs to the System**
  sources, types, formats, procedures

- **Outputs from the System**
  contents, format, layout

- **Constraints**
  hardware, software, environment and operational

- **Other Aspects**
  system start-up and shut-down, security procedures

Fig. 12.1 The functional specification

phase. Sometimes design may begin before the functional specification has been agreed by the client. In this case, it is unwise to proceed too far with the design in case the functional specification has to be changed in a way that has major implications for the designers.

12.6 SUMMARY

It is very important that the users of a proposed system read, understand and agree the specifications and that they are committed to their success. Securing this commitment requires that the specifications be presented clearly and that proper, controlled, methods are used to review them.

There are always options in the way in which a system may be developed, each one offering a different balance of functionality, cost and speed of implementation. Analysts and users must work together to identify options and to present a shortlist of these, together with their perceived costs and benefits to senior management for their decision.

Good communication skills, both in giving presentations and in producing written material, are vital to the analyst.

CHAPTER 12 CASE STUDY AND EXERCISES

Q1 The System Telecom LDS and DFDs which you have seen represent the selected business option, that is they are for the preferred required system. Consider which areas of processing could have been excluded for reasons of cost, and what are the two other options that would consequently be produced.

Q2 Identify the relative benefits and drawbacks of these options.

CHAPTER 13

From Analysis to Design

13.1 INTRODUCTION

We have looked in the last six chapters at the process of analysis, using the PARIS model to take us through the five stages of planning the approach, asking questions and collecting data, recording the information, interpreting the information collected and specifying the requirements. We will now turn our attention to the process of design, which was introduced in chapter 1 as the third stage in the model of systems development, redrawn here as figure 13.1

Fig. 13.1 Stages in system development

As we pointed out in chapter 7, although the model is useful because it indicates the relative position of the stages in the development process, there is frequently an overlap between analysis and design, with high-level design beginning during analysis and analysis continuing as part of design. In the following chapters, the design process will be described in detail. This chapter will focus on the transition from analysis to design.

The final deliverable from systems analysis is a document containing an unambiguous statement of the client's requirements for a new system. This document, called the *functional specification*, states what the development project will have to deliver in order to be considered a success. It is written by the analysis team and agreed and signed off by the client. The functional specification is the starting point for the designer, who will depend to a great extent on the accuracy and thoroughness with which the analysts have carried out their task. The analyst's understanding of the business, appreciation of the client's problems and documentation of requirements provide the foundation on which the designer will build a working solution. However, if the design is based on an inadequate understanding of the users' requirements, then the system delivered, however well it works, will not be a quality solution because it will fail to meet the real needs of the business. This can lead to a great deal of extra work for the developers, who then have to carry out further analysis, fix the design, and produce and test revised code in order to deliver a system which does conform to the client's requirements. This is a costly process, because, as Edward Yourdon has observed, it is easier to make a working system efficient than to make an inefficient system work.

In structured methodologies, a lot of emphasis is placed on the need to spend considerable time and effort ensuring that analysis is rigorous so that the design process, and later the implementation, will be straightforward. A key factor in this approach is the use of structured techniques, and in the next section of this chapter, we will discuss their value in 'bridging the gap' between analysis and design. We will then consider the objectives of design and the constraints on the process, before concluding the chapter with an overview of the stages in the design of a computer system.

13.2 BRIDGING THE GAP

In traditional approaches to system development, there was frequently a gap between the information about the system documented by the analyst, and the detailed technical task which lay ahead for the designers. While the existing physical system was described in detail, usually in the form of a lengthy narrative – sometimes described as a 'Victorian novel' – it still left the designer, who had to produce designs for files or databases, processes, interfaces and controls, with a lot of unanswered questions. Analysis ends with a description of what the new system must do, while design must specify how this will be done by selecting one of the many possible ways of doing it. The gap between the end of analysis and the beginning of the design is represented in figure 13.2.

All structured methodologies share the view that logical models must be integral to the process of systems development in order to ease the transition between analysis and design, and also to enable intermediate views of the system to be discussed with the users. As we described in chapter 7, a logical model describes exactly what the system does rather than how it is does it, by ignoring the constraints which determine how the current system is physically implemented. A required logical view is created by mapping the customer's requirements for a new information system onto the current logical model. The structured techniques used during analysis which provide this logical view were described in chapters 10 and 11, and include:

- Data flow diagrams
  representing the processes which manipulate the data as it passes through the system
- An entity model
  showing the relationship between the data items held within the system
- A data dictionary
  providing an overall consistent definition of the data used by the organisation. This definition can include the content of the data stores, data flows and processes shown on the data flow diagrams, and the entities which make up the entity model.

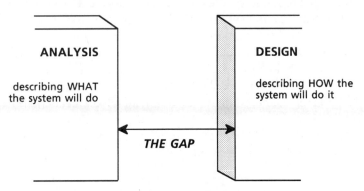

Fig. 13.2 The gap between analysis and design

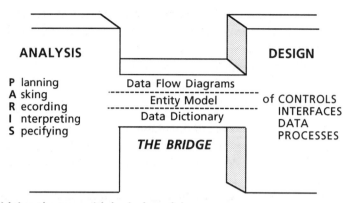

Fig. 13.3 Bridging the gap with logical models

These logical models provide a sound basis for the process of design, and as we shall see later, the designer will use, amend and develop these models to create design documentation. For example, the entity model is used in drawing *logical access maps* – also described as data navigation diagrams – which assist in the design of files or databases; while the data flow diagram is the starting point for designing inputs to the system such as an order form, outputs from the system such as a sales report and human-computer interfaces such as an on-line enquiry screen.

In this way the logical models support both the current system which is in operation, and the new system being developed, and can be used to overcome the analysis/design gap by providing a 'bridge' of techniques central to the processes of analysis and design. This bridge is represented in figure 13.3 which also shows the constituent parts of analysis and design as described in this book. Since another feature of structured methods is the signing off by the client of the deliverables at each stage of the development, the designer can have confidence that this 'bridge' represents a sound and accurate model of the system.

13.3 DESIGN OBJECTIVES AND CONSTRAINTS

It is important, right at the start of the design process, for the designer, or design team, to set clear objectives. The primary objective will always be to design a system which delivers the functions required by the client to support the business objectives of their organisation. For example, the system may be required to speed up the production of accurate invoices, so that the company's cash flow can be improved; or to provide up-to-date, detailed management information to improve the managing director's control over the business; or to help senior managers to make strategic decisions. In other words, to be a quality product – as we described in chapter 4 – the system must conform to the customer's requirements, and be delivered in a way which meets their expectations in terms of service. There are many ways in which these requirements might be met by a physical design solution, but there are a number of other objectives which must be considered if a good design is to be produced.

- *Flexible*
  The design should enable future requirements of the business to be incorporated

without too much difficulty. Often during the analysis phase, users may not be clear about exactly what they will require from the new system, for example which reports will be most useful to them. However during the evaluation period after the new system becomes operational, the real needs often emerge and a flexible design will be able to accommodate these new requirements. In addition, businesses change over time and a good design enables the system to reflect these changes.

- *Maintainable*
  This is closely linked to the previous objective because it is about change. A good design is easy to maintain and this reduces the client's maintenance costs, which usually represent a high proportion of the total lifetime cost of the system.
- *Portable*
  Still on the subject of change, a client who has bought a software system may wish to change the hardware on which the system runs. A good design is portable – in other words it is capable of being transferred from one machine environment to another with the minimum amount of effort to convert it.
- *Easy to use*
  With the increasing exposure of people to computer applications in the home as well as in the office, expectations of computer systems in terms of their ease of use are also increasing. A good design will result in a system which is 'user friendly' – easy to understand, not difficult to learn how to use and straightforward to operate.
- *Reliable*
  This objective is about designing systems which are secure against human error, deliberate misuse or machine failure, and in which data will be stored without corruption. While this is desirable in any computer system, for certain systems in the areas of defence, process control or banking, it will be a key design objective.
- *Secure*
  Security is another objective which must be considered by the designer. In order to protect the confidentiality of the data, particularly if it is commercially sensitive, it may be important to build in methods to restrict access to authorised users only, for example by introducing passwords.
- *Programmer-friendly*
  While the other objectives are mainly about delivering benefits to the client, the designer must also consider how easy it will be for the programmers to produce the code from the program specifications. By producing a programmer-friendly design, both the costs of production and the risk of building in errors are reduced.
- *Cost-effective*
  This includes a number of the other objectives, and is about designing a system which delivers the required functionality, ease of use, reliability, security, etc. to the client in the most cost-effective way.

Constraints In addition to thinking about design objectives, the designer must also be aware of the constraints which will have an impact on the design process. While the logical models of the system are the starting point for design, the physical realities of the environment in which the designer is working must be taken into account. In addition to time, which is always limited, and money – the available budget will limit the options available to designers – there are a number of other constraints which need to be taken into consideration. These are listed below:

- *Resources*
  An important constraint on any design solution will be the availability of resources – particularly the technology – to be used in delivering a solution to the client. For example, the system may have to be implemented on an IBM mainframe and developed using the Yourdon methodology.
- *The client's existing systems*
  A major constraint would be the need for a new system to interface with other systems – hardware, software or manual – which already exist and will continue to be used by the client organisation.
- *Procedures and methods*
  The final design might also be constrained by internal or external procedures, methods or standards. For example it might be a company standard to develop systems in SSADM, or the client might specify the Yourdon methodology. The CASE tool to be used in development might be specified, as might the programming language in which the system must be coded.
- *Knowledge and skills*
  This might be an internally or externally imposed constraint. The knowledge and skills of the development team may limit a designer's options, as might the competence or 'computer literacy' of the potential users. Also, financial considerations may be important here – 'how much does it cost to employ two trainees compared to the cost of employing one expert?'

As long as the constraints on the possible design solution are clear to the design team from the beginning of the process, allowances can be made or alternatives considered, in order that the stated objectives of design can be met.

13.4 AN OVERVIEW OF SYSTEMS DESIGN

In the next eleven chapters of this book we will be introducing a range of approaches, issues and techniques which, taken together, provide a comprehensive picture of the process of systems design. In this section we present an overview of the design process, pointing out the logic behind the sequence in which the chapters are written.

For the customer's business to survive in a competitive market place and also to fulfil legal obligations, the designer must protect the system's most valuable asset – the data it holds. We begin our description of the design process with an examination of system security and controls, since the requirements in this area must be established before beginning the design phase and these requirements might have a major impact on the design. In chapter 14 we consider the risks to the system, assess the need to protect data and control user access to it, and look at ways of achieving this using hardware and software solutions.

The design of input to and output from the system is discussed next because inputs, outputs and interfaces *are* the system to many users, and often the quality of the whole system will be judged by the quality of these visible products. We start with outputs, and then go on to consider the inputs required to provide these outputs, discussing technologies which can be employed, and the differing requirements of the customer, and the different physical environments in which they might be used. The main types of dialogue are considered, identifying key features, and special consideration is given to the WIMP interface. Chapter 15 ends with a discussion of ergonomics – the study of the relationship between people, machines and the environment – and its role in the design of workstations.

Having discussed human-computer interfaces, we move on in chapter 16 to discuss other interfaces: intra-system interfaces between processes within the same computer and protocols for communicating with peripheral devices.

Whilst data modelling is initially an analysis task, it is essential that the designer understands the technique so that the models can be questioned, added to or manipulated as the final design of the system evolves, to ensure the final data models accurately reflect the customer's needs. In chapter 17, we contrast a top-down approach – building an entity model – with a bottom-up approach – relational data analysis – and explain the value of the two techniques, used together, in producing a robust and reliable data model. The data dictionary is also described. This is followed by a detailed treatment of file design and database design, including database management systems, in chapters 18 and 19. Chapter 20 concludes our consideration of data design with a discussion of physical data design issues.

Having covered controls, interface design and data design, the only remaining area of systems design to examine is process design. Chapter 21 examines alternative approaches, beginning with guidelines on partitioning the system, and leading to an exploration of a number of different program specification models.

Under the heading 'choosing hardware' we describe how to select the most appropriate hardware, the factors influencing the selection, and the constraints and opportunities associated with the solution. The themes introduced in chapter 16 are developed further in chapter 23, which looks at inter-system communications. This area is becoming increasingly significant with the widespread use of Local Area Networks (LANs) and Wide Area Networks (WANs). These networks are described in chapter 23, which also explains how they are mapped onto the Open Systems Interconnection (OSI) model.

Finally, chapter 24 on 'Systems implementation' presents a view of the process which follows systems design. Topics covered include coding, testing, user training, data conversion and the maintenance cycle. The focus in this chapter is on the move from an existing system to a new one, and the implications of this process for the system designer.

13.5 SUMMARY

In this chapter we have looked at the interface between analysis and design. The value of starting with a set of logical models, produced during analysis, was emphasised and we described how these models bridge the gap between analysis and design. If analysis has been carried out thoroughly and documented in an unambiguous manner, the designer's chances of delivering a quality system are greatly increased.

It is important that the designer is clear about the objectives of design and understands the constraints which will affect any proposed design solution. With this in mind, the process of design, which includes the design of:

- controls
- human-computer interfaces – inputs, outputs and dialogues
- system interfaces
- data – logical models, files, databases, and physical storage media
- processes

can begin. And of course any design solution must be considered in relation to a number of factors, including the available hardware, software and middleware.

While analysis and design have been presented as two separate stages in systems development, we have emphasised that there may be considerable overlap between the two, and on large development projects the two processes may be occurring in parallel. The essence of a smooth transition between the analysis and design depends on the quality of the communication which takes place, and structured techniques provide an ideal means of communicating effectively.

CHAPTER 14

Systems Design: *Protecting the System*

14.1 INTRODUCTION

In the late 1970s, commercial businesses and their software developers became aware of the need to protect their software systems against three things: prolonged and unauthorised downtime, security breaches and loss of data or system integrity. Until this time these factors were really only considered for military, life-critical or leading-edge technology systems such as space exploration. Previously a commercial system was installed, then ran until it crashed or until a corruption was noticed, then the system maintainers worked to get it back up and running as quickly as possible. At that time protecting the system was very much an afterthought. Working with systems that are 10 – 15 years old, two things are very noticeable: trivial errors can make the system lock up or crash, and users frequently take high-risk corrective action like patching or zapping data areas. These activities are necessary when two related data items get out of step. The operator decides on which of the two is wrong and instead of going through the applications program the operator simply overwrites, or zaps, the data item. This has disastrous consequences if the operator changes the wrong data item, and since they are working with pages of numbers the probability of that is quite high. Organisations whose systems work like this are now weighing up the costs of retrospectively building in resilience and recoverability. Nowadays developers are much more proactive when it comes to building in protection. The once familiar apology 'sorry, our computer made a mistake' is no longer acceptable to the organisation who paid for the system or to their clients.

In this chapter we explore the possible risks to every system, and to each part of a system as detailed in figure 14.1. We examine their causes and consequences, and highlight key areas that must be discussed and agreed with the client before starting to define a system protection policy.

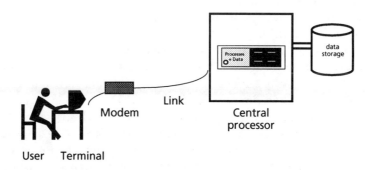

Fig. 14.1 System components that are at risk

14.2 DAMAGE TO THE PHYSICAL INSTALLATION

Computer systems perform best when housed in a dust-free, temperature-controlled environment, and most large installations operate in such conditions. However even when the processor is housed in such conditions, the terminals are likely to be on desks in a variety of offices. The most likely hazards in this environment are radiators, cigarette smoke, coffee cups and sometimes hand-held radio transmitters. System users may resent restrictions placed on them when using terminals, or when near to a terminal, but there are good reasons for them.

Terminals and personal computers themselves generate heat, so being in close proximity to a radiator may at best affect a computer's ability to respond to user commands and, at worst, may help burn out parts of the logic boards. Cigarette smoke, like dust, can easily clog filters and leaves a fine layer of dust inside the terminal. Eventually this will reduce its performance. This type of damage can be reduced by using a dust cover when the terminal is not in use, but in most offices this is the exception rather than the rule. Humidity and condensation will also affect performance. Spilled drinks are the most common office hazard. To have a keyboard professionally cleaned out after this happens costs almost as much as buying a new keyboard, especially if the drink is hot chocolate or soup!

On sites where key staff use radio transmitters to keep in touch, these transmitters have to be switched off when in the vicinity of certain types of computer. For example, the type of programmable logic units used to monitor and control a railway station signal system with large screen colour graphics showing points, train positions and signal status react to transmitters. They might not respond to operator commands or might show incorrect data when picking up interference and yet the operator would not necessarily recognise the cause of the problem. Any kind of electromagnetic source can corrupt data and displays in a similar way. The difference is that transmitters usually cause only temporary inaccuracy in the data, whereas magnetic fields can alter data permanently by changing the binary patterns. This code corruption is covered in more detail in section 14.3.

Flood damage can only occur if the system is located close to a large quantity of water: a river can burst its banks, dams can burst. So this type of natural disaster is less frequent than fire or storm damage and therefore in most cases it is easier to guard against, unless of course the system's raison d'être is water-related, such as controlling a water purification plant. The only other risk of water damage is where a computer system is housed on the floor below water storage tanks.

Storms may cause damage to the building where the system is housed. Falling trees, roofing or other debris may crash into the building, or electrical surges may damage circuitry. It is usually possible to protect the building from all but the most severe storms.

The natural disaster which most frequently affects computer installations is fire, whether caused by carelessness or storms. Burnt-out circuit boards and wiring can be difficult to trace and expensive and time-consuming to replace. Because it is relatively common, most computer rooms have a fire-fighting system installed: not water, of course, but halon gas. As soon as a fire is detected the gas is released, removes the oxygen in the room that feeds the flames, so the fire dies. Obviously this system cannot be switched on while people are in the room. Halon is only found protecting systems that are housed separately, not the smaller systems that are located in users' offices. It is only switched to automatic mode outside working hours. To date there are no records of whether the halon affects the hardware, but it is not thought to be damage free.

There are other risks to the system. Terrorist bombs have gone off in the City of London in an attempt to cripple the financial institutions and affect the British economy. The terrorists are not specifically targeting computing systems, although these are among the casualties. The effects are usually short-lived since financial institutions are security-conscious and will have back-up copies of the system and the data stored off-site. The affected hardware is usually part of a network of systems, so their business can be run from another office. Sabotage is the final risk to consider. One of the most recent involved a disgruntled employee who left 'timebombs' in the system. Every few weeks the computer would shut down and output a message requesting that the employee be brought in to sort out the problem and restart the system. On the third occurrence the company sued the employee. Most saboteurs do not leave their name!

14.3 DAMAGE TO THE SOFTWARE SYSTEM

The types of error we are looking at here are those inadvertently built in during coding and malicious damage. Code defects, code corruptions and code operating in inconsistent ways are caused by accident. Malicious damage is caused by viruses, hackers and fraudsters. The important issues in each case are how it happens, why it happens and how to prevent it.

Code defects are errors or bugs coded by programmers which have not been found during tests. These may occur for a number of reasons. In large complex systems, while many tests are run and rerun, it is virtually impossible to predict every possible combination of conditions. Bugs are therefore often left in the code simply because no-one thought to test that particular condition. Another reason bugs are overlooked is that program testers may not accurately predict the expected response to a set of test data, and therefore may accept a response that looks reasonable. Let's look at an example.

I want to list all staff in a department whose second name begins with the letter B.

Expected response	Actual response
Bill Bailey	Bill Bailey
Roy Bean	Roy Bean
Jim Bowen	Jim Bowen
John Brown	Joe Brown

The last two don't match so I check the test data file and discover that a Joe Brown does exist and does work for this department. I now believe that I simply wrote down the expected response incorrectly and accept the test has passed. I explain this to the test reviewer who accepts my explanation. Sometime later, let's say when the interface of this program to another is being tested, someone else spots the error. I check it again and I discover that John Brown also exists. I now also discover that this program loop always omits the last output. In this case both my expected and actual response lists are wrong.

This is a trivial bug and one which would be easy to trace and correct. However it's important to recognise that such simple errors can and do occur, and often this is due to the program coder/tester's environment. The programmer's excuse might go something like this: 'This test is only one of twenty tests run on this program; and this program has already undergone two major redesigns because the client keeps changing his mind so it will probably be changed again; and I am also working on three other programs; and I am working overtime because the project is behind schedule, and I really want to be doing design or teamleading instead of programming anyway.'

Accidental code corruption means that the binary pattern representing the code has been changed. Let's say a 16-bit word originally containing

 0011110110100100

is corrupted and now contains

 0000000100000001

When the program executes with this corruption, the system will do something. The problem is that the action the system takes is unpredictable. The certainty is that it won't do what it was originally coded to do. So corruptions occur when the memory area where code is stored is overwritten.

Corruptions have a number of different causes, but typically it happens because a memory area allocated to one program is also allocated to store a second program, or data. This type of corruption is shown in figure 14.2. It only happens if the memory management is faulty: it has not recorded that the area is already in use, so it allocates that memory area twice.

Sometimes the problem occurs when a program which has requested and been allocated a fixed quantity of memory does not stop writing to memory when it reaches the limit of its allocation. This may be because the exit condition of a loop is never met, so the program goes round and round the loop overwriting areas of memory that it should have no access to, similar to computer viruses. This is shown in figure 14.3. It usually comes to a halt only when it reaches the end of memory and the system 'hangs'. Alternatively, a user will recognise that the system is not responding and the system may be rebooted.

The final aspect of damage to the software system at the coding level – inconsistent code – is caused by having incompatible versions of programs resident on the system. This happens when a program is changed after other programs have already been installed using the original system design, or when the system is in the maintenance cycle.

For example, say a company decided to reformat their client account number. They want to replace the existing 8-digit sequential number with a 10-digit number where the first 3 digits represent a sequential area code number and the following 7 digits are the sequential client account number.

    Current format   Required format
    00000001 → 99999999 001 → 999, 0000001 → 9999999

Therefore a sample new account number might look like 1550000233, where 155 is the area code and 0000233 is the client number.

Many programs will read or write the client number and in all of these programs the data definition must be amended. Let's say that all these programs have been successfully amended, but that the client enquiry program is accidentally not reinstalled. This means the system is still running the old version of the program. A user wants to enquire on a client's account but they can't because the screen only accepts 8 digits which doesn't

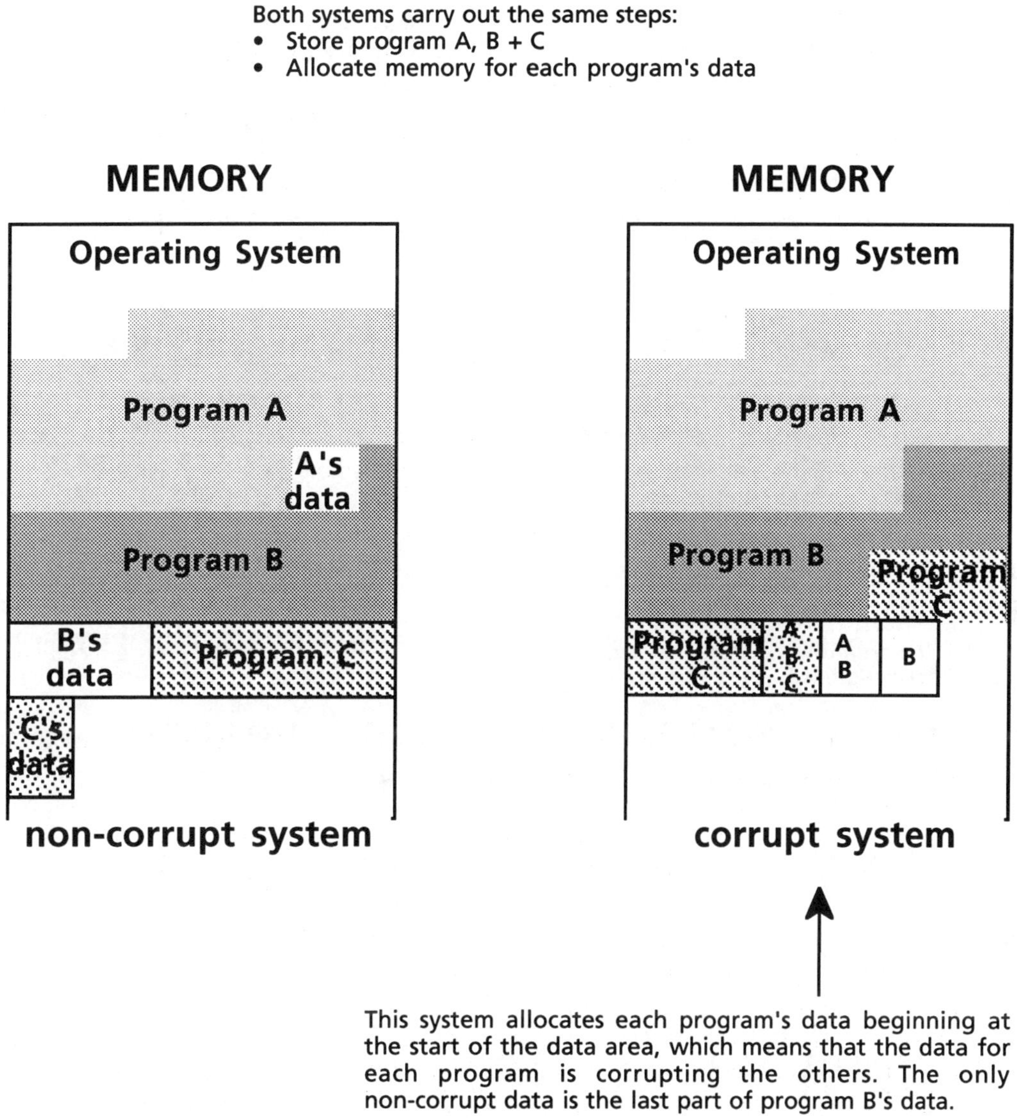

Both systems carry out the same steps:
- Store program A, B + C
- Allocate memory for each program's data

MEMORY

Operating System

Program A

A's data

Program B

B's data   Program C

C's data

non-corrupt system

MEMORY

Operating System

Program A

Program B   Program C

Program C A B C A B B

corrupt system

↑
This system allocates each program's data beginning at the start of the data area, which means that the data for each program is corrupting the others. The only non-corrupt data is the last part of program B's data.

Fig. 14.2 The difference between a corrupt and non-corrupt system

match the newly formatted client data area. This is a very visible example of the type of difficulty that arises when programs operate differently. Often the mismatch is much more subtle, for example when two data items that are used quite independently for most of the time are suddenly used together in one program. Only then would the problem be discovered.

Computer systems, particularly microcomputer-based systems, are affected by viruses. These are code corrupting programs often circulated on free demonstration disks. To encourage receivers to install the software immediately, the demonstrations will often use state of the art graphics. Viruses are so called because, once installed on a system,

208 Systems Analysis and Design

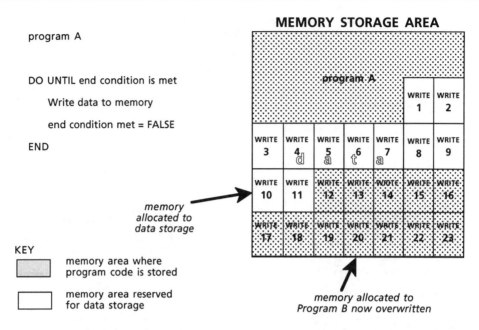

Fig. 14.3 A program dynamically overwriting memory

they lie dormant until triggered by some predetermined event. They then damage the system by altering parts of the software, sometimes even the operating system. Often simply inserting the floppy disk and running the demonstration program triggers the virus. Some viruses are relatively harmless and are installed by hoaxers. Many however are written for profit.

Receivers of these disks install the software and are then informed that their system will be corrupted unless they send a payment for another disk containing the software to eliminate the virus. Understandably, people are now suspicious of free demonstrations from organisations they don't know. Companies running networked PCs or fileservers are popular targets because the virus can be installed on one system and it will spread throughout the whole network. Such companies now have sophisticated virus guards which check each PC or fileserver on startup for viruses and will not allow the installation of software from a disk found to contain a virus. They also operate virus quarantine procedures where any disk suspected of containing a virus is run for a time on a standalone PC. When the quarantine period is over, the software is installed on the main system.

Hacking into a computer system is only possible if that system has links to the outside world, through a modem, or is part of a network of computers. It used to be a favourite pastime of home computing enthusiasts who were tired of the limitations of working on a standalone system and did it for the fun and the achievement. More often these days it is done by professionals who know exactly which system they want to access and what data they want to see or alter. Typically this might occur in the field of industrial espionage or when users of a system with only limited access seek access to the parts of the system that are currently off-limits to them.

Hackers write programs which dial one telephone number after another in sequence, and programs which enter one password after another in sequence. When they dial up a computer modem or the particular computer they want to access, the system login prompt appears on their screen and they run a password program which may gain access to the system. Most systems allow a number of retries before breaking the link, so the hacker must store the modem telephone number and know which passwords they've tried. They then redial and try more passwords, and so on until they find the right password. Figure 14.4 shows how hacking works. Hacking into a system you know nothing about is time consuming. This practice is illegal.

To defraud an organisation you have to know exactly how it works, what will be detected and what will go unnoticed and the length of time before the fraud will be detected. Therefore fraud is usually committed by staff who have the knowledge,

Hacking program
START
   CALL connect to system
   CALL enter password
STOP

Connect to system program
START
  initialise telephone number
  IF telephone number is known to be a computer system
  THEN
    dial number through modem
  ELSE
    DO UNTIL number connected
      increment telephone number
    dial number through modem
    END
  ENDIF
  store last telephone number dialled
STOP

Enter password program
START
  initialise password
  IF password is known to gain access to system
  THEN
    enter password
  ELSE
    DO UNTIL password accepted or link to system is broken
      increment password
        enter password
    END
  ENDIF
  store last password entered
STOP

Fig. 14.4 Hacking into a system

access, opportunity and inclination. Fraud is sometimes committed by writing fraudulent programs, more frequently however it is carried out by experienced computing professionals identifying a loophole in the system and exploiting it to their advantage.

For the first possibility to work the system manager has to be involved in the fraud or be negligent. This allows the fraudsters to install software which manipulates data to their advantage. A vigilant system manager will recognise that the software was not applications related and was unauthorised, so it would be short lived and the culprit might quickly be found.

Here is an example of how it is possible to exploit existing software. When half pences were legal tender, banks used to round down to the nearest one pence. Since this was a small amount to lose, clients were relatively happy with this practice. Some fraudsters seized the opportunity to credit these half pences to their accounts. The rounding down having happened, the fraudsters simply re-routed the extra half pence.

The most successful frauds have usually involved a system user and a computer support employee working together. This could work as follows. The user sets up a dummy supplier, and from time to time enters invoices from that supplier. The computer person alters the invoice processing to recognise that supplier and automatically clear any invoices for payment. The cheques are signed without question because the system has cleared them and they are sent to the supplier. If necessary the computer person can also suppress any information about that supplier so that all knowledge about it is removed from the system.

14.4 DAMAGE TO THE DATA AREAS

There are two main causes of data corruption: the actions of system users and the inadequacy of the original design. Users can be creative in finding ways to make the system do things it was not designed to do, and will make a habit of it if they think the system should be able to do it. Often this is done with the best of intentions, perhaps to save time or recover from a plant failure, and usually without realising that this will cause problems. These can only happen if the system design allows it.

For example, in an automated warehousing system, information about where the pallets were stored chained the records together. The first pallet of product A identifies itself as the head of the chain and points to the location of the second pallet and so on down the chain to the end. However the crane mechanisms were notoriously prone to failure. The crane sometimes picked up a pallet to be stored in the warehouse then stopped in the middle of storing it. The pallet was then manually recovered. Because manual recovery meant switching off the crane, this caused the crane to lose the data about that pallet. This in turn broke the product chain, which made retrieving any following pallets impossible without first fixing the chain. So if there were 50 pallets of product A to be stored and the crane failed on pallet number 32, pallet 31 pointed to a location that was thought to be empty and pallets 32 to 50 could not be found (figure 14.5).

Because the location for the 32nd pallet was empty the pointer to pallet 33's location was blank and therefore pallets 32 to 50 were not retrievable using the system. These broken chains usually went unnoticed until the warehousemen tried to retrieve pallets

Pallet number	Warehouse location	Data stored on the system
1	location full	points to next pallet
2	" "	" " " "
3	" "	" " " "
.	" "	" " " "
.	" "	" " " "
.	" "	" " " "
30	location full	points to next pallet
31	location empty	no pointer
32	location full	points to next pallet
33	" "	" " " "
34	" "	" " " "
.	" "	" " " "
.	" "	" " " "
.		
50	location full	points to next pallet

Fig. 14.5 Data chain broken due to hardware failure

of that product to load it onto a delivery lorry. The crane failures were not anticipated, so the software system was not designed to recover from them. The user's method of overcoming the problem then caused data to be inaccessible.

Another typical case can arise where the designed system is badly implemented; for example, where the value reported by the system does not match the value the user expects to see, or a crosscheck value held by the system. Let's say that at the end of each monthly billing period a cellular phone system adds up the cost of each call made on that phone for the month. This figure is then checked against a running total that was incremented each time the phone was used. At times, these two figures do not match. Which figure does the client accept as being accurate? This problem arises because in the design of the system, the part dealing with day-to-day processing of the calls and the part of the system dealing with the monthly billing processing are using the same data in different ways to produce different results.

14.5 DAMAGE TO THE CLIENT'S BUSINESS

Damage can be localised: one process may no longer be running; one personal computer may be stolen; one data area corrupted. Or it can be more widespread and bring the organisation to a standstill: the main processor or the whole installation may be inoperable. In the aftermath of any of these situations the client organisation has to know what to do next: whose services do they need? can they carry out business as usual? what alternative systems or procedures do they use?

Typical consequences of hardware damage. System downtime can only be avoided if the problem is due to a peripheral device, such as a printer. In these cases, the rest of the system will continue to operate. Until the printer is fixed or replaced the users can continue by rerouting print jobs to a different printer, or postponing them. Damage to the main processor, such as a burnt-out logic board, or a read/write head crash on the hard disk, will necessitate halting the system, installing the new hardware component, testing it, restarting the system and recovering the data. This assumes that the required hardware is instantly available. If it has to be delivered, or ordered, then the delay could be between half a day and two days.

The effects on the client's business will depend on the extent of the damage and the length of time the system is unavailable. The client might have to stop production, or go back to a manual system and record every event on paper.

Typical consequences of software damage. These could be relatively minor. Where a program 'hangs', it is sometimes possible to identify the program and simply restart it. However restarting one program can have an effect on those programs already running. For example, the restarted program may reset data flags to a value that the running programs won't accept as valid. Sometimes it is necessary to halt the whole system because one program has crashed. A more worrying scenario occurs if the code is found to be inconsistent when a number of new programs have been released onto a live system, and an error occurs after installation. Assuming the maintainers backed up the original system before installing the new programs, they can recover to a working system by saving the new system, reloading the old system, restarting the system and recovering the data areas.

The effects on the client's business will again depend on the length of downtime and the time it takes for the damage to be noticed. For example if fraud continues for six months, more will have been lost than if it was noticed after only six days.

Typical consequences of data area damage. The effects of a damaged system can be anything from inconvenient to life-critical. Consequences will vary depending on the data problem. If a charity's mailing list has been corrupted and the postcodes are missing the consequence might be that some mailshots are delivered late or not delivered at all. If a formula for a new type of oil was stored on a system and fell into the hands of competitors, the originators would quickly lose their competitive advantage. If a military aircraft wrongly identified a friendly aircraft as an enemy it could be shot down. This occurred in 1991 when an American warship stationed in the Arabian Gulf identified an Iranian airliner as an Iraqi fighter plane and shot it down.

However it would be wrong to assume that simply because an organisation operates in a commercial marketplace all security problems should be treated as profit-threatening, or that all inconsistencies in a defence establishment threaten lives or political stability. Alongside the technical problems that have to be resolved are the personal issues. If bad design or inadequate advice about the consequences of decisions was the cause, the client may lose confidence in the system and its developers and this will take months to rebuild. Two issues must be investigated: how is the relationship between the software maintainers, client organisation and system users affected and how is the relationship between the client organisation and its clients affected?

Consider what happens when an on-line flight booking system goes down. Prospective travellers might phone back later but they are as likely to phone another airline. For that airline to decide whether it is cost-effective to prevent this happening, they might use a simple algorithm:

Bookings per hour x hours to recover x probability of risk occurring

However this takes no account of the subconscious effect on their clients, who might start thinking along the lines of 'if their computer system is unreliable how reliable are their planes?' or 'if they can't give me instant information, they don't value my custom'. These are less tangible, and the impact lasts much longer than the system downtime.

14.6 DISCUSSIONS WITH THE CLIENT

It is important to enlist the help of the client in identifying risks and suggesting appropriate security measures. The software developers bring to these discussions their expertise in protecting systems, and the client brings business knowledge about most probable risks and their consequences. Just as the business and performance requirements were specified and agreed between the two parties, so too must be the risks and the security aspects.

Protecting the system and building in recoverability will cost extra time and money. It may also slow down system response times and require extra memory space, or even more processors. They are not part of the application system and clients may be understandably reluctant to pay out thousands of pounds to protect their system against something that may never happen. If the worst should happen however then they will hold the software developers accountable. For this reason analysts have a responsibility to themselves to persuade the client to take this issue seriously. Outside bodies such as institutional shareholders can be relied upon to influence the client where the risk is such that it may threaten the profitability or commercial survival of the organisation. Analysts also have to help define protection measures wherever the organisation needs to comply with legislation such as the Data Protection Act, and perhaps remind the client of the penalties for failing to do so. It may be appropriate to consult organisations who specialise in contingency planning and disaster recovery to ensure that all eventualities have been covered.

In an ideal world clients and developers would discuss in turn each data item, data area, program, and hardware component, identify all the possible risks to each and assess the cost of preventing an error against the cost of correcting it. This doesn't happen however because the cost of such investigation and specification of security requirements would far outweigh the payback. What actually happens is that three types of issues are discussed. These are system not available, security breaches and loss of data/system integrity.

System not available. The maximum acceptable downtime is determined by how long the client estimates they can maintain a high enough level of service or productivity without the system. Organisations running widely used systems that have to be taken down, for routine housekeeping or update activities, like to be able to warn their clients well in advance. For example cashpoint machines within a geographical area might display this type of warning, when in their idle state, for one week before the event. The message would also suggest alternatives, such as directing customers to a nearby associated outlet, to save the bank from a deluge of complaints.

Lack of availability is particularly worrying when the system crashes and is down for a long period. This might still be acceptable if the maintainers can tell the users a definite restart time, but if the maintainers don't know what is wrong, they can't estimate how long it will take to put right. This puts the client in the difficult position of trying to cope without the system for an unspecified period. This in turn makes it difficult for them to decide whether to wait for the system to be restarted or to continue without the system. For organisations where any downtime is too long, say a nuclear power plant, more than one processor is required to ensure that this situation never arises (see section 14.7.2).

Security breaches. The client has to consider threats from inside and outside the organisation. If, for example, the organisation has formulae that only a few trusted staff know, then encrypting that data would significantly reduce the possibility of that data getting into the wrong hands. This might be especially useful to organisations that have high staff attrition rates.

The client also has to decide how much they trust people outside their organisation who might have legitimate access to their system. In the early 1990s, Virgin Airways shared a system with British Airways until some members of BA misused access to Virgin's data. It would have been possible to forecast the risk, but unlikely that Virgin could have prevented it.

Finally the client has to estimate how likely it is that people external to the organisation will want to damage it. If the organisation is politically sensitive, a military or nuclear organisation, or seen as damaging the global environment, there are real risks. Financial institutions are also popular targets for security breaches that enable fraud or theft to take place. Even if the client is not in any of these high risk groups, they will still have to estimate how their business would be affected if someone installed a virus, stole a workstation or hacked into their system and printed off a copy of all of their data.

Loss of data/system integrity. This means loss of accuracy or any action that causes data or software to be missing or wrong. Since errors and omissions account for over 50% of risks to the system, the developers must ensure their system design is not at fault. Again there are external risks to the integrity of the system from viruses and hackers.

Having identified the risks most likely to affect the system, the client now has to decide which of these they need or want to prevent, those they want to be able to recover from, and lastly those risks which they are prepared to accept and do not believe it is necessary to protect against.

14.7 BUILDING IN PROTECTION

There are a number of ways of counteracting possible risks to a system. We will look at three alternatives:

- Protection that depends on software alone, whether the code is part of the manufacturer-supplied operating system or written by the developers
- The protection we can develop using a combination of software and hardware
- Hardware-only protection.

The costs of these options must be balanced against each other and against the consequences of the risks they are designed to prevent or cure.

14.7.1 Using software only

Where a data area is shared between many users and all of them can access the data at the same time, there is the risk of the data getting out of step. Let's look at an example; refer to figure 14.6.

The problem occurs because the data was not write-protected and the second user read the data before the first user had updated. The stock file thinks there are 60 Mars bars available for allocation, but actually there are –10.

Using a *data lock* will ensure that this can't happen by making the second person wait to read the quantity in stock until after the first person has updated the quantity. See figure 14.7.

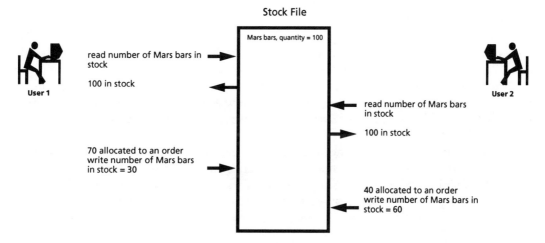

Fig. 14.6 Data corruption due to lack of data lock

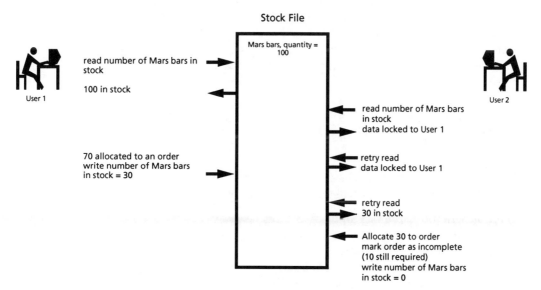

Fig. 14.7 How a data lock works

processing	data
DO UNTIL change counts are equal	
read data	0102DIR00230010SW000
	18409LG09003201ACK09
reread change count	0102
END	

Fig. 14.8   How change counts work

In today's multi-user environments, where processors are becoming increasingly faster, it would be unthinkable not to use some data access control mechanism of which data locks is the main one.

An alternative is to use *change counts*. These operate in a similar way to data locks. The change count is stored as part of the data. The count is read first, then the data, then the count is read again. If the count is not the same on both reads, the data is read again, and the count is reread and checked. Counts are used in systems where there is an unstoppable writer: a device where if the data is not written to shared memory at exactly the time it occurs, it will be lost forever. Such systems are rare. Change counts work as shown in figure 14.8.

Change counts are useful when using a data lock would be over-engineering and would significantly reduce the performance by requiring more processing time.

Other software devices can be used to protect data against intrusion, for example data encryption prevents data being readable by the human eye, but the encryption algorithm could be decoded by another system. This device is used if sensitive data has to be transmitted across public networks.

One way to restrict access to a data or code file is to allocate user privileges. Most operating systems will provide this as an option which system managers can enable or disable. The decision to use privileges will be made in the early stages of development, often even before the hardware is purchased. When a file is created, privilege flags are set. These will identify the file's owner, and what type of access each user has.

Fred creates a file FRED.1, and the privilege flags look like this:

    owner = true
    read = true
    write = true
    update = true
    delete = true

Jim, on the other hand, has read-only access to the file so when Fred copies FRED.1 into Jim's directory, the privilege flags look like this:

In Fred's directory	*In Jim's directory*
owner = true	owner = false
read = true	read = true
write = true	write = false
update = true	update = false
delete = true	delete = false

Users are often only aware of privileges when they are prevented from doing something such as deleting a file which was created by someone else.

There are also levels of privileges such as 'user' and 'superuser'. The difference between the two is that someone who only has user-level privileges must work within the constraints imposed on them by the superuser, but within these constraints and within their own directory area they have all they need to do their job. Superuser privileges are usually only assigned to system managers. This enables them to

- increase or reduce the amount of memory allocated to users
- to reset a user's password in case the user forgets it
- to create or delete system users
- to access user's files
- to assign and change system task priorities. For example if there is a long queue of jobs waiting for the printer, the superuser can move an urgent job to the head of the queue.

It is important to recognise that privileges are not tied to who the user is, but to how the user identifies himself or herself to the system. Fred might do most of his work logged in on his personal identifier at the ordinary user level, but he may also know how to log into the system as a superuser and he may use that knowledge to move his jobs to the head of the queue.

Passwords allow access to the system, or parts of the system, that may be confidential or dangerous if accessed inappropriately. The password format is specified by the system manager and this would be something like 'a minimum of 6 characters, a maximum of ten, must be alphanumeric and start with a letter'. The system manager will also enforce update rules to ensure that passwords are changed regularly, perhaps every two months or after 100 system accesses. Passwords are used to give the user sole access to that part of the system containing their files. This could contain programs they are coding, or diagrams, reports, letters or memos.

Unless they know another person's password the user will not be able to access that person's files. This can give users a somewhat blinkered view of the system. A typical user's view of the system is shown in figure 14.9. The directories in the dotted boxes are the areas that Fred does not have access to. One way for system managers to give users read-only access to files written by other users is to designate common access directories.

Fred cannot see the files contained in the dashed boxes, only those in his directory and the general directory. This diagram also shows that the file fred.1 is now available to anyone who can log on to the system. Software developers might use this approach to release tested programs. Each developer has their own directory, access to which is controlled by a password, and when the program has been signed off as working, the program is released to the general project directory.

There are also levels of password and these can be used to effectively 'lock' areas that are more private. This would be the case where a user logs onto a system at the beginning of the day and logs off at the end of the day, and from time to time during the day wants to read or send electronic mail. Electronic mail is treated in the same way as postal mail, it is private and only intended for the addressee. This means that the user keys in one password to access the system and a second password to use the electronic mail facility. Using passwords, and levels of passwords, ensures that sensitive material can only be accessed by authorised personnel.

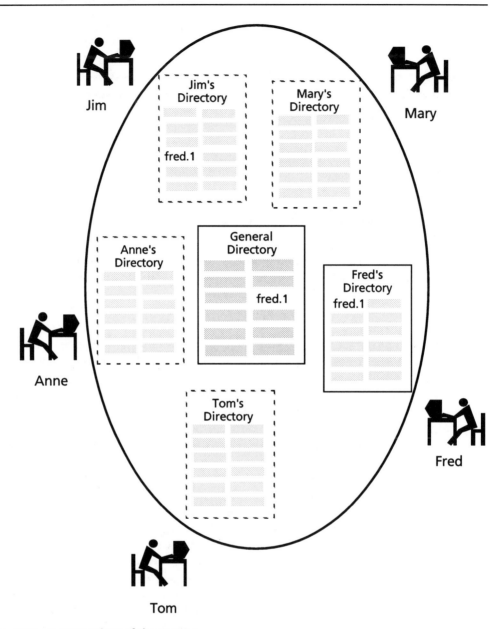

Fig. 14.9 A user's view of the system

System access logging is a piece of manufacturer-supplied software. This is where every time a user logs on to the system, information about the log on is recorded: user-id, terminal-id, log on time. It will even register attempts at entering an invalid user-id or password. This type of information can be useful for tracking down would-be hackers. It helps the system manager to spot which terminals or modems are potentially vulnerable so that steps can be taken to make them less vulnerable by, for example, changing the telephone number of the modem.

It can also help track down fraudsters. This is because whenever a data file or program is updated the system timestamps the new version. So if the fraud is spotted, the system manager can check back through the system access log to discover who was logged on at that time and identify the suspects. Version control is part of updating a program or data definition. A programmer carrying out maintenance work on a program registers in the file header: the version number, date, their name, and the reason for the update, change request or error report where applicable. As such it is a manual procedure, but the code listings reflect the manual procedure. Version control ensures that the latest version of an updated program is installed in the new release of the system. It also allows the maintainer to go back to an earlier version of the system if there are errors in the latest release.

Configuration management is a similar process to version control, but at the system level rather than at the program level. Let's consider a system that is being installed in phases. The phase 1 system configuration contains programs A, B, C and D. Phase 2 adds programs E and F and replaces A with A2. Phase 3 removes B which was a temporary process and replaces it with programs G and H. Configuration management identifies at any point in the system life cycle which programs are in the current system, and which versions of those programs.

14.7.2 Using a combination of software and hardware

User identifiers, sometimes called logons, can be used to restrict the type of access a person might have. This works in a similar way to passwords and privileges except that these are not set up on creation of the file but are coded by the developers into the applications software. Let's assume you work in the accounts department and want an update on how a certain product is selling. You logon as 'accounts', the process validates your user-id and your are given enquiry-only access to the sales files. If you had logged on as 'sales' you could have amended the files. The logon you used prevented you from changing data you didn't own. The hardware equivalent of this is where the system restricts access based on the user's terminal. So if the terminal you are using is in the accounts office you will again be given 'read-only' access to data outside the control of the accounts department. The system uses the terminal-id instead of a user-id to decide access rights.

Standby hardware is used for critical systems so that if the main processor crashes, the standby hardware takes over with as little loss of data as possible. A good example of the need for this would be in a process control system. The computer monitors and controls every piece of equipment – valves, motors, switches, robots etc. – so if the system crashes, production stops. Dual processors, sometimes called master and slave processors (figure 14.10), eliminates or reduces loss of production by running on the master processor until that system crashes and then switching to the slave with little or no loss of data.

For two processors to run as one there has to be a software handshaking device. This is a simple message sent between the two system at defined time intervals. The message simply indicates 'I'm still running'. If the time interval is 1 second, and the slave doesn't receive this handshake after 2 seconds the slave assumes the master has crashed and takes over as the master.

This instantaneous changeover is known as a hot standby because the two systems are always running and as well as the handshakes the master is keeping the slave's data areas in step with its own. The slave is doing very little processing and is mostly waiting for the master to break down.

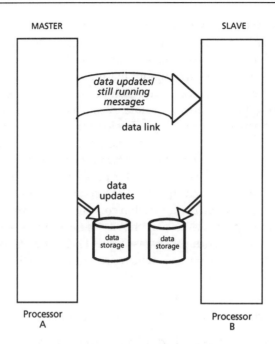

Fig. 14.10 Dual processors providing a hot standby

When a system crashes some data is usually lost because the system was doing something at the time of the crash. To be able to recover the data areas after a system crash, the data has to be stored on a non-volatile medium such as a magnetic tape or a disk or another computer as in the dual processor example above. Periodically saving data to a transaction log, or dump file as explained in chapter 18, allows the user to recover the data areas on restarting the system.

14.7.3 Using hardware only

It is possible to set up a screen to mimic what is happening on a nearby screen without the two being connected to the same system. This clearly allows illegal access to a system. The only hardware which helps protect the system and does not require software assistance to do so is a Faraday shield. The shield creates a physical barrier which prevents the second terminal from eavesdropping on the first.

14.8 FORMULATING A PROTECTION POLICY

The amount of effort spent on securing the system has to be based on an estimation of the consequences of the risks to that system. There would be little point in spending thousands of pounds to prevent an error occurring if that the error would probably occur once a year, would be a minor inconvenience and would take five minutes to correct. The client must have the final say when deciding which risks they need protection against since they are paying for it, but they must also have a very clear picture of risks they will not be protected against, and accept that responsibility.

Preventative measures to consider are:

- *Siting.* Housing the system in a separate room or building. In some military installations and other politically sensitive applications such as the Olympic Games, systems are located underground for maximum protection.
- *Redundancy.* Running more than one processor, which requires complex software to manage the handshakes, the data transfer and the changeover.
- *Power supply.* Using separate power sources for each processor; otherwise if the power source fails, all processors go down.
- *User access.* Restricting each user's access using logons, passwords and privileges can be used to limit the damage, whether by accident or design, that users can do to data or code they don't own.
- *Fault tolerance.* Allowing a number of retries on data entry or transmission, or of defaulting to a defined value, keeps the system up and running. A system that is not fault tolerant would simply report an error and stop.
- *Reviews.* Using review methods such as desk checking, walkthroughs or Fagan inspection on both code and design documents picks up more errors than code testing alone. The human brain makes connections that are not picked up in testing. For example it recognises that although something might not be technically wrong, it is handled differently in this program than in others.

Curative measures to consider are:

- *Spares.* Carrying spares if vulnerable components have been identified, and ensuring that these spares won't be damaged in the meantime or outlive their shelf-life.
- *Hardware warranty contract.* When hardware breaks down, a hardware support engineer is on site within an agreed time. The shorter the response time required, the higher the cost of the contract.
- *Error reporting.* The system reports the error and waits until the operator fixes it before continuing processing. Alternatively, if the program is processing a number of records such as monthly bills to clients, it might report the error to a logging file and continue on to the next client. The operator investigates the accounts that are in error at a later time.

Finally, the following steps are helpful when preparing a protection policy:

- Identify all possible risks.
- Decide on the likely consequences of each risk, both tangible and intangible: envisage the worst possible scenario. Optimism at this time could prove dangerous later.
- Assess the cost of each preventative measure and use the client's experience to make the decision between preventing the risk and curing it.
- Specify any required preventative action: take into account the effect on the organisation of the consequences, and the cost of minimal and maximal solutions both financially and in terms of system performance.
- Investigate what is already available: hardware options, user procedures, manufacturer-supplied software, packages.
- Consider subcontracting parts of the system development to companies that specialise in the security area.
- Explore how easy it would be to amend any already existing and therefore well-tested software to meet your requirements.

- Write your own software. Developers are often tempted to see this as the first option because bought-in software may not exactly match their needs. We suggest that the developers' main task is to produce the applications code. Security software has been written for almost every other system produced: why start from scratch when you can use already designed, already tested code from somewhere else?
- If it is necessary to write bespoke security software, use only experienced and dedicated career programmers. Do this to ensure high-performance, thoroughly tested, quality code; errors in this type of software can be very difficult to spot.

14.9 SUMMARY

In this chapter we have looked at why software systems must be secured against everything from incompetent operators to acts of God. For this to be done successfully, we need to base the system protection policy on the client's business experience and the analyst's experience with similar systems. Developers must continually ask themselves and their clients 'what are the consequences if this goes wrong?' Armed with this information the software developers and their client can assess the probability of each risk occurring and estimate the cost of preventing and curing it, then make decisions about which risks they will protect the system against and how.

This seemingly over-cautious approach will ensure that when accidents or disasters happen, the system is resilient enough to meet the user's performance requirements. All of this must be carried out while bearing in mind that the system has a finite lifespan and the client has a finite budget.

CHAPTER 14   CASE STUDY AND EXERCISES

Q1 System Telecom relies very much on having its systems up and running when they are needed. How would you formulate a protection policy for the system described in the level 1 DFD? Remember to consider the risks to the system and the risks to the business.

Q2 Field sales representatives in System Telecom all use laptop PCs to send call reports or orders to their local office. They also receive sales information directly from the local office minicomputer. What security measures should be taken to protect this activity? Remember that System Telecom's style of operation is 'lean and mean', but equally it regards competitive information as having a high value.

CHAPTER 15

Systems Design:
Human–Computer Interfaces

15.1 INTRODUCTION

In this chapter we shall look at the design of the interfaces which provide the means of communication between the computer system and the people who use it. Such an interface can be described as the point at which the user and the computer system meet. We will consider the output the users receive, the input which they enter, and the dialogues through which the user and the system talk to each other. The inputs and outputs which are referred to are those which involve people. Inputs to, and outputs from, other systems or other processes in the same system are described in the next chapter.

The specification of outputs, inputs and dialogues forms a key part of the designer's task because of the high visibility of these interfaces to the users of the system. The output produced by the computer is the main reason for developing the new system. If the users are clear what they want from the system, the inputs needed to produce this output can then be identified. Should the output fail to meet the requirements of those who have to use it, the system can be seen as a waste of time by those users, who may have little appreciation of the internal, and therefore invisible, sound design or elegant code. The layout of a screen and the consistency of the dialogue will be important to those users who have to spend a lot of time sitting in front of terminals, and who will base their evaluation of the whole system on these interfaces. If the method of entering data is difficult, because of a poorly designed input form for example, then the chances of inaccurate or incomplete data entering the system will be greatly increased, which in turn will adversely affect the value of the output produced.

Some designers see the design of forms, screens and reports as a less important or less interesting part of their task than grappling with the logical design models. However the time and effort spent getting the interface design right will enable the designer to meet the design objective, described in chapter 13, of producing a system which is easy to use. In this chapter, we first consider how interfaces are identified as a result of drawing the system boundary, and then go on to discuss the design of outputs, inputs and dialogues. Finally we describe one other area of interface design – the design of the workstation. This is important for the designer because it can significantly affect the performance of the end user, it can be critical in their perception of the system, and it also has health and safety implications.

15.2 AGREEING THE SYSTEM BOUNDARY

In order to identify where in the new system the interfaces will occur, a useful starting point for the analyst or designer is the data flow diagram for the required system. This is a logical view which was described in detail in chapter 11. The example shown in figure 15.1 is a simplified version of a required DFD for an order processing system. The line around the edge represents the system boundary and shows those parts of the system to be automated – inside the line – as well as those parts which will remain outside the new computer system. This boundary must be determined in consultation with the users of the new system and in the end it is their decision as to which processes are automated, and which require human judgement or intuition and are therefore best done by people. In discussing this with the client, there are three questions to be asked about any process:

- Can it be automated?
- Is it economic to do so? The answer to the first question may be 'yes', but will it be worth the money in terms of the benefit it will bring to the business?
- Is it operationally possible to automate it? In other words will the environment allow it? There may be resistance to the automation, for example from trade unions.

The designer will present options, but the final decision rests with the client. When the system boundary has been agreed, both the client and the system developer will be able to see clearly what is in and what is out of the new system. In the data flow diagram in figure 15.1, inputs are represented by data flows entering the system and outputs are those data flows leaving the system. The points at which one of these lines enters or leaves a process box mark the actual interfaces with the system, and procedures have to be designed to handle these. This includes the design of screens and dialogues which allow the users to give instructions or to ask questions about the stored information. The system boundary can also be used to illustrate the phases in the development of a system by extending the original boundary with dotted lines on the DFD to represent the future stages of the project.

15.3   OUTPUT DESIGN

Having identified from a logical model of the new system where the outputs will be, by listing those data flows which cross the system boundary as they leave the system, the next stage in the process is to determine their content. Again structured techniques play a useful role here, because the designer can turn to the data dictionary to find the content of each data flow which represents an output. Let's consider an example from the DFD in figure 15.1 – the output described as *Stock Valuation Report*. According to the data dictionary, the contents of this data flow are as follows:

DATA ITEM	TYPE	LENGTH
Inventory group	Character	25
Stock code	99999	5
Product description	Character	30
Unit cost	999.99	6
Quantity in stock	99	2
Unit selling price	9999.99	7
Expected sales	999	3

Systems Design: Human–Computer Interfaces 225

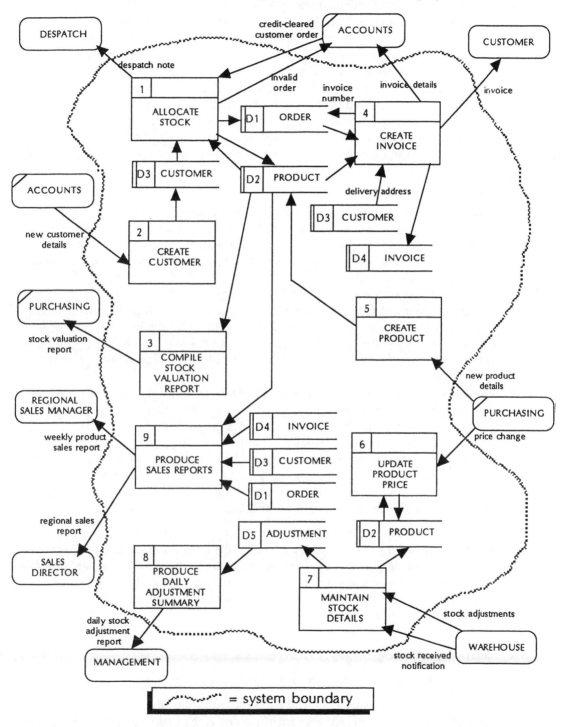

Fig. 15.1 The identification of outputs, inputs and dialogues on a data flow diagram

Once the content is known, the designer must select the appropriate method or technology to present the information, and then create the document, report or display which contains the required information. It should be noted that not all of the outputs required by the user may be shown on a data flow diagram. For example, a number of ad-hoc reports may be required on demand from data stored in the system, or exception reports, which show only data items which are unexpected, or above or below a predefined limit, may have been identified as a requirement in the requirements catalogue produced by the systems analyst. These should not be overlooked by the designer.

A quality output is one which meets the requirements of the end user, and which presents the information in a way which is clear, easy to read and visually attractive. In order to decide on an appropriate method of presentation, and a suitable format, a number of questions need to be asked:

- *Who receives the output?*
  What profile can you create of your target population and their needs? The output may be received, for example, by users within the company, eg. a listing of accommodation available, or by people outside the organisation, such as customers, eg. an invoice, or government departments, eg. inland revenue returns, or by management, eg. a monthly sales report.
- *Under what circumstances will the output be received?*
  Does the environment place constraints on the technology which can be used? For example if the output is to be generated on a factory floor, the device used may have to operate effectively in dusty or dirty conditions, or in a noisy environment where an error message 'beep', which would be perfectly audible in an office setting, would be inappropriate.
- *What use will be made by the receivers of the information on the output documents or screens?*
  If the design of the output is to result in a quality product, then its purpose, from the user's point of view, must be understood by designers.
- *When and how often is the output needed?*
  What implications will the required frequency of information have for the selection of an output method? For example, a warehouse supervisor may require a daily stock report, whereas senior management may only need to receive reports once a month (although they might also require the facility to make ad-hoc enquiries).

The answers to these questions will help the designer make decisions about whether a display or a printed copy of the output is required, the type of device needed to produce the output required, and the layout of the information which would best meet the needs of the users. Let's now consider some of the options available to the designer, firstly a brief review of the technology available, and then an examination of the alternative ways of presenting information.

15.3.1 Output technology

When considering output devices, there are two main alternatives:

1. Printing, using any of a variety of different types of printer, or
2. Displaying the information on the screen of a mainframe terminal or a microcomputer.

There are two main types of printer – *impact printers*, which work like a typewriter where a hammer strikes an inked ribbon to produce a character on paper, and *non-impact printers*, which work in a number of different ways to produce a character on paper, without making physical contact with it. Examples of the former are the dot matrix and daisy wheel printers, while examples of the latter include the laser and ink-jet printers. While impact printers have traditionally been more commonly used in computer systems because they are less expensive, new systems are increasingly making use of laser printers to produce printed output because of the better print quality, the possibility of integrating graphics with the text, and because the cost of the technology is decreasing.

A visual display unit (VDU) or computer screen is the other main technology for producing output. This method is appropriate where only a transient image, rather than a printed document, is required, although typically there would be a facility within the system for making a 'hard copy' of the information on screen. Traditionally, the display would be monochrome, but colour monitors are now being used more in developing systems, and high resolution graphics enable sophisticated presentations of data to be displayed on screen.

In addition to the above two methods, a number of other alternatives are available. Plotters can be used to produce coloured line drawings, such as graphs, diagrams, and maps; output can be transmitted as a facsimile or a digital message; output can be in the form of sounds, such as beeps or clunks for example, music or video images; or if the output is to be stored, rather than being read straight away, it can be recorded onto magnetic media, optical disks or onto microfilm or microfiche. (Microfilm is a strip of film along which individual frames are arranged, and microfiche is a rectangle of film, the size of an index card, containing frames arranged in a grid. In both cases one frame corresponds to a sheet of paper.) Speech output is also possible now, either as *concatenation*, by chaining together pre-recorded units of human speech (this is used on telephone services such as the speaking clock), or by a method called *synthesis by rule*, in which artificial speech is produced through the use of rules of pronunciation, intonation and the physical characteristics of human speech. A number of software packages use this method, which although less limited than concatenation can still sound rather artificial. The use of speech output for error messages, instead of the familiar beep, is an attractive idea for designers, because it can not only attract the user's attention to an error, but also state the nature of the error.

15.3.2 Presenting information

Whether the output is printed on paper, or displayed on screen, three important principles apply:

- The first and most important point is about content. The message here is to keep it simple. This is achieved by presenting only that information which is needed by the user. Clearly, if this is to be achieved, it is important to know the answers to the questions suggested earlier, particularly the first three. The designer can be tempted to include on a report all the available information but this usually proves to be unhelpful to the end users.
- The second principle is to ensure that the page or screen is uncluttered and easy to read. Often reports which are poorly designed have so much crammed onto one page that it is very difficult for the reader to make sense of it. The use of white space on a printed page significantly increases its readability.

```
                        Management Reports - VAT Return
                                                              Date from : 010180
                                                                    to : 3111299
                                                                        Page : 1

Code:      T0       T1       T2       T3       T4       T5       T6       T7       T8       T9
Rate:     0.00    15.00     0.00     0.00     0.00     0.00     0.00     0.00    13.04     0.00
-----------------------------------------------------------------------------------------------
                             Sales Tax Analysis (by Invoice)
Sales Invoices - Nett
       8218.39  38707.02     0.00     0.00     0.00     0.00     0.00     0.00     0.00   603.00
Sales Invoices - Tax
          0.00   5806.07     0.00     0.00     0.00     0.00     0.00     0.00     0.00     0.00
Sales Credits - Nett
          0.00      0.00     0.00     0.00     0.00     0.00     0.00     0.00     0.00     0.00
Sales Credits - Tax
          0.00      0.00     0.00     0.00     0.00     0.00     0.00     0.00     0.00     0.00
-----------------------------------------------------------------------------------------------
                            Purchase Tax Analysis (by Invoice)
Purchases Invoices - Nett
       2868.60   4933.98     0.00     0.00     0.00     0.00     0.00     0.00   244.41 19882.35
Purchase Invoices - Tax
          0.00    740.12     0.00     0.00     0.00     0.00     0.00     0.00    31.88     0.00
Purchase Credits - Nett
          0.00      0.00     0.00     0.00     0.00     0.00     0.00     0.00     0.00     0.00
Purchase Credits - Tax
          0.00      0.00     0.00     0.00     0.00     0.00     0.00     0.00     0.00     0.00
-----------------------------------------------------------------------------------------------
                                  Nominal Tax Analysis
Bank Receipts - Nett
          0.00      0.00     0.00     0.00     0.00     0.00     0.00     0.00     0.00     0.00
Bank Receipts - Tax
          0.00      0.00     0.00     0.00     0.00     0.00     0.00     0.00     0.00     0.00
Bank Payments - Nett
          0.00      0.00     0.00     0.00     0.00     0.00     0.00     0.00     0.00     0.00
Bank Payments - Tax
          0.00      0.00     0.00     0.00     0.00     0.00     0.00     0.00     0.00     0.00
Cash Receipts - Nett
          0.00      0.00     0.00     0.00     0.00     0.00     0.00     0.00     0.00     0.00
Cash Receipts - Tax
          0.00      0.00     0.00     0.00     0.00     0.00     0.00     0.00     0.00     0.00
Cash Payments - Nett
          0.00      0.00     0.00     0.00     0.00     0.00     0.00     0.00     0.00     0.00
Cash Payments - Tax
          0.00      0.00     0.00     0.00     0.00     0.00     0.00     0.00     0.00     0.00
Journal Debits
       3260.27      0.00     0.00     0.00     0.00     0.00     0.00     0.00     0.00     0.00
Journal Credits
          0.00      0.00     0.00     0.00     0.00     0.00     0.00     0.00     0.00     0.00
-----------------------------------------------------------------------------------------------
                                  Tax Analysis Summary
Inputs - Nett
       2868.60   4933.98     0.00     0.00     0.00     0.00     0.00     0.00   244.41 19882.35
Inputs - Tax
          0.00      0.00     0.00     0.00     0.00     0.00     0.00     0.00     0.00     0.00
Outputs - Nett
```

Fig. 15.2   A cluttered report

- The final point is to arrange the information logically on the page or screen, in such a way that it can be easily and quickly used and understood. Every output screen or report should include a main heading or title which identifies the purpose of the output, and subheadings which identify the various sections within it.

In addition it is helpful to the reader if reports and screens from the same system show consistency in the way that the information is arranged, so that users will know where to look to find the information which is important to them.

15.3.3 The use of tables and graphics

Reports are commonly arranged as tables, especially those containing financial information, and this is appropriate if the detailed information is arranged in discrete, labelled groupings, if totals are included and few explanatory comments are needed. However the report in figure 15.2 is an example of poor information design. While all the data required is present, the layout is cluttered and there is too much redundant information which reduces the impact of the key data items. All the zeroes, for example, are unhelpful, and the reader needs some way of deciphering the codes on the top line.

Figure 15.3 on the other hand shows a report in which the layout is clear and the information unambiguous. This weekly product sales report is one of the outputs from the order processing system which was represented by the DFD in figure 15.1, and is clearly labelled with a title, makes good use of white space and does not include redundant information

Graphics can be used effectively to present data in ways other than tables, and low-cost technology is now available to produce high-quality diagrams and charts. However, these graphics need to be used with care if they are to be effective, and much will depend on the designer's judgement and the requirements of the users, as described by the analyst.

The types of graphics used to present numeric information – pie charts, bar charts and line graphs – are illustrated in figure 15.4.

GALAXY
Electrical Products Limited

WEEKLY PRODUCT SALES REPORT

PRODUCT CODE = 01714   DESCRIPTION = BRAUN UK12 FOOD PROCESSOR

INVOICE	INVOICE DATE	CUSTOMER NO	NAME	QUANTITY	NET VALUE
5403	9/11/90	143	VFM ELECTRICAL	4	200
5406	9/11/90	204	STROUD ELECTRICAL	6	300
5409	9/11/90	524	GALAXY, READING	10	500
5413	9/11/90	508	GALAXY, CARDIFF	10	500

Fig. 15.3   Weekly product sales report

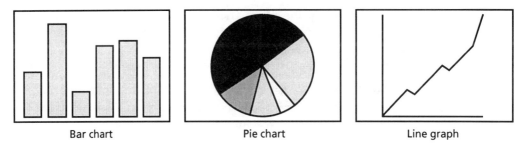

Fig. 15.4 Types of business graphics

15.3.4 Specifying outputs

Having decided on the content of the output, the technology to be used and the format in which the information will be presented, the designer must then undertake the detailed design work involved in specifying the report. This will usually be done on a *print layout chart* for a printed document – an example is printed as figure 15.5 – or as a *display chart* in the case of an output to a VDU screen – shown in figure 15.6.

A print layout chart illustrates the content and format of a printed output produced by a computer system. The proposed column headings, subheadings and some representative data and totals are written on the chart in their correct positions. The headings will predominantly consist of constants, but guidelines published by the NCC suggest that if any variables (which are retrieved or computed from stored or input data) are included, they should be enclosed in square brackets. However if data item lines consist of all, or nearly all, variables, they need not be bracketed. While the print layout chart is a detailed design document, it can be difficult for the users to get a clear picture of what the actual reports will look like. To ensure that any reports produced meet the clients requirements it is helpful to show the users a representation of the finished report before any coding begins.

The display chart is used for specifying the visual layout of output which appears on the screen. The main convention in filling in display charts is to enter all the information as plain characters except the variable fields which are enclosed in square brackets. As with reports, a simulation of the output is useful, as well as the paper specification, to help ensure it is what the user requires and this can be achieved using prototyping.

One final point about outputs from the system. In addition to the standard printed reports, special forms may also be produced. These include *preprinted stationery*, where the system will only add the details which are variable; *multipart forms*, the most common type of which consist of NCR (No Carbon Required) paper which has a special chemical coating on the back so that, when pressure is applied to it, the image appears on the copy underneath; and *turnaround documents*, for example gas bills, which are outputs produced by the system and sent out to customers, to return later to be re-input to the system. This type of document will usually contain characters which can be read by special equipment such as an optical scanner, so it can be identified when it returns.

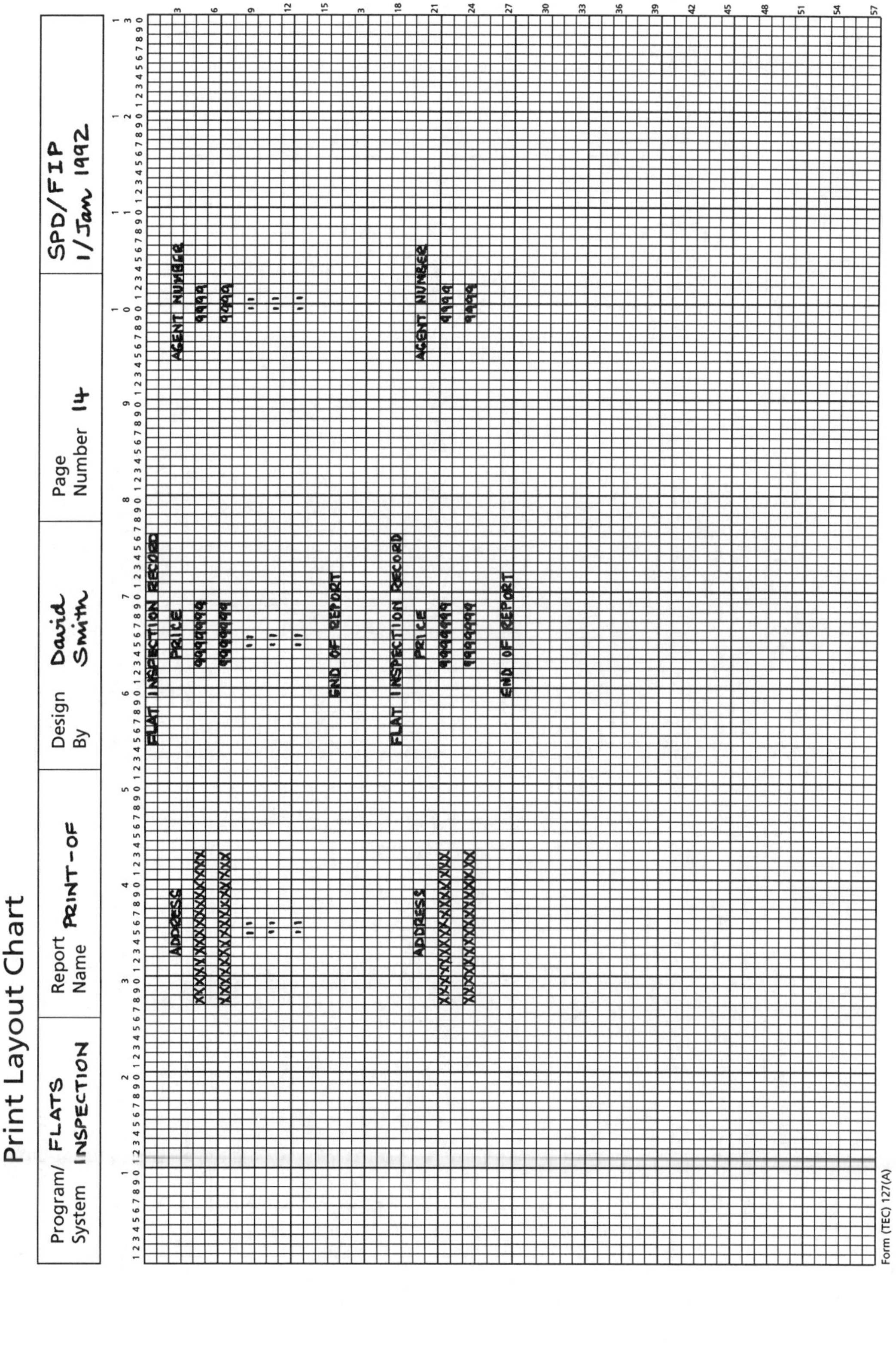

Systems Design: Human–Computer Interfaces 231

Print Layout Chart

Program/System	Report Name	Design By	Page Number	
FLATS INSPECTION	PRINT-OF	David Smith	14	SPD/FIP 1/Jan 1992

```
FLAT INSPECTION RECORD
    ADDRESS            PRICE        AGENT NUMBER
    XXXXXXXXXXXXXX    9999999       9999
    XXXXXXXXXXXXXX    9999999       9999
        ''              ''           ''
        ''              ''           ''
        ''              ''           ''
                  END OF REPORT

FLAT INSPECTION RECORD
    ADDRESS            PRICE        AGENT NUMBER
    XXXXXXXXXXXXXX    9999999       9999
    XXXXXXXXXXXXXX    9999999       9999
        ''              ''           ''
                  END OF REPORT
```

Fig. 15.5 Print layout chart

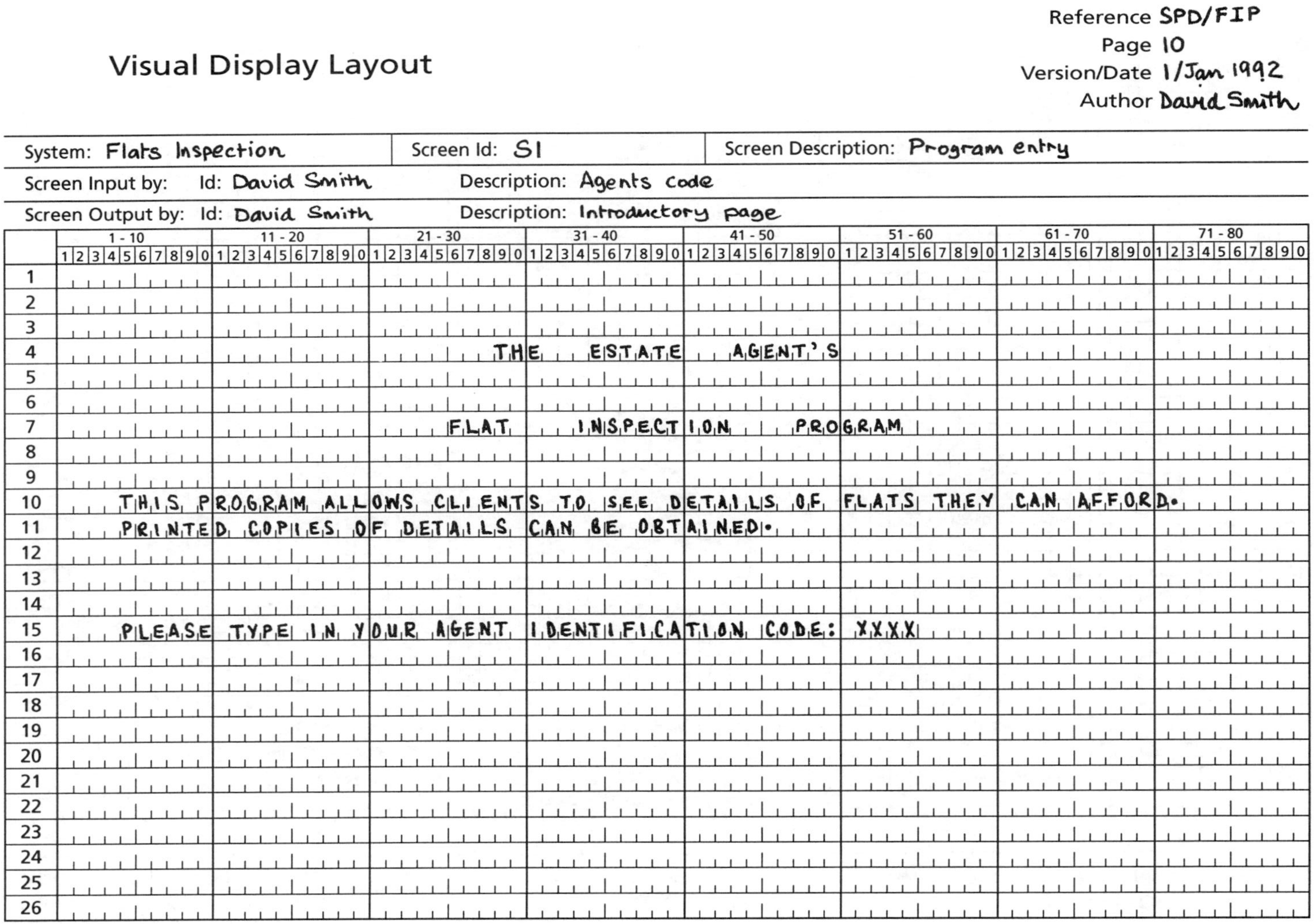

Visual Display Layout

System: Flats Inspection	Screen Id: S1	Screen Description: Program entry
Screen Input by: Id: David Smith	Description: Agents code	Reference SPD/FIP
Screen Output by: Id: David Smith	Description: Introductory page	Page 10 / Version/Date 1/Jan 1992 / Author David Smith

Row 4: THE ESTATE AGENT'S
Row 6: FLAT INSPECTION PROGRAM
Row 10: THIS PROGRAM ALLOWS CLIENTS TO SEE DETAILS OF FLATS THEY CAN AFFORD.
Row 11: PRINTED COPIES OF DETAILS CAN BE OBTAINED.
Row 15: PLEASE TYPE IN YOUR AGENT IDENTIFICATION CODE: XXXX

Fig. 15.6 Display chart

15.4 INPUT DESIGN

When designing input, the objective is to ensure that the data which will be processed by the system is collected and entered into the system efficiently, according to the specified requirements, and with the minimum of errors. In discussion with the client, the designer will choose a method of input which is cost effective and acceptable to the end users. The process of input design, like output design which was described earlier, consists of four stages:

- firstly, identifying the inputs into the system, by listing the data flows on the required logical data flow diagram which cross the system boundary on their way in;
- then determining the content of these inputs by inspecting the data dictionary;
- next choosing an appropriate input device to change the user's data into a form which can be read and processed by the computer system;
- and finally completing the detailed design work involved in specifying forms, input screens and any other data collection documents.

We will examine each of these four stages in turn and look at the design options available and the steps which can be taken to ensure that the objectives of input design are achieved.

The first two steps are linked together, and involve the designer in looking at the data flow diagram for the required logical system, and the data dictionary. If you look back to the DFD in figure 15.1, a number of inputs for the order processing system can be identified: *new customer details, credit-cleared customer orders, new product details, price changes, stock adjustments* and *stock-received notifications*. If the data dictionary for the system were to be inspected, the contents of the data flows which correspond to these inputs could be determined. For example the data items which make up the input called *new customer details* are:

    customer number, customer type, sales region, discount code, name, address, and delivery instructions,

while *new product details* contains the following data:

    product number, description, manufacturer, origin, inventory group, product class, despatch unit, re-order level, discount prices, vat code and VAT rate.

In this way the contents of each input can be confirmed, as can information about the source, volume and frequency of the input.

The next stage is then to choose the appropriate technology for introducing the data contained in each of these inputs into the system. There are a wide variety of different ways of entering data, and the choice of the appropriate method will depend on a number of factors, the two most important of which can be summarised in the following questions:

- Which method will be most suitable to the needs of the users who have to enter the data? While a keyboard will be appropriate in many situations, it may not be appropriate to a checkout operator in a supermarket, where speed and accuracy are important.
- Which method will be most suitable for the format and volume of the data to be entered? If an educational institution has a large number of multiple choice answer sheets completed during examinations, and wishes to put these results onto the system, then an optical mark reader – which can read the input directly from the students answer sheets – would be the most suitable.

Whichever method is chosen, it will include some or all of the following steps: (1) the initial recording of significant data by the user; (2) the transcription of data onto an input document; (3) the conversion of the data from a 'human-readable' into a 'computer-readable' form; (4) verification of this data conversion to pick up any errors; (5) the entry of the checked data onto the computer system; (6) the validation of the data by the system to ensure it is logically correct; (7) the correction of any errors highlighted by the data validation program. As passing through all these steps makes data entry a costly process, the key guideline for the designer is to make the process as simple as possible. If this can be done in a way which minimises the cost to the user, minimises the chance of data transcription errors occurring and minimises the delay in entering the data onto the system, then the designer will once again be building quality into the system.

While there are a large number of different input devices, they can be grouped according to the way in which data is entered: either by keyboard transcription from clerical documents, by direct input onto the computer system via a peripheral device, by direct entry through intelligent terminals or by speech or body movement.

15.4.1 Keyboard transcription from clerical documents

This is still the most common way of entering data – from an order form, a time sheet, or an application form for example. The device used is the keyboard or keypad. There are a number of different types including the *QWERTY keyboard*, which has the keys arranged in a layout invented by Christopher Scholes and used on conventional typewriters; the *Dvorak keyboard*, which allows greater speed and lower error rates as a result of frequently used keys being located together; the *alphabetic keyboard*, which is very similar in design to the QWERTY keyboard, except that the keys are arranged in alphabetical order; the *chord keyboard*, which requires several keys to be pressed simultaneously to form letters or words, and can be extremely fast when used by an experienced operator – an example is the Micro Writer keyboard, which consists of five keys which fit the fingers of the right hand, as well as two shift keys, and these seven keys, in various combinations, can produce upper and lower case characters, special characters and numbers; and the *numeric keypad*, which consists only of keys marked with numbers, decimal points, and arithmetic operators, as well as an enter key, and is used for the rapid keying of numeric data. In addition to the keys found on a typewriter, keyboards may also have a number of special-purpose *function keys*, which are programmed to perform a particular action – such as 'escape' – when pressed, as well as *cursor control keys*, which are four keys, marked with arrows, used for moving the cursor around the screen. The layout of keys on a typical QWERTY keyboard is shown in figure 15.7.

If the number of transcription errors are to be minimised when this method of data entry is used, then a number of steps need to be taken. Some are the responsibility of the user organisation, such as the training of the keyboard operators, while others rest with the designer. In the latter category, these steps would include: first of all, designing an input form which is easy to complete; minimising the amount of data which the user needs to record on this form (on a customer order form, for example, the customer's name, address and account number can all be preprinted); and ensuring that the design of the input screen matches the input form – with the data fields in the same order and the same relative positions.

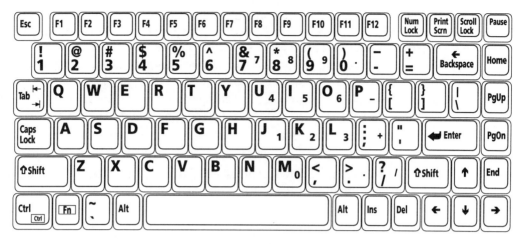

Fig. 15.7 Layout of a keyboard

15.4.2 Direct input onto the computer system via a peripheral device

Another approach to minimising the errors introduced into the system is to reduce the possibility for human error by using input devices which require little or no action by the user after the data has been initially recorded. These peripheral devices include:

- *Bar code readers*, which read the input contained in bar codes, can take the form of pens, as used by librarians when issuing books, or the gun-like versions used in shops. Bar code readers can also be embedded into work surfaces, as is the case at many supermarket checkouts, and read bar codes which are passed over them.
- *Optical character recognition (OCR) devices*, which read printed characters directly from a page, and are widely used to accept input where the volume of data is high, and the input needs to be entered quickly. For example, turnaround documents, such as gas and electricity bills, are produced by a computer system, sent to the customer and then re-accepted as input, when they are read in by OCR devices.
- *Magnetic ink character recognition (MICR) devices* are similar to OCR devices, but are sensitive to the magnetic rather than optical qualities of the characters and are therefore more reliable. This system is used by banks, and characters printed using a special font can be seen at the bottom of every cheque in a cheque book. The ferrite impregnated ink with which these characters are formed is magnetised by the MICR device, so that the characters can be recognised and accepted by the system
- *Optical mark recognition (OMR) devices*, which detect the position of hand drawn or machine printed marks on documents. These marks are read from specially designed forms so that the marks will be in the correct positions, and these devices are therefore best suited to stable applications where the production of large quantities of these input forms will make the method cost effective.
- *Automatic scanners*, which accept input either as pages of text which can be read into the computer system at high speed, or in the form of pictures or diagrams.
- *Touch-sensitive tablets or screens*, which detect the position of a finger touching them, and interpret this as a particular item of input; for example, in on-screen telephone

dialling facilities, touching the telephone number on the screen will cause that number to be automatically dialled. Such devices are very easy for people to use with little training, but the surfaces must be kept clean.
- *Pointing devices*, for example the *mouse*, which can accept continuous input when it is rolled over a flat surface to move the position of the cursor around the input screen, or discrete input when a button on its surface is pressed to select an on-screen object. The *trackball*, which is a ball embedded in a fixed socket, and the *joystick*, a small stick which can be moved in any direction within a fixed socket, are similar input devices which can be used to move the cursor.

Other direct entry devices include machines which read magnetised plastic badges, or which accept input from the magnetic strips on the back of credit cards.

15.4.3 Direct entry through intelligent terminals

An 'intelligent' terminal is one which has processing power and the ability to store data, as opposed to a traditional terminal which does not have these facilities and can only be used to input data directly into the system. There are many examples of this type of input, which eliminates the need for input forms. One example is the point-of-sale (POS) equipment used by many shops. Data is directly entered via a keyboard into an intelligent terminal, often a personal computer of some kind, and validated. It is then either sent directly to the main computer – in an *online* system, such as an airline ticketing operation – or stored on disk, and entered onto the main system at some later stage – a *deferred online system*. This type of direct entry is often a feature of distributed computer systems, where local processing and storage of data are seen as an advantage to the business needs of the organisation. A number of retail outlets carry out stock checking using small, hand-held computers, which can then be connected to the main computer system and the data which is stored in the memory downloaded.

15.4.4 Input by speech

Input using speech recognition, whilst a desirable method of data entry because it is a natural method of communication, leaving the hands free to carry out other tasks and requiring very little training of data input staff, still has a long way to go in its development. It has only been used in certain specialised applications, and even then is limited to single word or short phrases, and the system has to be 'taught' to recognise the speech patterns of the people whose voices will be used to provide input. This is because speech recognition devices have problems recognising where one word ends and the next one starts, adapting to the voice patterns of different people, distinguishing between similar-sounding words and expressions (figure 15.8) and can be confused by background noises. However, some predict that speech input will one day replace the keyboard as the primary method of data input.

Input can also be accepted from devices responsive to head or eye movements of the user, using methods such as the detection of reflected light from the eye, the tracking of head movements or the recording of muscle contractions through electrodes.

The final stage in input design, having identified the inputs and their content and

Fig. 15.8 The problems of speech recognition
(from the *Usability Now Guide*, 1990)

chosen the appropriate technology, is to do the detailed design work. Many of the principles which were outlined earlier, when discussing output design, are applicable here. The designer should be concerned with keeping the amount of data that has to be input to a minimum by limiting the collection of data to information which is variable, such as the number of items of a particular product ordered, and data which uniquely identifies the item to be processed, for example an indication of whether payment is being made by cash or credit. Data which is constant, or which can be stored or calculated by the system, does not need to be entered. To ask the user to do so wastes time, increases the possibility of transcription errors and makes the job of data input to the system seem like an additional overhead rather than an integral part of an individual's work. When designing input forms and screens the key, as with output design, should be to keep things as simple and uncluttered as possible. A well designed input form will have a logical sequence to the information requested, will read from left to right and from top to bottom of the page, and will contain sufficient 'white space' to make it easy to read.

In specifying the inputs required by the system, the designer will need to ensure that they meet the requirements of the users, and that they can be clearly understood by the programmers who will write the programs to handle the input. Any form to be filled in by the user should be accompanied by an explanation of its purpose and instructions on how to complete it in a user manual.

15.5 DIALOGUE DESIGN

For most users of computer systems, the main contact with the system is through an on-screen dialogue, and their perception of how 'friendly' the system is will depend on the characteristics of this dialogue. Ambiguities and difficulties in the dialogue cause problems in training and operation and lead to systems underperforming. For example it is important that developers use *style guides* to ensure that screens which are part of the same system, but which are designed or programmed by different individuals, have a consistent layout. Because the dialogue involves a two-way exchange of information, this area of system development is also referred to as *human computer interaction* or HCI. It is an important area for the developer who, in seeking to design an effective dialogue, will be concerned to ensure that the conversation between the system and the user flows freely. This depends on understanding the characteristics and requirements of the users, as well as the type of hardware and operating system available, and on having an appreciation of the principles of screen design and the different ways in which the dialogue can be structured. This is particularly important in the design of interfaces in a real-time system, where the objective is to minimise the human interaction with the computer system, while ensuring that the interaction which does occur is effective. In this section we will concentrate on screen design and the different types of dialogue, as well as describing a number of guidelines for the design of an effective graphical user interface, and types of support available to the users.

15.5.1 Screen design

The quality of the screen design can have a direct impact on the performance of the users of the system, and the designer needs to consider the format as well as the content of the screens on which the dialogue, or interaction, between the user and the system is based. A number of features of screen design are worth discussing here:

- *Text* – must be easily readable. In addition to choosing an appropriate font and size for the characters, readability can be improved by using lower and upper case letters, rather than the approach sometimes adopted in screen design of using all upper case. Evenly spaced text, with an unjustified right margin, is easier to read than right justified text which has spaces of varying sizes between the words. The use of concise phrases, familiar vocabulary and appropriate abbreviations make it easier for the reader to understand the text. The most visible section of the screen is the upper left-hand corner, and it is a good idea to locate important messages in this area. Again it is important that the designer understands the characteristics of the end users in order to deliver a quality product. Beginners, who are usually looking at their fingers, will notice error messages which appear on the bottom line of the screen, whereas the top right corner of the screen is a more appropriate location for experienced keyboard operators,
- *Highlighting* – can be used to make parts of the text stand out from the rest. There are a number of different ways of doing this. The text can be

    in UPPER CASE
    underlined
    in a **bold** typeface

    | enclosed in a box |

    | or in a shadowed box |

    or it can be larger than the rest.

    In addition, because the medium is a VDU screen, rather than paper, other techniques can also be used to highlight the important information. It can be made to *flash*, to be

brighter than the surrounding text, or *reverse video* can be used – creating an effect rather like a photographic negative, with white text on a black background.
- *Colour* – another set of design options is possible, in addition to those described above, if the display is in colour. Text can be highlighted by being in a different colour to the rest or being enclosed in a coloured box. Background colours can be changed or a design convention can be used in which different types of information are displayed in different colours. The consistent use of colour on screens within the same system is important, and the designer must be wary of using too many colours or creating lurid combinations as these will work against the effectiveness of the screen design. In addition, the designer must also be aware of avoiding colour combinations which could cause problems for those users who are colour-blind.
- *Graphics* – can be used to good effect for displaying information, especially trends in numerical data. They can be coloured, solid, three-dimensional or animated, and the designer must decide on what is appropriate to the purpose. Another use of graphics is as an integral part of the structure of the dialogue – known as a *graphical user interface (GUI)*. A key feature of GUI design is the use of small pictures called icons which are used to represent entities such as files, documents and disks. An icon should communicate exactly what it represents to the user without the need for accompanying text. Where they are used, each icon should be easily distinguishable from any others, and used consistently to represent the same thing. Examples of icons used on Apple Macintosh applications are shown in figure 15.9.
- *Animation* – although this is little used in screen design, it can be a powerful technique for attracting the attention of the user, because the eye is always drawn to a moving object; to mark the position of an object, for example, a blinking cursor can be used; or to communicate a message, a clock with a moving hand, or an hourglass with moving sand, indicate to the user that they have to wait while some processing is carried out by the machine.

15.5.2 Dialogue types

There are a number of different approaches which can be taken when designing the conversation or dialogue between the user and the computer system. Essentially a dialogue consists of the user responding to a prompt from the computer by providing input. This

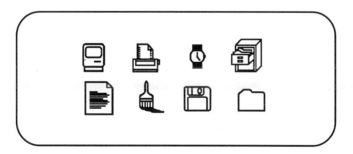

Fig. 15.9 A selection of icons

input is processed by the computer and a response is output to the screen, which in turn may prompt the user for the next input.

The main dialogue types are summarised below. It is up to the designer, having considered the alternatives, to decide which of these is most appropriate, based on the requirements and characteristics of the end users.

- *Menus* In this approach, the user is presented with a set of alternatives from which one has to be selected. These alternatives can be displayed on screen using text or graphics, and usually each alternative has a letter or number associated with it. When the user selects an option, the letter or number is entered via the keyboard, or the arrow keys are used to move the cursor between the options. Alternatively if a mouse is attached to the computer, an option can be selected by pointing at it. An example of a menu is shown in figure 15.10. An item from this menu can be selected by typing either the appropriate number or the intial letter of the word.
Menus are widely used in screen design because they require minimal effort, and skill, on the part of the user. This in turn reduces the training requirement when preparing individuals to use the system. A common approach is to structure the menus hierarchically in a 'nest': selecting an option on the first menu takes the user to a second menu from which another option is chosen, and so on. This allows the number of alternatives on any one screen to be kept to a minimum. When designing nested menus, it is important to build in short cuts to allow experienced users to move quickly through the system and avoid the frustration of having to go through several menu screens for each transaction, and to allow an exit which doesn't involve travelling back through each of the previous menus. The principles of good design should also be applied by keeping the menus simple, with an upper limit of ten choices per screen; minimising the number of levels of menu through which a user has to navigate; and ensuring consistency in the way options are selected from each of the related menus.

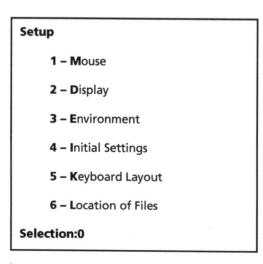

Fig. 15.10 Example of a menu screen

Systems Design: Human–Computer Interfaces 241

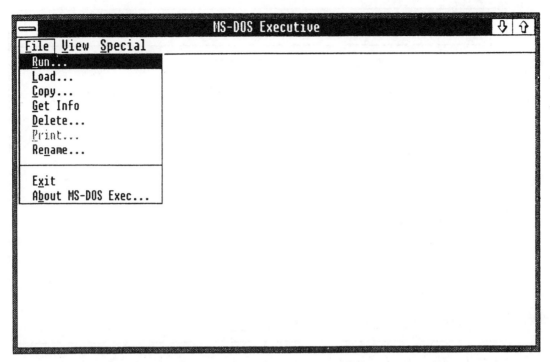

Fig. 15.11 Example of a pull-down menu

In some applications a menu is permanently displayed in a bar across the top of the screen from which other menus can be pulled down as required. This is known as a *pull-down menu*, and is represented in figure 15.11.

- *Question and answer* This type of dialogue involves the system prompting the user by asking a question, and then carrying out the processing associated with the user's answer. The questions usually require one of a limited number of answers, e.g. yes or no, which eliminates errors and reduces the amount of validation of answers required. It is another dialogue type which is useful for beginners, but which can be frustrating for experienced users. It is commonly used when checking instructions given to the system, for example

    Do you really want to delete this file?   Type 'Y' or 'N'
    (where the default is 'N').

- *Form-filling* This is particularly appropriate for the user who has to enter information into the system from a paper input form, and a good design will ensure that the layout of the screen resembles the original document as closely as possible to minimise the risk of errors. This type of dialogue is also called a *template* and an example is shown in figure 15.12, the screen design for the input *new product details* shown on the DFD in figure 15.1, the content of which was described earlier.
  The areas in which data has to be entered will usually be highlighted in some way to stand out from the background text – using colour shading or graphics for example.

```
┌─────────────────────────────────────────────────────────────────────┐
│                    Galaxy Electrical Products Limited                │
│                           SCREEN LAYOUT                              │
├─────────────────────────────────────────────────────────────────────┤
│                  NEW PRODUCT INPUT/PRODUCT AMENDMENT                 │
│                                                                      │
│     PRODUCT NO. [_ _ _ _ _ _ _ _ _ _ _]   DESCRIPTION [_ _ _ _ _ _ _ _ _ _ _ _ _ _ _ _ _]
│                                                                      │
│         MANUFACTURER [_ _ _ _ _ _ _ _ _ _ _]   ORIGIN [_ _ _ _ _ _ _ _ _ _ _]
│                                                                      │
│         INVENTORY GROUP [_ _ _ _ _ _]    PRODUCT CLASS [_ _ _ _ _ _ _ _ _ _ _]
│                                                                      │
│         DESPATCH UNIT [_ _ _ _ _ _ _ _]   RE-ORDER LEVEL [_ _ _ _ _ _ _ _ _ _ _]
│                                                                      │
│         DISCOUNT PRICES A [_ _ _ _ _]                                │
│                         B [_ _ _ _ _]                                │
│                         C [_ _ _ _ _]                                │
│                         D [_ _ _ _ _]                                │
│                                                                      │
│     VAT CODE [_ _ _ _]         VAT RATE [_ _ _ _ _ _]                │
└─────────────────────────────────────────────────────────────────────┘
```

Fig. 15.12 Example of a form-fill screen

When data has been entered, the cursor moves to the next input area. Form-fill is useful when a large amount of similar information has to be entered quickly into the system. A variation of form-filling is panel modification. In this type of dialogue, a screen of data is presented to the user who can then move around it to amend specific items. This is appropriate if existing records held on the system are being amended by an experienced user.

- *Command language* This type of dialogue can appear very 'unfriendly' to the inexperienced user as the prompts are usually very limited, and response by the user has to be syntactically correct in order to be accepted by the system. However it is a very fast and efficient dialogue for the experienced user and short commands or abbreviated forms of these commands will initiate specific processing. There is a great deal of freedom to move around the application, as the user rather than the system is structuring and ordering the conversation. The designer should ensure that there is a visible output on screen to show that the user's input has been accepted by the system and should endeavour to make error and help messages in such a dialogue as meaningful as possible. An example of a command language dialogue which is very widely used is MS.DOS, a sample of which is shown in figure 15.13.

The earliest and still the cheapest form of screen-based interaction uses a set of menus to implement hierarchical selection, but this technique effectively constrains the operator to only making a limited number of selections. From the operator's viewpoint, this can be a slow, cumbersome and tedious way to work, effectively providing a very narrow interface to the system and menus can be particularly irksome

```
              Type EXIT to return to Menu
              C:\>A:DIR

              Not ready reading drive A
              Abort, Retry, Fail?R

              Not ready reading drive A
              Abort, Retry, Fail?R

              General failure reading drive A
              Abort, Retry, Fail?F
              Invalid drive specification
              Bad command or file name

              Type EXIT to return to Menu
              C:\>
```

Fig. 15.13 Example of a command language dialogue

for the experienced operator. A command language, with defaults, is generally preferred by experienced operators. However, there is more to remember and more can go wrong - the operator is not so constrained as by menus and there are opportunities for 'finger trouble'. Hybrid systems use a combination of menu and command techniques for interaction. For example, menus may be displayed in 'novice' mode, or by request from the operator, with command specification available as an override or default.

- *Natural language* In contrast to the command language dialogues, natural language conversations are much closer to the way way people would speak. This type of dialogue relies on artificial intelligence and expert systems which will recognise key words or phrases input by the user and translate them into instructions which can be followed by the computer. The problem with natural language is that most natural languages, like English, are characterised by ambiguity and inconsistency in their syntax. These dialogues therefore require the use of a limited vocabulary and syntax, with which users need to be familiar before they use such a dialogue.

In practice a designer may use a number of these dialogue types in combination when designing the screens for a system, depending on the requirements of the system and the needs and experience of the users.

15.5.3 WIMP interfaces

This is an increasingly common interface used in many applications. The WIMP interface is so called because it incorporates:

WINDOWS, several of which can be opened on screen at the same time,
ICONS which represent entities such as files and documents,
a **MOUSE** to select and move icons around the screen and
PULL DOWN MENUS.

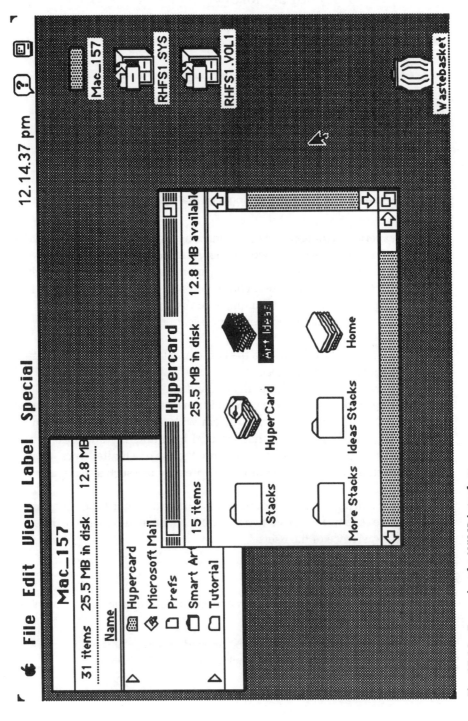

Fig. 15.14 Example of a WIMP interface

An example of this type of interface, from an Apple Macintosh machine, is shown in figure 15.14.

Apple Computer Inc have published a number of design principles to help designers who are developing products for Macintosh computers. Ten of these principles, from the *Macintosh Human Interface Guidelines* (1992), are summarised below, and are useful for any designer creating a WIMP interface for a computer system.

Metaphors   This involves using representations of familiar, concrete ideas from the outside world to help users understand concepts and features which are part of the application. For example, using graphic representations of files and folders to simplify the saving and storing of information.

Direct manipulation   One of the key concepts in a WIMP interface, direct manipulation enables the user to directly control the objects represented on screen. So, for example, a folder can be dragged into a wastebasket in order to delete information held on the system or a menu pulled down from the bar at the top of the screen.

See-and-point   If the user can see what they are doing on screen, and have a pointing device – the mouse – they can select options from a menu, or initiate actions by pointing at objects on screen. In the Macintosh desktop interface, the user will point first at an object of interest, called a noun, and then at an action from a menu, called the verb, in order to initiate a piece of processing. For example, using the mouse, a file can be opened by pointing at the icon for the file (the noun), pulling down the appropriate menu, and then pointing at the 'open' command (the verb).

Consistency   Consistency is just as important when designing WIMP interfaces as it is when designing any other interface – a point which we have emphasised throughout this chapter. It allows users to transfer their skills and knowledge from one application to another, because of the consistency in appearance and behaviour of the interfaces.

WYSIWYG   Another important principle is WYSIWYG, meaning 'what you see is what you get'. This involves, for example, using meaningful names rather than codes or abbreviations on menus and making sure that there are no major differences between a document as it appears on screen and as it appears when printed on paper.

User control   This means that the designer of an interface ensures that the user, and not the computer, is always in control of the dialogue and is able to initiate chosen actions. This allows the user to be more actively involved in the interaction than if the computer were guiding the user through it.

Feedback and dialogue   It is important to keep users informed about their progress through an application, by providing feedback as soon as they have given a command. This might involve the designer building a visual or aural response into the dialogue, or a message which explains exactly what caused an error. Again this will ensure the user remains actively engaged throughout the dialogue.

Forgiveness   If actions carried out by the user of a computer system are reversible, the designer has built in 'forgiveness'. This encourages the user to experiment and learn how to get the best out of an application. Messages which notify the user of potentially destructive actions, such as deleting an application or data file, give the user a chance to change course.

Perceived stability   In order to help users to learn to use new applications within a system, familiar, understandable and predictable features of the interface are desirable. The designer should consider the consistency of layout, graphics, colour, etc. which increase the user's perception of stability. For example, when familiar commands on a pull-down menu are unavailable, they can be dimmed rather than deleted from the display.

Aesthetic integrity This means that information displayed on screen is well organised and visually attractive. As we discussed in the section on screen design, the usability of an interface will be enhanced if layout is simple and clear and the graphics used are appropriate.

15.5.4   User support

As well as the important principles of simplicity and consistency, which we have stressed throughout this chapter, the designer also needs to consider how to provide help if a user has problems in selecting an option from a menu or in continuing a particular dialogue. Such help may be off-line, through the use of documentation or help desks, or on-line, using screens which explain particular functions to a user.

When designing off-line help, which is usually paper-based but which could also be in the form of audio or video tapes, a key principle is to think about the needs of the users who will be referring to them. They will be keen to find a quick solution to their problem and will frequently be anxious and in a hurry. Manuals should therefore be kept simple and not overloaded with information, and should be relevant to the common types of problems encountered by users. The most useful way of developing such material is by working closely with end users and modifying the text in the light of the difficulties they experience when using the system.

On-line help can be basic, usually in a concise form with reference to a manual where more detailed information can be obtained; it can be context-sensitive, providing appropriate information about the particular function being used at the time the help is requested; or it can be intelligent, responsive to the route a user has followed while navigating through an application and their needs when the help screen is called.

User support will also take the form of training for those who will be interacting with the computer system. Again, this can be provided off-line, in the form of a training event (either a taught course or an open learning programme), or on-line, as a tutorial which takes the user through the functions of the system and then through a series of structured exercises.

It must not be forgotten, when designing a computer system, that users are part of this system, and can be considered almost as an extension of the computer. Their role must be defined and designed like any other component of the system, and thought must be given early on as to how they can best be supported in carrying out this role.

15.6   ERGONOMICS AND INTERFACE DESIGN

A final area to address in this chapter on interface design is the subject of ergonomics. Ergonomics is the study of the physical and mental reactions that people have to their working environment, and computer ergonomics have been the subject of a lot of discussion in recent years. It is an applied science which draws on the findings of three other areas of study: anatomy, which provides information about the dimensions of the body (anthropometry) and the application of forces (biomechanics); physiology, which helps explain the expenditure of energy and the effects of the physical environment on the body; and psychology, which contributes an understanding of how people process information, make decisions and solve problems (cognitive skills), how they produce and control changes in the environment (perceptual-motor performance), and how they behave and interact with others at work. An ergonomist also needs to understand the computer system and the working environment in which the two will come together.

The findings of ergonomists are useful to system designers, who recognise that computer systems can only be effectively implemented if human factors are taken into account from the beginning of the development process. This is because the reason for developing the system is not just to develop more powerful technology, but to meet a business need by improving communication, providing effective tools and minimising the chance of errors being made.

In addition the design, implementation and use of computer systems is also governed by law, and all video terminals and personal computers must now be designed to good ergonomic standards. With effect from 1st January 1993, a new set of regulations has been in force to meet the requirements of EC Directive VDU 90/270. The *Display Screen Equipment Regulations* requires employers to perform an analysis of workstations in order to evaluate health and safety conditions to which they give rise for their workers, particularly as regards possible risks to eyesight, physical problems and problems of mental stress, and to take appropriate action to remedy the risks found; and there are stiff penalties for non-compliance.

In designing an effective and ergonomically safe workstation, attention must be paid to a number of points, highlighted in the Display Screen Equipment Regulations. The diagram in figure 15.15, and the accompanying text, summarises these points.

1. The body should be upright and the body weight fully supported on the chair.
2. There should be good lumbar support, and the seat back should be adjustable.
3. The height of the seat should also be adjustable.
4. When the height of the seat is correct, the forearms should be horizontal, the shoulders relaxed and not hunched up and the angle between forearm and upper arm about 90 degrees. In this position the hands will be in line with the home run (the ASDF row) of the keyboard, and there will be minimum extension, flexion or deviation of the wrists.
5. Feet should be flat on the floor, and a footrest should be provided if required.
6. Thighs should be supported by the front of the chair, but there should be no excess pressure from the chair on the underside of the thighs or the back of the knees.
7. Space should be allowed for changing the position of the legs and obstacles removed from under the desk.
8. Space should be provided in front of the keyboard to support hands and wrists during pauses in keying.
9. The distance between eye and screen should be in the range 350 to 700 mm, so that data on the screen can be read comfortably without squinting or leaning forward.
10. The height of the VDU screen should be adjusted according to the user's typing proficiency, and the needs of the task.

A. For those with touch typing skills, or whose work is purely to screen without referencing written material, the top of the screen should be level with the eyes, and should be angled approximately 15–20 degrees to the horizontal.

B. Users who are not trained touch typists, and switch between screen, keyboard and reference material, should have the screen set in a lower position to avoid eye and neck strain in constantly switching back and forth. The screen plane should be as close as possible to 90 degrees from line of sight.

11. Contrast and brightness controls should be set to produce a display which is comfortable to use, and should be adjusted to take account of varying light conditions during the day.
12. The screen should be positioned to avoid glare from either direct sunlight or artificial light. Where possible blinds should be used to shield the screen.

The layout of a workstation should be designed to allow the user to maintain an upright body posture without having to repeatedly twist to access supporting material. This involves maintaining a balance between the individual's position, the VDU screen and the keyboard. Figure 15.16 shows two working arrangements which conform to this recommendation (A and B) and another (C) which does not.

15.7 SUMMARY

We have explored a number of different aspects of interface design in this chapter, which is intended to be an introduction to the subject. Much has been written about interface design and human–computer interaction, particularly about the cognitive approach to HCI, which is outside the scope of this chapter.

The starting point for interface design is the drawing of a system boundary, which allows inputs, outputs and dialogues to be identified. The designer can then determine the content of each input and output, using the data dictionary, and agree appropriate technologies with the user.

To be effective, a dialogue has to be both functional and easy to use, and we discussed issues of screen design, dialogue type and graphical user interface design which would contribute to this objective. It is important to state that dialogue design, like any other part of systems development, is an iterative process, and there may be a number of changes before the design is finalised. The involvement of the user at an early stage is vital to ensure that this aspect of the system meets their requirements.

A key goal of interface design is to maximise the usability of the system by ensuring that users can carry out their tasks effectively, efficiently, safely and enjoyably and in order to do this a designer has to be aware of the available technology and the needs, tasks and characteristics of the eventual users of the system. Technical developments have enormously widened the range of human–computer interaction possibilities at the hardware level. The major developments include:

Larger, flatter, colour screens.

New CRT controller chips which are oriented towards high-speed windowing displays, with 1024 x 768 pixels taking over from the lower-resolution CCTV-derived displays

Systems Design: Human–Computer Interfaces 249

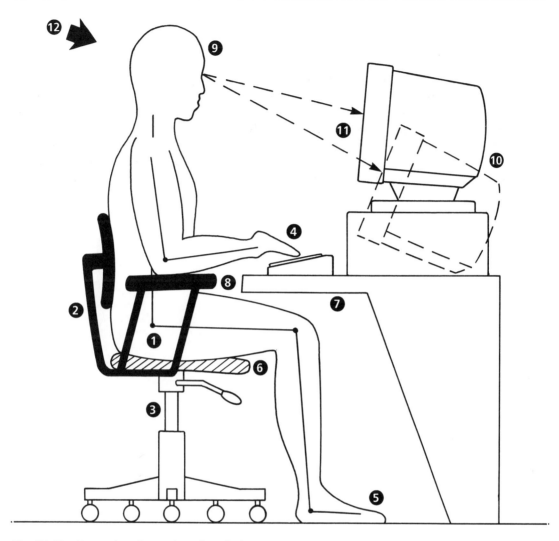

Fig. 15.15 Key points in workstation design

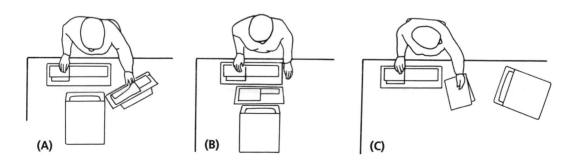

Fig. 15.16 Workstation layout

Liquid crystal displays
Coloured LEDs
Speech synthesis and recognition chips.

The technology now exists to take some account of the psychology of the human being who is operating, and contributing to, the system. While it can be argued that all this new technology costs money, it must also be remembered that new technologies in the IT industry fall rapidly in price once they have been around and adopted for a year or two, and also that the user interface is the most prominent view of the system the customer has. There are many examples of poor interface design, and it is clear that the time invested in designing interfaces which meet the real needs of the user contributes significantly to the effectiveness of the system delivered to the client.

CHAPTER 16

Systems Design: *System Interfaces*

16.1 INTERFACES DEFINED

The first step in analysing a complex system is to divide it into a number of smaller, more manageable objects and then analyse each of these smaller objects separately. So, for example, a simple on-line banking system might be divided into:

- Database containing information about the bank's customers and their accounts
- Interaction with a customer. This is shown in figure 16.1. On the right is a bank customer using a terminal to make a balance request, and on the left a central database.

What is also shown is a set of messages passing between the terminal and the central database. First, a password is sent for validation. Then, after this has been accepted, a balance request is sent and the reply returned.

This set of messages and the sequences in which they occur is the interface between the terminal and the central database. If the interface is not well understood, then any attempt to analyse the subsystems will be inadequate. For example, suppose in our example, that details of the password validation were not defined. The designers of the terminal might assume that they had to send the password to the central database for validation. The designers of the central database, on the other hand, might suppose that the validation is performed locally by the terminal. Both designs might work, but they won't work together.

So far, this example has imagined that interfaces are simply ways of dividing a system into manageable parts, but there are other interfaces which are important: interfaces which are concerned with the way the system works. The system must, for example, interact with the outside world. In figure 16.2 we see that:

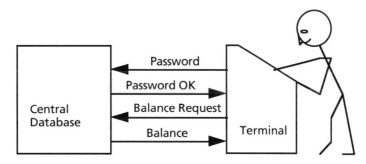

Fig. 16.1 A bank terminal

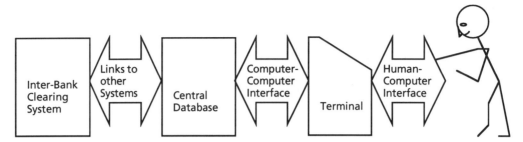

Fig. 16.2 External interfaces

- There is an interface between the user and the terminal, called the Human-Computer Interface (HCI)
- There could be another interface which links the central database and the inter-bank clearing system.

As we saw in the previous chapter, the HCI is concerned with the design of screens, the way messages are displayed, the way the user is expected to respond, the use of passwords and cards with a magnetic strip, and so on. The links to other systems are usually thought of under the heading of data communications. Both of these are discussed in more detail in other chapters of this book and are not discussed further here except to emphasise that they are just as much interfaces as a procedure call or shared memory, and can be analysed in just the same way. The complexity of data communications protocols, for example, often leads to the false assumption that there is something qualitatively different about data communications. This is not true.

In this chapter we shall consider the analysis of the boundary that enables the system to be divided into well defined subsystems, and the mechanisms which might be used to implement an interface. First of all we have to identify the interfaces. There are three ways in which we can do this.

Partitioning by organisation Organisations naturally divide themselves into different parts either by function or by geography. There may, for example, be a head office, a sales office and some factories, and these may be located in different parts of the country, or simply on different floors of the same building. These functions and locations represent natural divisions within a company. There will be information flows between them. Where the divisions are geographic, the flows of information are normally grouped under the heading Data Communications. Where different functions exist in the same place, there may well be common use of computers. For example, the accounts department and the sales office might have access to the same database.

Where there is a real distinction between uses being made of the computer system, rather than between the departmental function, then there is a natural boundary across which information may flow. This should be reflected in the design of the computer system. If, on the other hand, there is little distinction in the need for computer services, then there is an opportunity for savings by providing a common service. The sales office and the accounts office have very different functions but if their main use of computers happens to be word processing, then there is no distinction in the use and, therefore, no sensible system boundary.

Partitioning by data flows Partitioning by organisation works well at a high level but there is also a need to find ways of identifying interfaces between individual computers, between processes within a single computer, and even between the different parts of a single process. The general idea when analysing data flows is to partition the system to localise complex interfaces. Think again of the bank terminal we discussed earlier. Figure 16.3 shows three parts of this system with arrows to indicate the complexity of the interaction between each. The question is, where should the functions provided by the bank terminal be implemented? Should the terminal hardware provide relatively few functions directly, with most functions being implemented as part of the central banking functions? Alternatively should the terminal be given more processing power?

Suppose the terminal functions and the central database functions were to be located together as shown in figure 16.3. In this case, large numbers of messages pass across the interface, and the closely coupled terminal and database communicate in a much simpler way!

A better solution would be to place the support of the bank terminal functions with the customer as shown in figure 16.4. In this case the majority of the communication takes place locally between the customer and the terminal, with relatively little communication over the longer journey to the central services.

Partitioning by data ownership The partitioning which we have discussed so far is on the basis of the flows of data round a system. Increasingly, the use of Object Oriented Analysis and Design (OOA and OOD) has introduced an alternative view based on what is known as 'Data Encapsulation'.

Consider an example. Suppose you are a member of your local Squash club. You want to arrange a game with another member, Mary, but you don't know her phone number. So, you do the obvious thing and ask the club secretary to tell you the number. What you don't do is ask where the secretary stores the members list and in what form – and then go to look for it yourself! This may sound silly, but that is effectively how thousands of computer systems over the years have been designed. Data was considered public. So,

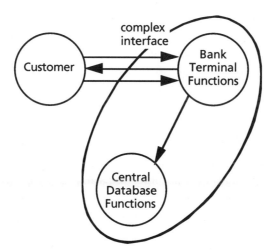

Fig. 16.3 Poor partitioning which produces a complex interface

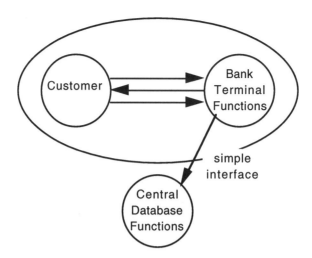

Fig. 16.4 Good partitioning which produces a simple interface

every part of the system had to understand the structure of the data. If the data structure had to change, to include a new field for example, then every part of the system that accessed the data would have to be rewritten. Back at our Squash club, you don't care about how the secretary stores the members list so long as she can answer the question. At some stage she could convert an aged card index system into an online database and you would never know. In programming terms, if the data is 'encapsulated', its structure is known only by the program which owns it. Other programs can only gain access to the data by using access functions provided by the program which owns it. In a system designed in this way, the key feature is the identification of the data and the partitioning of the system into areas of 'data ownership.'

Having identified interfaces, the next step is then to understand the different kinds of interfaces. You should be wary of assuming, however, that this is a simple two-stage procedure involving first identifying the interfaces and then analysing them. As a designer you have some choice over where the interfaces go. If your first attempt to partition the system produces interfaces which turn out to be too complex, you'll need to revisit the partitioning process: move step by step towards partitioning so that interfaces are simple enough to be defined in a clear, unambiguous way. This iterative process is shown in figure 16.5.

16.2 ANALYSING INTERFACES

The next problem to be tackled, then, is how to analyse an interface.

A point that will be emphasised again in chapter 24 dealing with Data Communications is the importance of separating what an interface is used for from the physical form that the interface takes. That is, to separate the reason for having an interface from the choice of mechanism.

The purpose of an interface might be to carry accounts data from the sales office to the accounts office. The forms in which this transfer takes place could vary widely. The two

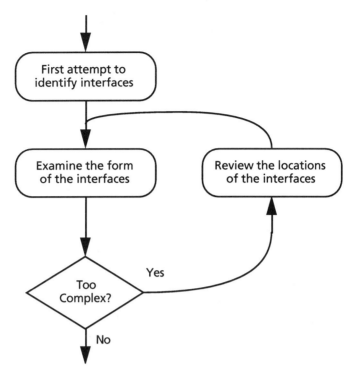

Fig. 16.5 Looking for interfaces

offices could, for example, share access to a computer on which is a database into which the sales office place data and from which the accounts office read it when they want it. Alternatively, there might be a data communications link between separate computers used by the two offices. The same interface could even be implemented by sending a fax from one office to the other. The analysis of the interface will suggest what form the interface should take. So, the decision illustrated in figure 16.5 can be understood to mean 'Is the partitioning acceptable in the light of the ways we have to implement the interfaces?' We'll examine later the different mechanisms which are available, but here we'll consider the interface itself in relatively abstract terms.

There are three diagramming methods which are widely used to describe interfaces:

- State Transition Diagrams
- Time Sequence Diagrams
- Data Flow Diagrams.

They are not alternatives, but are different tools with different capabilities.

Imagine a system which waits for an event, acts on it and then waits for another event. Each time it waits there may be a different set of events on which it can act. Interfaces can be described in this way using what is called a *state transition diagram* .

Consider, for example the interface between you and your television. A possible form of the interface is shown in figure 16.6. The circles represent the waiting for an event: these are called the 'States'. Arrows between them show the events which move the interface from one state to another.

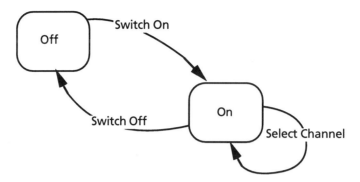

Fig. 16.6 State transition diagram of the control of a television set

The interface starts in a state called Off in which there is only one event which can have any effect: Switch On. Having switched the TV on, the state changes to On. In this new state there are two events to which the interface can respond: Select Channel, which does not cause a change of state, and Switch Off which causes the state to change back to Off.

A more complex interface might have many more states. An analysis of the interface to a video recorder, with all of its different controls, could have well over 50 states. In such complex circumstances, using a table is often clearer. The events are drawn across the top of the table, and the states down the side. Text in each box in the table then describes the associated processing. In our TV set example this would look something like figure 16.7. In this example nothing happens in the empty boxes. In a more complex interface they might cause error processing because an event has occurred in the wrong state.

State transition diagrams don't tell the whole story, however. They show the events which control the interface, but they don't show any information passing across it. In the case of the TV set, for example, there is no information about a picture being displayed which, after all, is the whole point. In such cases a time sequence diagram can be useful.

A *time sequence diagram* shows 'things', whether data or control, passing across an interface. Figure 16.8 shows the television set example in this form.

The line down the centre represents the boundary between the user and the TV. The arrows crossing it represent 'things' crossing the interface. An arrow has also been added to the central line to indicate that time is to be thought of as flowing down the page. At the top of the diagram a Switch On signal passes to the TV which responds by sending a picture back. The user then changes channel and in response the TV sends a different picture. Finally the user sends a message to Switch Off. This form of diagram captures

Events States	Switch On	Switch Off	Select Channel
On	• Switch on the power • Change state to 'On'		
Off		• Switch off the power • Change state to 'Off'	• Select another programme

Fig. 16.7 State transition table showing the control of a TV set

Systems Design: System Interfaces 257

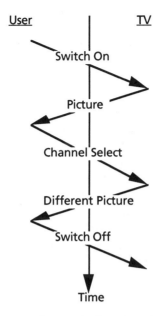

Fig. 16.8 Time sequence diagram of control of a television set

both the control and data aspects of an interface, but cannot show the more complex sequences. For example, it does not show that having switched off the TV, the user can switch it on again, nor does it show that the user can repeatedly change channels.

A data flow diagram captures a static view of the communication between parts of the system. It looks superficially similar to a state transition diagram. However, where the state transition diagram shows the control of the system, the data flow diagram, as in figure 16.9, shows the movement of data. It does not represent state changes or changes in time.

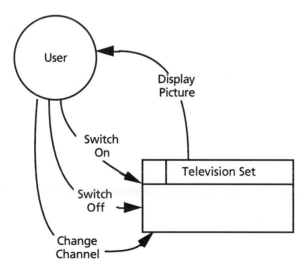

Fig. 16.9 Data flow diagram for the control of a television set

16.3 PHYSICAL FORMS OF INTERFACES

This section will examine the mechanisms which are available to implement an interface. It concentrates on communication between processes within a computer. The first mechanism to be discussed, the procedure call, is the simplest in that only one program is running. Later methods such as mailboxes and shared memory assume that more than one program is running at the same time. Finally, to give a taste of the variety of solutions which are available, two special mechanisms are mentioned: the Ada 'Rendezvous' and UNIX 'Standard Input' and 'Standard Output'.

The commonest way to pass information from one part of a program to another is a *procedure call*. In figure 16.10, the call and return arrows show the flow of control which may or may not be accompanied by data.

The two pieces of communicating code are so closely linked that a subroutine call is almost more a part of the program structure than an interface. However, simple though it is, it does provide a boundary between two processing worlds. In most languages there is careful control over the information which can pass between the two worlds.

There are two ways in which information is passed:

- Pass-by-reference
- Pass-by-value

If a parameter is passed by reference, then the subroutine is given access to the same piece of memory as the calling program. This means that if the subroutine changes the value, then the calling program will see the change. This is the mechanism which was used by default in older languages such as the early versions of FORTRAN. It works perfectly well, but the calling routine has to be able to trust that the subroutine will behave itself and not corrupt things which it should leave unchanged. More recent languages such as C pass parameters by value. That is, a copy of the parameter is made and the subroutine is given access only to the copy. The subroutine can corrupt this as much as it likes because it will be thrown away when control returns to the main program. In C, a programmer wanting to pass a value out of a subroutine must explicitly pass into the subroutine an address in memory. This means that the programmer has to be explicit about what the subroutine is allowed to change. In the Ada language, every parameter must be marked as 'in', 'out' or 'in–out' to identify the direction in which changes to the parameters are allowed to flow.

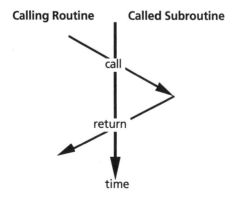

Fig. 16.10 A subroutine call

Older computer systems often ran as what were called batch systems, a form of processing which is becoming increasingly rare. Such systems were written as a series of standalone programs. The first would read input, perhaps from punched cards, and write to a file. The next program would take this file as its input and write its output to another file and so on. The interfaces between the programs were these shared files. There are some disadvantages with this approach, not least of which is the fact that one stage must be completely finished before the next can start.

This idea of using an intermediate file still exists. In the MS-DOS operating system widely used on personal computers you can enter a command such as :

dir | more   use the 'dir' command to list the contents of the current directory, but instead of displaying the list, feed it into the 'more' command. This will display the output a screenfull at a time.

This is implemented in MS-DOS by writing the output from 'dir' to a temporary file. Only when 'dir' has completed its work is 'more' allowed to read the file. When 'more' has finished the temporary file is deleted.

Instead of sharing a file, two programs running on the same processor can *share a data area within memory*. One program writes to the shared area, the other reads from the same area (see figure 16.11). This is a particularly useful solution if there is a lot of data passing and processing speed is important, but it is important to prevent clashes between sender and receiver. The reader must be prevented from reading before the writer has completed writing the data, otherwise only part of the data will be read, and the message will be corrupt. This problem did not occur under the older file-based batch systems because the reading program was not run until the writing program had finished writing to the file. The same idea of not allowing access by the reader while the writer is writing still exists. However, the suspension of the reader now takes place for each message, not for the whole file of messages.

Mechanisms are provided by many modern operating systems to pass messages from one program to another through a mailbox. This is rather like one program sending a

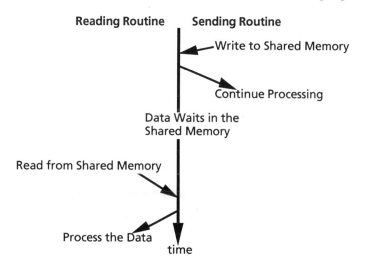

Fig. 16.11 Communication by shared memory

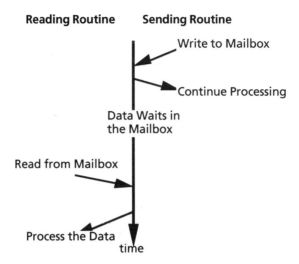

Fig. 16.12 Asynchronous communication by mailbox

letter to another program. You can visualise the mailbox as a kind of pigeonhole maintained by the operating system where the message is retained until the recipient reads it.

The kinds of services which could be expected by the user from the operating system are:

- Send a message
- Read from my mailbox
- Tell me if there are any messages waiting
- Advise me immediately when a message arrives.

Messages can be sent synchronously or asynchronously. In synchronous communication the sender cannot proceed until the receiver has received the message, but more often asynchronous communication (figure 16.12) is used because the programs involved do not keep in step with each other and are usually doing completely different things.

Some modern languages have communications features built in. For example, the Ada language supports its own built-in synchronous communication method called a *Rendezvous* (figure 16.13). This is rather like accessing a subroutine in one program from a call in another. The functions used are CALL and ACCEPT. The calling program issues a CALL which must be matched by an ACCEPT in the called program. The programs can issue these calls in any order: the program which issues its call first is suspended until the other catches up and issues its own call. Code specified as part of the ACCEPT is then executed using parameters supplied by the caller. During the Rendezvous, data can pass in both directions.

The UNIX operating system tries to make all interfaces look like files. A program has an input channel called STANDARD INPUT and an output channel called STANDARD OUTPUT. It can read from one and write to the other without knowing whether the messages are coming from a file or from another program. There is, for example, a UNIX command 'ls' which lists the contents of a directory, rather like the 'dir' command in MS-DOS that was mentioned earlier. Some of the ways in which it can be used are:

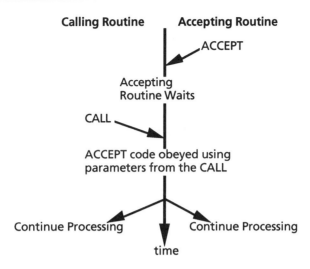

Fig. 16.13 Ada Rendezvous

    ls   display the content of the current directory at the terminal
    ls > file   create a file called 'file' and write to it a description of the content of the current directory
    ls | more   make description of the content of the directory and pass it to the program 'more' which will display it one screenfull at a time.

The point is that in all these cases the 'ls' command is identical. In each case the 'ls' command thinks it is writing to a file called STANDARD OUTPUT. The last example 'ls | more' is similar to the MS-DOS 'dir | more' command. In the case of UNIX, however, there is no intermediate file created. The data is passed directly from one program to the other: it just looks as if a file is being created.

Most of these forms of interface require *synchronisation* between the reading and writing processes. It would be unwise, for example, for a process to read from shared memory before the message has been completely written. Processes also often need access simultaneously to resources which are by their nature non-shareable. Imagine what would happen if two processes could write to the same printer at the same time! This implies the need for some form of communication between the processes. The communication on this occasion is concerned with the control of those processes, not with passing data between them. Take the example of access to a printer mentioned earlier. If there was a flag which was set to 1 if the printer was available, and to 0 if it was in use, then access could be controlled so long as every process which wanted to access the printer performed its processing along the following lines:

```
IF flag=0
   THEN
      suspend until flag=1
   ENDIF
set flag to 0
access the printer
set flag to 1
```

While one process uses the printer, all other processes are suspended. Then when the flag is reset to 1, all the suspended processes are given the opportunity to access the printer. What actually happens at this point depends on the operating system. Possibilities are:

- The process which has been waiting longest gets the printer
- The highest priority process gets the printer
- Chance! – the first process to run gets the printer.

You should notice, as well, that the problem is not quite as simple as has been implied. Consider what happens if, between testing the flag and finding it set to 1, and trying to reset it to 0, another process runs and sets the flag to 0.

One form of control that can be used to prevent this is often called a *semaphore*. This is a more general version of the idea of a simple flag. Suppose in this case that there are several printers so that several processes could print at the same time. The flag used in the earlier example could become a count of the number of printers available. The logic then becomes:

```
IF count=0
   THEN
       suspend until flag>0
   ENDIF
   decrement the count
   access a printer
   increment the count
```

Many operating systems provide functions, often called wait and signal, to support semaphores:

```
wait(s)    IF s = 0
           THEN
               suspend until s>0
           ENDIF
           decrement s

signal(s)  increment s
```

16.4 INTERFACES TO PERIPHERALS

Historically, computers had relatively few types of peripheral devices attached to them. In the 1960s and 70s, common computer peripherals consisted of magnetic disk and tape units, printers, terminals, teletypewriters, card readers and not much more. Most of these were relatively slow devices with a simple interface. Today there are many more ways to get data into and out of computers. Mice, tracker-balls, light pens, CD-ROM drives, keyboards, colour VDUs, magnetic disks, optical fibres and laser printers spring to mind immediately, and these are the 'standard' devices. Move into industry and the computer may well be controlling a robot. In your own kitchen there may well be a computer which controls your washing machine. Some modern cameras have two computers: one in the camera body and the other in the lens.

When discusing interfaces in general, the importance of separating the use of an interface from its implementation has already been noted. Such a distinction is even more important when widely varying peripherals are to be used. It is bad design to have to completely rewrite a piece of software because the user wants a different printer! To achieve this, the operation of the program needs to be separated as far as possible from the real characteristics of the peripherals. It might, for example have available to it a command such as PRINT which prints a line of text on the printer. However, its knowledge of the behaviour of the particular printer to be used would be limited to such things as the length of the print line. The more it knows about the printer, the more it is likely to be incompatible with some other printers.

At the lowest level, the *device driver* supports the peripheral directly and is aware of the detailed control sequences needed to support it. If you change the printer you will need to change the device driver, but in most cases that is all you would have to change.

The user expects a very high-level interface, but the device driver supplies a very low-level interface. If these are just too incompatible, an intermediate *device handler* could be introduced. This can supply buffering and queueing services as well as providing a more convenient access to the device driver. The device handler exists to bridge the gap between the application and the device driver. It makes the device easier to use.

While the device handler can hide many of the characteristics of a peripheral, it cannot hide them all. Because peripherals vary so much, it is not possible to give hard and fast rules. The following, therefore, is a list of issues to be aware of:

- What *speed* does the device operate at? Few peripherals can operate as fast as a computer processor. Some, indeed, are particularly slow. A keyboard, for example, can only deliver characters as fast as someone can type. If the application is doing nothing but processing these characters then it might be acceptable to wait for each character. In general, however, it is better to separate responsibility for the collection of characters into a device handler which can then either raise an interrupt when a character arrives, or be inspected at regular intervals. Similarly, if characters are to be written to a printer, there is no point in the application waiting for the device after every PRINT statement. A device handler which accepts the text and holds it until it can be printed is a better solution. This mechanism by which an intermediate store holds information until a slow device can accept it is called *buffering*. In general, whenever two devices of significantly different speeds are to be linked, there will be a need for buffering.
- What is the *unit of transfer*? That is, what is the form of what is passed into the application program? Some devices communicate in terms of single characters, others in terms of kilobytes of data at a time. A magnetic disk, for example, may well be able to read and write 512 or even 1024 bytes at a time. Here, the device handler is again important. It would be poor design to allow every application direct access to the structure of the disk: every disk is different. An application is much more likely to want to talk in terms of 'Files' and 'Records within those Files'.
- What is the *data representation*? IBM mainframe computers represent the letters of the alphabet with a code called EBCDIC. Most other computers use a code called ASCII. These two representations are incompatible. These incompatibilities are made even worse when more detailed differences are examined. The files produced by two different

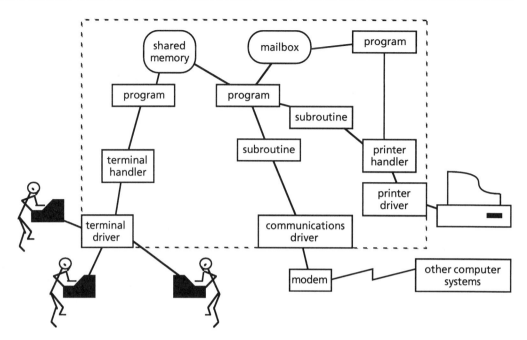

Fig. 16.14 The variety of interfaces

word processors are unlikely to be compatible. The codes used for 'paragraph indentation', 'reset page number' and so on will be completely different. Beware, then, of interfaces that complicate the data representation. We said earlier that interfaces should be as simple as possible.

16.5 SUMMARY

A complex system must be partitioned in order to make it easier to understand. In human terms we'd describe this by saying: 'divide and rule'. There are many different kinds of interfaces. Figure 16.14 shows some of them.

When deciding where to make the partitions, care must be taken that:

- Geographically separate processing is not unnecessarily combined
- Wherever possible, the more complex the interface the more local should be its implementation: you can get away with transferring several megabytes per second between tightly coupled processors, but you won't manage it over a telephone line to the far side of the world.

The other key to good interface design is to remember that an interface implies that data is crossing a boundary between two physical entities. Using diagrams to represent graphically the structure of the interfaces helps other software developers and users to picture how it works.

CHAPTER 17

Systems Design:
Logical Data Design

17.1 INTRODUCTION

Data is something which an organisation invests in but which only has value to the organisation when it is accurate and properly controlled. Business processes change frequently but the underlying data is relatively stable and unless the core business of an organisation changes, the data it uses will remain unchanged. For example, the processes needed to control a warehouse stock retrieval system may change from being completely manual where paper records are kept and where stock is stored and retrieved by forklift truck, to being fully automated. In the new system incoming goods are identified to the system by barcodes, the system decides where to store the item and it controls cranes and automated goods vehicles to transport items to and from locations. The core data, item numbers and locations, remain unchanged, even though the processing has changed dramatically.

During the 1960s systems were usually built to follow the current processes, and little attention was paid to data analysis and data design. This approach suited the technical limitations of the times where many applications used serial magnetic tape files used in batch processing systems. Because no thought was given to the fact that data has a right to exist on its own, separately from any processing that used it, data came to be stored in the system only when a process required it. This also resulted in each user holding their own customised version of the data. This meant a proliferation of different names, sizes and formats for the same data item. One piece of data could be held many times in different files in different formats.

With the introduction of structured methods and database techniques, greater attention was paid to data analysis: a method which considers data in its own right, independent of processing limitations or hardware and software constraints. It also ensures that there will be only one version of the data, and any duplication is explicitly agreed and controlled. The resulting data model provides a complete picture of the data used by the organisation. It consists of

- Data entities
- Key fields for entities
- A list of attributes for each entity
- Relationships between entities.

17.2 THE TOP-DOWN VIEW: ENTITY MODELLING

In chapter 10 we considered entity models in the context of recording information about the existing system. The same principles apply here and are restated with additional examples as part of the overall approach to data design. We begin then with some revision about entities, their properties and the relationship between entities. A data entity is something about which an organisation needs to hold data. Data entities are not only tangible and concrete, such as 'person', but may also be active such as 'accident', conceptual such as 'job', permanent such as 'town', and temporary such as 'stock item'. Entities are always labelled in the singular: 'student', never 'students'. Although the system will hold information about more than one student, the analyst is interested in identifying what the system has to know about each occurrence of 'student'. The entity label 'student' is a generic description. An entity must have the following properties:

- It is of interest to the organisation
- It occurs more than once
- Each occurrence is uniquely identifiable
- There is data to be held about the entity.

Entities have attributes. An attribute is a data item that belongs to a data entity. For example, a bank may have a data entity 'customer' which could include the attributes 'account number', 'type', 'name', 'address', 'phone number', 'account balance', 'overdraft limit' and so on. So a data entity is really a group of data items and in concept is closer to a record rather than a data field.

An entity must have a key that gives each occurrence of the entity a unique reference. For 'student' we could try to use 'name' as the key but there could be more than one John Smith, so 'name' is not unique and therefore can't be used as the key. To find a unique identifier therefore we could make one up and choose the student's admission number because it uniquely identifies each student in the college. This is a simple key because it

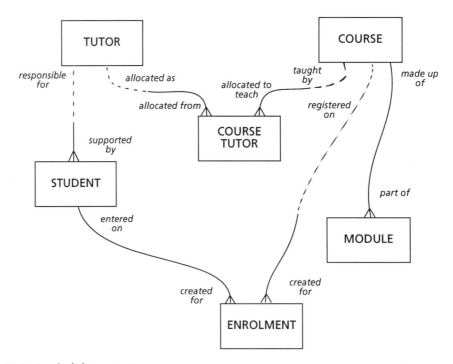

Fig. 17.1 Logical data structure

Student	**Tutor**	**Course**
admission number	tutor	course code
name	tutor name	course title
address	address	course cost
telephone number	telephone number	
tutor	grade	
year of entry	skills area	
career intention	salary	

Fig. 17.2 Entity descriptions

consists of only one data item. If we had not assigned a unique number to each student, we would have to use more than one data item as the key, such as 'name, date of birth, and address'. A key like this consisting of more than one field is called a complex key. Where possible it is preferable use simple keys.

There are three possible relationships between entities: one-to-one, one-to-many and many-to-many. Only one-to-many relationships are modelled on a data structure. The logical data structure for a college enrolment system is shown in figure 17.1.

For each entity shown on the data model the analyst compiles a list of attributes, and checks that the attributes support the relationships defined on the model. Figure 17.2 shows some of the entity descriptions which support the data structure.

Producing the logical data model is an iterative process which begins with a basic outline model of the current system. This first attempt can be made without having defined the data content of each entity in detail. Gradually, with further fact-finding interviews and cross-checking the system models, this initial data model will be expanded and refined. The data model(s) which are produced should not be used in isolation. They are just one of the techniques used to understand and define the new system and should interact with the other models if successful systems development is to take place. The data model is produced in parallel with the process model, and is used to check the data stores accessed by the process model.

17.2.1 The entity relationship matrix

An entity matrix is drawn to establish the 'direct' relationships between entities. It shows that each relationship has been considered and resolved in a way that satisfies the requirements for the user's new system. To illustrate this, let's take a few candidate entities from the case study system and draw up an entity relationship matrix for them. Normally the matrix is drawn for all entities in the system but it is possible, and helpful, to show the relationships between entities within one functional area. Optional and exclusive relationships are entered as direct relationships, which means the link must exist on the physical system.

The diagonal axis (figure 17.3) acts as a mirror with every relationship on one side of the axis reflected on the other side. That means it is only really necessary to complete one side of the matrix, but it is good practice to complete the whole matrix, and then check it for consistency. An 'x' is placed in the box when the entity has the same name. This is because it is not helpful to say 'student has a 1:1 relationship with student'. Where the relationship is indirect, this is indicated by '-'. For example each student attends one or more courses, but the relationship is not direct because it is routed through 'enrolment'.

Entities	tutor	student	enrolment	course	course tutor	module
Entities						
tutor	x	1:m	-	-	1:m	-
student	m:1	x	1:m	-	-	-
enrolment	-	m:1	x	m:1	-	-
course	-	-	1:m	x	1:m	1:m
course tutor	m:1	-	-	m:1	x	-
module	-	-	-	m:1	-	x

Fig. 17.3   An entity relationship matrix

If there is a query over whether this is a direct or an indirect relationship, then the customer is consulted to resolve it. It may be necessary to assume it is a direct relationship until data modelling is complete and then question the relationship again. The relationship highlighted reads as 'student has a many-to-one relationship with tutor'. The main benefit of drawing the matrix is to validate the data model.

17.2.2   Summary of entity modelling

We can summarise the main points about entity modelling as follows:

- On an entity relationship diagram each entity is represented by a rectangular box.
- A 'direct' relationship between a pair of entities is that which can be described without reference to some other entity on the matrix, and is shown as a line on the diagram.
- The line between two boxes is a relationship, or access path, and corresponds to an entry on the entity relationship matrix.
- A 'crow's foot' at the end of this line denotes the 'many' part of the relationship.
- The 'crows foot' end of the line denotes a 'member'; the single end denotes an 'owner'.
- An owner may have different types of members.
- A member may have more than one owner.
- A member cannot exist without an owner, unless the relationship is optional.
- Many-to-many relationships are converted to two one-to-many relationships by the addition of junction data.

Logical data modelling provides a solid foundation for any system to be developed, whether file based or using a database. However, the logical data design is an ideal model, so before data is eventually stored in the system the data will be quantified and the model will be flexed to meet the constraints of the physical system.

17.3   THE BOTTOM-UP VIEW: THIRD NORMAL FORM ANALYSIS

Normalisation of data is a process of removing duplication, and grouping related data to minimise interdependence between data groups. The less interdependence that exists between data groups, the less impact a data modification has, such as increasing the size of a data item. This approach to data analysis allows the analyst to build a data model starting with the attributes and subsequently grouping these together to form entities.

The analyst then identifies the relationships between those entities. If the proposed system is not replacing an old system, copies of existing input and output will not exist – therefore third normal form analysis cannot be carried out but there are very few systems in development that do not have predecessor systems.

To take data through third normal form analysis, you first need access to all data the organisation stores in the system. Typically this will be done by collecting one copy of every type of form and report. If there is an existing computer system, screen printouts can also be used. Forms provide the input to the system, reports identify the output, and screens are a combination of the two. From these system inputs and outputs every data item that appears is listed. This gives the complete picture of all of the data that is important. The analyst must then identify how these data items relate to each other. Unlike entity modelling, third normal form analysis is a procedural method of modelling the data. So let's start by listing the steps involved.

1. Identify all system inputs and outputs.
2. For each of these
   - list all data items and identify a unique key (unnormalised form)
   - remove repeating groups (first normal form)
   - remove part-key dependencies (second normal form)
   - remove inter-data dependencies (third normal form)
   - label the relation.
3. Merge entities with the same key.
4. Apply third normal form tests.
5. Draw a logical data model showing the relationships between entities.

Each successive step refines the structure of the data further. To illustrate how this works, we'll normalise the data items found on the student enrolment form shown in figure 17.4.

Student Enrolment Form

Name .... Tutor .... Year of Entry .... Admission No. ......
Address: .... Tutor Dept ....
....
....
Telephone ....

GCSE Results	
Subject	Grade
GCSE results validated	
Signed ....Tutor	

Enrolment: Examined Courses				
Course Code	Course Title	Module Code		Course Cost
			Total Cost:	

Career Intention

Payment Method: V☐ A☐ CH☐ INST☐
AMOUNT OUTSTANDING: ....

Fig. 17.4 Student enrolment form

The next step is to list all the data items and identify the key data items. There are a number of rules to follow when selecting the key. Firstly, it must be unique so that it uniquely identifies all of the remaining data items. It can never be blank, otherwise the remaining data items are not accessible. A short simple numeric key is preferable since this makes searching and sorting easier. Figure 17.5 shows the enrolment form data in its unnormalised form. The key that uniquely identifies this set of data is 'admission number' since each student will have a unique admission number.

To move the data into first normal form we remove any repeating groups of data. Looking back to the layout of the enrolment form above, the repeating groups are easy to identify. Subject and grade appear several times, as do module and module code, which themselves are within a repeating group of course code and course title. Having separated out the repeating groups we must identify a key for each group, and this key must maintain the link to the original data set as in figure 17.6.

We now need to remove part-key dependencies, thus structuring the data into second normal form. This means that if the key to a set of data has two or more data items in it, every data item in the set must be tested against the individual parts of the key. For example, we can find the description and cost of a course simply by knowing the course code. We don't need to know the student's admission number to find these data items. So we separate it out as shown in figure 17.7.

```
admission number
name
address
telephone number
tutor
tutor department
year of entry
subject
grade
results validated signature
course code
course title
module code
module title
course cost
total cost
payment method
amount outstanding
career intention
```

Fig. 17.5 Enrolment form data in its unnormalised form

```
admission number
name
address
telephone number
tutor
tutor department
year of entry
results validated signature
total cost
payment method
amount outstanding
career intention

admission number
subject
grade

admission number
course code
course title
module code
module title
course cost
```

Fig. 17.6 Enrolment form data in first normal form

This leaves some relations containing only key data items. These maintain the links between data groups, and will go on to become link entities, also it is possible that after normalising other inputs and outputs we may find attributes for such an entity. We now move the data into third normal form by removing interdata dependencies. This is where data items within a set which is already in second normal form are tightly related to each other, but are not directly related to the key. In this case, the tutor's department is dependant on who the tutor is, not on the student's admission number. The data item 'tutor' now appears in one data set where it is the key, and in another data set where it is just an attribute. In the student entity we label tutor as a foreign key to show that it is the key to another data set. Foreign keys are marked with an asterisk (figure 17.8).

We now label the relations; in this case appropriate entity names are 'student', 'tutor', 'GCSE subject', 'course', 'enrolment', 'course module', and 'module'.

<u>admission number</u>
name
address
telephone number
tutor
tutor department
year of entry
results validated signature
total cost
payment method
amount outstanding
career intention

<u>admission number</u>
<u>subject</u>
grade

<u>admission number</u>
<u>course code</u>

<u>course code</u>
course title
course cost

<u>admission number</u>
<u>course code</u>
<u>module code</u>

<u>module code</u>
module title

Fig. 17.7 Enrolment form data in second normal form

<u>admission number</u>
name
address
telephone number
*tutor
year of entry
results validated signature
total cost
payment method
amount outstanding
career intention

<u>tutor</u>
tutor department

<u>admission number</u>
<u>subject</u>
grade

<u>admission number</u>
<u>course code</u>

<u>course code</u>
course title
course cost

<u>admission number</u>
<u>course code</u>
<u>module code</u>

<u>module code</u>
module title

Fig. 17.8 Enrolment form data in third normal form

What started out as one long list of interrelated data items has now been structured into a number of discrete entities. This same process is followed for each of the forms, screens and reports in the system. When all system inputs and outputs have been normalised in this way, we combine duplicate entities. Because in our example we've normalised only one form we don't have duplicate entities. Our next step is to apply the third normal form tests. These tests can be summarised as taking each data item in turn and asking: 'does finding the value of this item depend on knowing the key, the whole key and nothing but the key?'

The final step is to draw the data model. This is a less complex data model because it represents only the direct relationships between entities, and does not explore optional or exclusive relationships. The data model for our example is shown in figure 17.9.

Since this method of data analysis represents a considerable amount of work, it is not recommended as a preferred starting point. Data analysts often start by using the more intuitive entity modelling approach, then third normal form analysis on those forms or reports the client identifies as key system inputs or outputs. Third normal form analysis may also be used where information from users is contradictory, so that the data actually used and stored is identified. It only models the current physical system, whereas entity modelling also takes account of what the client wants in the new, required, system. However, for a small system, or to be certain that no data has been missed, data analysts would want to carry out both entity modelling and third normal form analysis. This provides them with two different views of the same data.

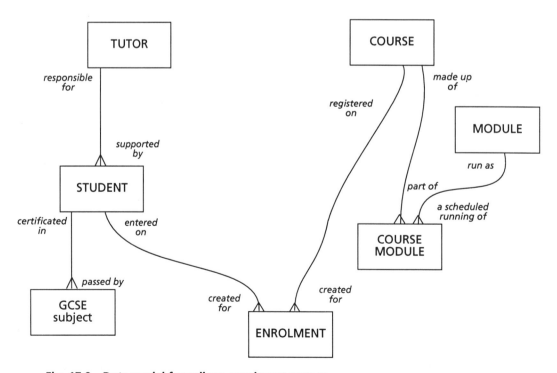

Fig. 17.9 Data model for college enrolment system

17.4 MERGING THE DATA MODELS

The final stage in logical data design merges the top-down view of the data and the bottom-up view. The aim is to identify any candidate data entities or relationships that need further investigation. We may find that an entity was not identified in the top-down approach but is shown by the results of normalisation to be needed. We may also find that some conceptual entity which was not identified during normalisation is required for the new system. We use this dual approach because if we treat data design as a totally intuitive task we can miss things and similarly, if we carry out normalisation and don't reflect on the results with a business eye, we can also miss things. Again it is the combination of the logical, thorough approach and the business knowledge that produces the most appropriate solution. Figures 17.1 and 17.9 illustrated the two views of the student enrolment system, and as you can see there are significant differences. Merging these two results in the data model is shown in figure 17.10.

In the merged model we have added the entity 'GCSE subject' from the bottom-up data model, and examined the relationships between 'student' and 'GCSE subject' to test for optionality. We did not normalise any documents that gave us information about the 'course tutor' entity so, although it did not exist on the bottom-up data model, it remains as part of the model.

This approach will duplicate effort of course, but double checking at this stage ensures that the data design is accurate.

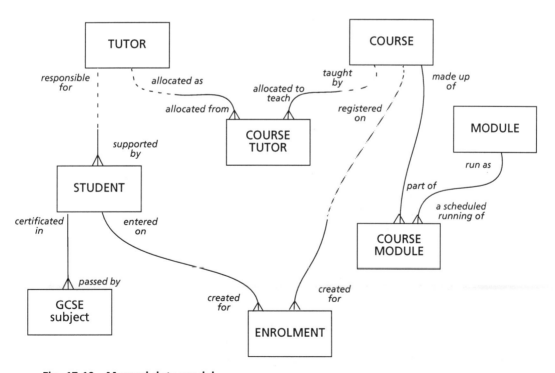

Fig. 17.10 Merged data model

17.5 TESTING THE DATA MODEL

Having built this idealised form of how the data entities in the system relate to each other, we must now check that this structure will support the business needs of the organisation. To do this we take each input to and output from the system and a copy of the data model, and then draw onto the data model the route through the data structure required to produce each report, and each input. To draw the route for every input and output on the same data model would result in an unreadable diagram. It is therefore simpler to copy the data model several times and draw one access map on one copy of the data model. James Martin's version of this model is called a 'data navigation diagram' (DND), while the SSADM equivalent is called a 'logical access map' (LAM). So to produce a class list for a given course for our example, figure 17.11 is what the logical access map looks like.

In this example, only three read accesses are needed to produce the class list, which implies that this would be a simple report for the system to produce. Each read access is labelled sequentially, and the initial read data is shown, in this case the course code. It may be that when drawing an access map, the analyst discovers that a relationship has been missed from the model, or that the route is too circuitous which would mean reading a lot of data would not be shown in the report. This would prompt the designers to reconsider relationships, and possibly add some new ones. This is not a problem, indeed it is the reason for using logical access mapping.

17.6 THE DATA DICTIONARY

To be certain that each relationship shown on the data model is accurate, we have to look behind the model at the entity descriptions showing all the attributes. Then we must look more closely at these attributes to ensure that we have not defined the same attribute under two different names. This is where the data dictionary is invaluable. It provides a detailed view of the data, and how that data is used during processing. This gives us, at a very low level, a link between data and processes.

A data dictionary may be used to support a simple file-based system, or a more complex database system. A data dictionary contains meta data; that is, it contains data about data. In its simplest form a data dictionary will hold basic information about each data item such as: name, size, validation rules, records, files, programs. A simple example of this might be student enrolment number. In our example the enrolment number is made up of three data items: the college number, the year of entry and the admission number. Typical data we might store about this data item are shown in figure 17.12.

Everything we need to know about each data item is stored in the dictionary because

- It is easier to trace and maintain data since each data item is defined once only.
- Productivity is increased as a result of the re-use of previously defined data items with consequent avoidance of errors.
- Central control allows data to be traced through the system, making maintenance easier.

The dictionary can be a manual one, with one page of data per data item, as shown in figure 17.12. This means that the data dictionary, even for a small system, can be over a hundred pages. However there is only one dictionary in the system, regardless of how many databases and/or files there are. The major drawback however when using a

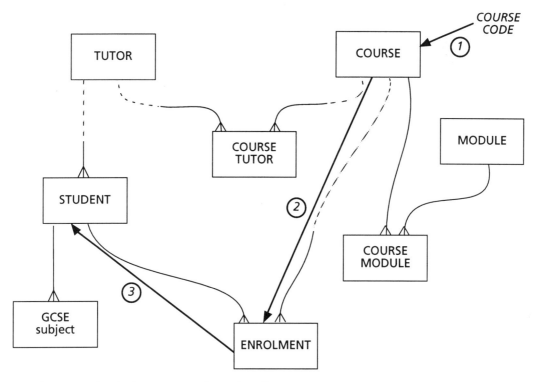

Fig. 17.11 Logical access map for a class list

manual data dictionary is that developers can forget to complete the cross-referencing, or can get it wrong. This causes confusion or errors later. Sometimes developers may use this as an excuse to avoid having a data dictionary at all, especially if the system is small, but the errors and difficulties caused by not having a central data reference usually outweigh those caused by its incorrect use.

The dictionary can of course be automated, and to a large extent this solves the problem of developers forgetting to follow through dictionary updates, because the dictionary automatically updates cross-references and related entries. They do however have a number of drawbacks. The quantity of data stored and the time taken to check cross-referencing can mean that updating the dictionary can be a slow process. Where a dictionary is associated with a particular DBMS, such as Ingres, it is usually implemented based on the use of that DBMS, so that building the database from the dictionary becomes a simpler task. The best automated dictionaries use database management systems for their internal data storage, and thereby support other database facilities, such as full recovery and online multi-user updating. Another benefit of automated dictionaries is that changes made to one screen, automatically update any related information. So that if we amend the 'where used' section to add the name of a common routine, the system automatically accesses the data held about that common routine to say that this data item is also used in the routine. This prevents the confusion which might occur if one part of a manual data dictionary is out of step with another.

Data Element Name: ENROL NO

Short description: Student enrolment number, this is made up of COLLNO, YEAR, ADMIT

Aliases: STUDID, APPLICNO   Type: positive integer, plus check letter

Format 999-9999999-A
        A B C   D

— Values —

Discrete

Part A is a discrete value assigned to identify this college. This is the same for all students attending this college.

Part B is the year of enrolment, e.g. "93"

Part D is a check-letter.

Continuous

1. Part C is continuous, within year 00000 - 99999
2. Positive values only
3. Last enrolment number assigned is stored in LASTENR

Validation

1. Check college id matches COLLNO
2. Check ENROLNO does not appear on X ENROL list.
3. Use algorithm to validate checkletter (part D)
4. Check it is < LASTENR

Editing

1. Once assigned, ENROLNO is never changed
2. When a student leaves ENROLNO is not assigned to another student, it is added to the XENROL list of unused enrolment numbers

Where used

Programs
Enrol student (ENROL)
Record exam results (RECRES)
Assign tutorial groups (ASSTUT)

Common routine
Validate enrolment number (VALENROL)

Fig. 17.12   Data dictionary entry for student enrolment number

17.6.1 Advanced features of a data dictionary

As was suggested previously, developers might want a central store holding not just data, but information about how that data is used. For this reason some data dictionaries allow the user to enter the data definitions, data flows, common processing routines, and layouts for screens, forms and reports.

The information held about a data flow includes:

- names of data items contained
- the origin and destination of the flow
- and what event triggers the flow.

Data held about common processing is usually

- a description of what it does
- its inputs and outputs
- a list of processes it is called by
- and detailed processing such as would be found in a program specification.

If common processes are defined in this way in the data dictionary, there is no need to write program specifications for the routines. Screen layouts, sometimes called skeletons or proformas, may be defined in the dictionary, but not if they are included in the functional specification. The reason for including them here is because the dictionary is a working document used all the way through development, whereas the functional specification, written in the early stages of design, might become more of a reference document.

17.7 SUMMARY

Since data is the key to an organisation, and the techniques for constructing the definitive logical data design are many, it is worth summarising the process step by step.

- Identify as many candidate entities as possible. During the fact-finding phase of analysis you should document entities in the existing system and add others identified in discussions with the client about their requirements for the new system.
- Rationalise the entities. Check whether the same entity has been identified twice but under different names. Combine duplicate entities. Check whether an entity exists outside the boundary of the system. Examine the data content of each junction entity to ensure that it is in fact required.
- Draw an entity relationship matrix as an aid to identifying the direct relationships between the entities.
- Represent each of the identified entities as rectangles on a sheet of paper.
- Initially, only draw the definite relationships. Later, having constructed the logical data model, any relationships which were shown on the matrix as being questionable can be reconsidered to decide whether they are in fact required.
- For each relationship decide which entity is the owner and which is the member. If both appear to be members a many-to-many relationship exists. Identify the junction entity. An entity can be the owner of one relationship and the member of another.

- We recommend an average of between eight and twelve entities, and a maximum of 15 on any one model, simply because this makes the model more readable. Large systems will typically have anything between fifty and two hundred entities but to model all of these together would be difficult to do and cumbersome to use.
- The initial diagram may be untidy with relationship lines crossing each other. To aid understanding and communication, redraw the diagram to produce a clearer representation. Also to aid clarity, draw the member entities below the owner entities on the diagram.
- Test the data structure against the existing system, then against the required system. Check that each form screen and report required can be produced by following a path through the data model.
- With the client, reconsider any relationships you are not sure should be implemented, and make a final decision. It is important that the true significance of each of the relationships on the model is understood to ensure that the data being retrieved is precisely that which is required. It is important too to exercise care when rationalising the entity model since it is easy to over-rationalise as a result of having inadequate information about the data content of the entities or the true nature of the relationships between them.
- Revalidate the rationalised data structure against the user's requirements for the new system. Once the data structure is complete, walk the client through it, and once they have agreed it ask them to sign it off!

CHAPTER 17   CASE STUDY AND EXERCISES

Q1 Perform third normal form analysis on the following list of unnormalised data.

Subscriber no
Telephone no
Rental start date
Bill start date
Bill end date
Subscriber name
Service type
Call no
Call date
Number called
Call period
Call charge
Total amount payable

Q2 Name the resultant relations and draw a partial data model.

Q3 Identify the types of keys found during this exercise.

CHAPTER 18

Systems Design:
Files

18.1 INTRODUCTION

The way in which data is organised and accessed can be crucial to the effectiveness of a computer system. Data can be stored in files or in a database. In this chapter we look at file design and in the next chapter we consider databases. Much of what the computer does is data processing. This involves taking input data, doing some processing on that data and producing output data. Physically this data consists of alpha-numeric characters grouped into data items or fields; for example, a customer name or address. Related fields are grouped into records. A customer record might contain the fields 'customer name', 'address', 'telephone purchase date', 'telephone number' and a 'customer reference number'. A file is an organised collection of related records. System Telecom's customer file would, for example, contain a customer record for every one of System Telecom's customers.

This chapter describes the most common types of file and the ways in which files can be organised, and how the records within the files can be accessed. We also look at many of the factors which determine the optimum file organisation and access method for a particular application. Finally we consider how the file design can be documented.

18.2 TYPES OF FILE

Eight different types of files are described in this section.

Master files contain records which are critical to the system and its users. The records in a master file store permanent information of long-term value to an organisation and are used regularly in the organisation's key systems. System Telecom, for example, might have a master file containing details of customers, another with information about the company's own employees, and another with details of call logging stations. The records in a customer file could contain the following fields :

Customer number	Name	Address	Phone purchase date	Telephone number
0001	M. Jones	9 Uxbridge Road Pinner, Middx HA9 7RD	24.2.88	081 866 3147

The records in the logging station file might contain these fields:

Log station reference	Location	Reliability	Last service date	Network service date
LON 0051	Hyde Park Corner Underpass	93%	9.8.92	4.2.93

Transaction files contain transient data relating to business activities, such as telephone calls logged. They are used mainly to update master files. Transaction files usually contain records relating to a particular period of time. For example, a new file may be created each day and written to between system start-up to system shut-down. The next morning, that file will be used to update master files, and a new transaction file will be created for that day's transactions. Transaction files are sometimes known as transaction logs or log files or update files or change files.

Output files contain information for output from the system; such as data for printing as a report. They are usually generated by processing one or more of the system's master files and a transaction file.

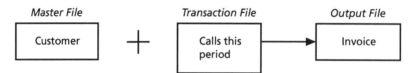

Transfer files carry data from one stage of processing to another. A transaction file for example may be the input to a sorting process so that the subsequent sorted file can then be used to create an output file.

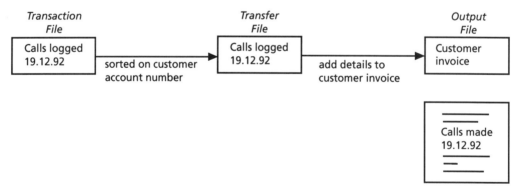

Security or dump files contain copies of data held in the computer at a particular moment. Their purpose is to provide a back-up, to permit recovery in case data is lost or damaged. Therefore security copies are stored off-line, on a magnetic tape, disk or other computer.

Archive files contain archive information for long-term storage. System Telecom might want to archive details of payments made by customers over past years. Archiving is often required by law for tax and audit purposes.

Library files contain library routines such as utility programs and system software. The term can encompass any file containing any compiled computer program. For example, our mobile telephone system will need a program which works out which logging station to use. Such a program will be coded and compiled and may be used by several applications in the system. Therefore it could be said to be one of a library of files.

Audit files are used by a computer auditor to check that the programs are functioning correctly, and to trace any change to master files. Such a file contains copies of all transactions which have been applied to the permanent system files.

18.3 STORAGE MEDIA

Files are stored on backing storage until they are needed for processing. The media most commonly used for backing storage are hard disks, floppy disks and magnetic tape. Disks are similar in concept to compact disks or to long playing records. Magnetic tape is analogous to audio cassettes.

18.3.1 Magnetic disk

Hard disks are the main file storage device of mini and mainframe computers. Hard disks are made of metal that is coated with a thin layer of magnetisable oxide. They are often mounted in packs (figure 18.1).

A typical disk pack contains six disks which have ten magnetisable surfaces. The outermost two surfaces are not used for recording data. Each of the ten surfaces has its own read–write head. Each surface is divided into tracks (figure 18.2) and there are typically 200 tracks on each disk.

A track is subdivided into equal-sized blocks, and these blocks are the smallest addressable units of data. This means that when data passes between the disk and the central processor, a whole block of data is transferred. There are gaps between blocks – inter-block gaps – which contain the block's address. The disk pack revolves continuously while the computer is switched on. To read files from a disk the read–write heads move to the required tracks and data is sent down a data bus to main storage. The read–write heads all move in unison and are always positioned over equivalent tracks on each disk

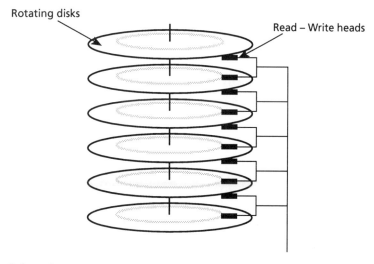

Fig. 18.1 A disk pack

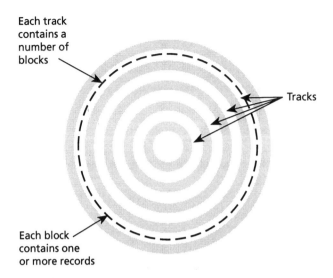

Fig. 18.2 Tracks on a disk surface

surface. The tracks that can be accessed for any particular read–write head position form an imaginary cylinder. So we can treat a disk pack as consisting of 200 cylinders, each with 10 recording tracks. Each block contains a number of records. If the block is 256 bytes and the record length is 240 bytes, the block contains one record. Records on disks can be accessed directly using their address, defined in terms of the cylinder, track and block numbers.

To write a record to disk, data passes from main memory through a read–write head, onto a track on the disk surface. Records are stored, one after another, in blocks, on each track.

The time taken to read from or write to disk is made up of the following:

- *Head movement (seek time)*
  This is the time to reach the correct track. It tends to be of the order of 30 ms.
- *Rotational delay (latency)*
  This is the time taken for the disk to rotate so that the correct record is under the head once the correct track has been found. It is likely to be roughly 10 ms.
- *Head switching*
  This is the time taken to activate the head to read or write, and is usually negligible.
- *Channel transfer time*
  This is the time taken to transfer information between the processor and the storage device.

Average access times are approximately 0.1 seconds and the data transfer speed to and from disk is of the order of 200,000 characters a second. The capacity of a disk pack is typically 20 megabytes or approximately 20 million characters.

With microcomputers, as well as small hard disks, floppy disks (diskettes) are also used for storing files. These are similar to hard disks but are lighter and non-rigid. They have only one (single-sided) or two (double-sided) recording surfaces. Floppy disks are more prone to contamination by dust or heat or magnetic corruption. They have a smaller capacity than hard disks and a slower data transfer rate.

18.3.2 Magnetic tape

Magnetic tape is held on a reel. A tape deck is used for writing data to magnetic tape from the processor, and for reading data from tape to internal storage. As with disks, reading and writing is performed via read–write heads.

The tape is wound from one spool to another, like audio cassette tape in a tape recorder. It is not feasible to read from one tape reel and to write to the same reel. Rather, to update a file a new version has to be created by writing to another reel (figure 18.3).

The magnetisable surface of the tape contains rows of magnetic spots, each spot representing a binary digit or bit. There are either seven or nine spots per row, depending on the convention used, and each row represents a byte or character, using a particular code. In this way each character in a field, each field in a record and each record in a file can be stored. Records can be of fixed or variable length. Records are stored in blocks and the number of records within a block can also be fixed or variable. Between each block is an inter-block gap, to allow the tape to be stopped and started (figure 18.4).

The rate of transfer of data between the central processor and a magnetic tape unit depends on the speed with which the tape can be spooled, the packing density of the magnetisable spots, the block size, the inter-block gap and whether the tape is stopped and started between each block. Data transfer rates are of the order of 100,000 characters per second. Magnetic tape is a cheaper form of storage device than disk and is more secure against corruption. However it can be more cumbersome to use and there are limitations in the way data can be accessed.

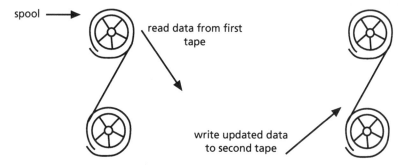

Fig. 18.3 Updating a data file using two magnetic tape units

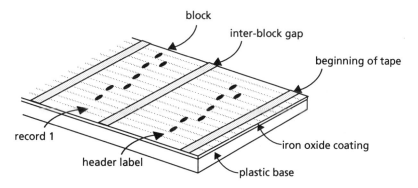

Fig. 18.4 An example of a seven-track tape

18.3.3 Other storage devices

There are other media which can be used to store files but these are less common. Historically, punched cards and paper tape were used (figure 18.5). These have patterns of holes punched to represent data, using a particular code. The idea is similar to the raised dots of braille. Their major limitation is that they cannot be re-used because the punched holes cannot be closed again. Paper tape and punched cards are serial recording media, like magnetic tape.

Another storage device is the magnetic drum. It consists of a cylinder in which the outer curved surface is coated with a magnetic material. Magnetic drums are direct access media, like magnetic disks.

18.4 FILE ORGANISATION

Whatever organisation is used, a file will generally have two special records, referred to as labels, at the start and end of the file (figure 18.6). The header label is the first record in the file. Its main function is to identify the file. It contains a specified field to identify

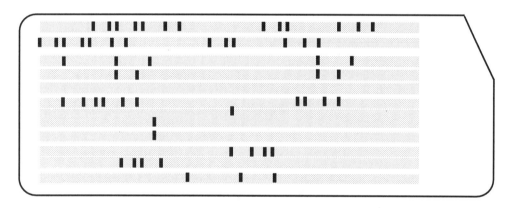

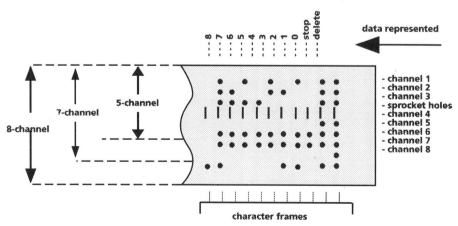

Fig. 18.5 Examples of punch card and paper tape

Fig. 18.6 File organisation

this record as a label, the file name, the date it was written and perhaps a date on which it can be deleted. It may also state whether access to the file requires certain privilege rights and whether the file is for enquiry or updating. This header label is checked by a program before the file is processed to ensure, for example, that the correct reel of tape has been mounted, or the correct disk mounted on the disk drive.

The trailer label occurs at the end of the file, and also has a field to identify it as a label. It generally contains a count of the number of records on the file which can be checked against the total processed by the program. If the file takes up more than one reel of magnetic tape or one disk the trailer will contain the reel or disk number too. It is useful as an end-of-file marker. There may be a record preceding the trailer label which contains control totals or similar information, again used for verifying the integrity of the file data.

The remainder of this section looks at a variety of ways in which the records in files may be organised. The first two methods can be implemented on either disk or tape. The remaining organisations require direct access and therefore cannot be used with tape.

18.4.1 Serial organisation

This is the simplest way in which records can be organised in a file. The records are placed one after another each time a record needs to be stored in the file. No particular sequence is followed – records are not sorted according to the value of a key field, for example. Each record goes into the next available storage space. This method gives maximum utilisation of space but no room is left for inserting records. It is similar to the way you might record songs on to a tape. You would record song 1, then song 2, then song 3. You would not record songs 1 and 3 and then go back and insert song 2.

Serial organisation can be used for files to be stored on magnetic tape or magnetic disk. In either case, records are placed on the storage device one after the other with no regard for sequence. Examples of the files which might be organised serially include transaction files, output files, security files and archive files. The main disadvantage of serial organisation is that it does not cater for direct access to records. If the required record is in the fifteenth position in the file, the first 14 must be read prior to accessing record 15. If access is required in any order other than that in which the file was written, the file must be sorted.

18.4.2 Sequential organisation

Like serial organisation, sequential organisation is also appropriate for files which are to be stored on either tape or disk. In this case, records are sequenced on the value of one or more key fields. For example, the records in a customer file might be sequenced in alphabetical order, or in ascending order of some unique customer number (figure 18 7).

Customer number	Name	Address	Telephone number
0001			
0002			
"			
"			
3219			

Fig. 18.7 Records organised sequentially, using customer number as the key field

Sequential organisation is appropriate for master files, or for sorted transaction files in a batch processing environment. It is not generally used for on-line systems demanding fast response, other than for recording data for later analysis 'off-line'. This is because it takes too long to find each individual record if the file has to be read from the beginning each time.

The advantages of sequential organisation are:

- It is a simple method of writing to a file.
- It is the most efficient organisation if the records can be processed in the order that they are read.
- It can be used for variable length records as well as fixed length records.

18.4.3 Indexed sequential organisation

An indexed sequential file has records stored in sequence like sequential organisation but, in addition, an index is provided to enable individual records to be located directly after reaching the index. Indexed sequential organisation is used with disks but not with magnetic tape. Data held by Telephone Directory Enquiries could be held in the form of an indexed sequential file. The index could be used to locate all data for a particular town. The data for that town would be held sequentially, with names sorted into alphabetical order.

When such a file is first created the records must be sorted so that the value of the key field increases on consecutive records. It is also necessary to think about how much extra space may be needed for record insertions to cater, for example, for new customers being added. Spare space or overflow areas may be needed on each track, and on each cylinder in case more records need to be inserted. Also additional cylinders may be required for overflow of the file, to cater for further record insertions. Initially, additional records would be inserted into empty space between existing records on a track in the appropriate position, according to their key fields. When there is no space for a record insertion at the sequentially correct position, the record would be put into track overflow space and when this is full the record would be put into cylinder overflow.

When the file is written to the disk, indexes are created to enable groups of records to be accessed. The indexes may include, for example, a cylinder index and a track index. The first cylinder on the disk pack typically contains the cylinder index and the first track on this cylinder contains the track index. These are limit indexes so the highest key value

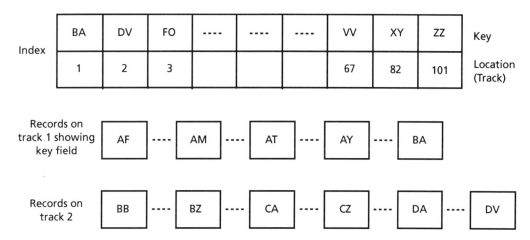

Fig. 18.8 Limited index organisation

for each track is held in the track index and the highest key for each cylinder is held in the cylinder index. Figure 18.8 illustrates the idea of the track index.

To access a record, the indexes are inspected to determine which track the record is on. This track is then copied into memory where the records are inspected to obtain the one which is required. If the records on the track are in sequence, a binary chop search can be used to reach the desired record. Using this method the designer knows the average and maximum access time for any record. This is described in section 18.5.2. Limit index organisation can also be used when the records on a track are not strictly in sequence (as long as the last record on the track still has the highest key value).

The main advantage of indexed sequential organisation is its versatility. It combines direct access to a group of records with rapid sequential scanning of the group to obtain the record required. It is a widely used method of organising files. The problem with indexed sequential organisation, however, is that when records are inserted, they have to go into overflow areas. This can slow down access since if a record cannot be found on the track where it would be expected from its key and the limit index, it will be in the overflow areas so these then have to be searched. Depending on the volatility of the file, it may need to be reorganised fairly often, because otherwise too many records will be in the overflow areas and access times will become unacceptably slow.

18.4.4 Random file organisation

A randomly organised file contains records stored with no regard to the sequence of their key fields. Like indexed sequential files, random files can only be used with direct access methods and therefore require direct access media such as magnetic disks, rather than serial media like tape.

A mathematical formula is derived, which, when applied to each record key, generates an answer which is used to position the record at a corresponding address. The records are retrieved using the same formula. This process of address generation is also known as *hashing* (figure 18.9). Typical algorithms for deriving record addresses are based on division taking the quotient as the starting point for the search or division taking the

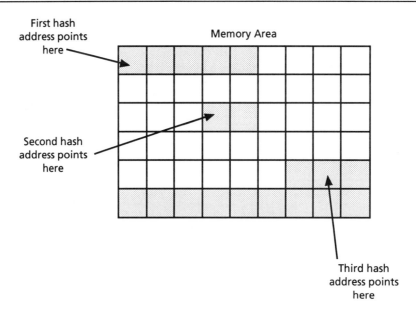

Fig. 18.9 Hashing

remainder. The main problem is to devise an algorithm that achieves a fairly uniform distribution of records on the disk. Also, variable length records are difficult to deal with, and 'synonyms' can occur, where several keys result in the same address because the algorithm does not produce a different address for each different key. In these cases, records are put into the next available space.

If we wanted to store data and the hashing algorithm had produced a remainder of 1, the system goes directly to the start of that sector and reads until it finds a null character, then it stores the data in that location. Notice that each sector in the memory area has a different number of locations used.

The main advantages of random file organisation are:

- No indexes are required.
- It permits the fastest access times.
- It is not necessary to sort master files or transaction files.
- It is suitable for volatile files: records can be inserted or deleted indefinitely without reorganising the file.

18.4.5 Full index organisation

With a full index (figure 18.10), there is an entry in the index for every record. This entry gives the address of the record. The index is arranged in ascending order of the keys although the records themselves can be in any order. There can be an index for each of several key fields enabling records to be accessed by more than one key. For example we could access a customer record by customer name or geographical location.

The drawback with full indexing is the size of the index as it can be large. Space is required to store the index and it can take a relatively long time to read the index and

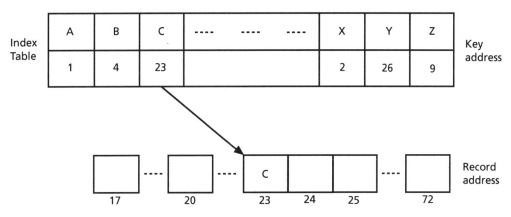

Fig. 18.10 Full index organisation

then access the record from the address. Full index file organisation is only appropriate for disks or other direct access storage media, not for magnetic tape.

18.4.6 Chained files

The scope of a file can be extended by using pointers. This is often done when a combination of direct and sequential access is required. To illustrate the idea, consider a warehouse system used by a food processing company (figure 18.11). Their warehouse has a number of aisles, each with a number of columns and levels. Each batch of, say, cake mix is coded with a date and pallet number. Samples of the cake mix are tested, which takes up to one day. The batches of cake mix are stored in the most convenient space, not necessarily in physically adjacent locations. If a sample fails the test, the batch must be destroyed so the locations of each pallet of that particular batch must be known. This can be done with a chain or linked list of addresses.

The system stores a pointer to the start of the chain, in this case 030403 (i.e. aisle 3, row 4, level 3). From the start of the chain it can follow the chain of pointers to the end. The pointer is actually a field within the record containing the address of the next record. In

LEVELS 4	CHOCCMIX 9302001 0003				CHOCCMIX 9302001 0002
LEVELS 3		CHOCCMIX 9302001 0006		CHOCCMIX 9302001 PALLET NO.1	
LEVELS 2			CHOCCMIX 9302001 0005		
LEVELS 1				CHOCCMIX 9302001 0004	

Fig. 18.11 Warehouse aisle 3

this way the records are organised as a linked list. A similar approach is used in CODASYL databases which are discussed in chapter 19.

In the processed food example the chain file would be as shown in figure 18.12.

Some chain files only use forward pointers; we have shown a file with forward and backward pointers. The final record in the list is assigned a forward pointer of zero. It is easy to insert or delete records using chaining, simply by modifying the pointer in the preceding and following records: if we removed record 3, we would amend the forward pointer in record 2 to point to pallet number 4, and the backward pointer in record 4 to point to pallet number 2.

Figure 18.13 shows the chained file showing where the chocolate cake mix is stored after pallet 3 is removed. So the chained file maintains a link from one pallet to the next. The two data items which had to change are highlighted in bold.

Using chains in this way allows for flexible use of the storage area. If there was no chocolate cake mix in the warehouse, no storage space is reserved for it. If there are 600 pallets of cake mix the system handles it in exactly the same way as it handles 6 pallets. Data designers have to make a decision about whether deletions can be made only at the head or tail of the chain or whether mid-chain data can be deleted. For example, if a pallet of cake mix was badly damaged due to a crane breakdown you would want to remove the pallet and delete any reference to it in the data files.

record 1	Ø	CHOCCMIX 93020010001	03 05 04
record 2	03 04 03	CHOCCMIX 93020010002	03 01 04
record 3	03 05 04	CHOCCMIX 93020010003	03 04 01
record 4	03 01 04	CHOCCMIX 93020010004	03 03 02
record 5	03 04 01	CHOCCMIX 93020010005	03 02 03
record 6	03 03 02	CHOCCMIX 93020010006	Ø
	Pointer to previous location	Product details: description, batch number + pallet number	Pointer to next location
	A Ø pointer indicates the head or the tail of the chain.		

Fig. 18.12 A chain file

record 1	Ø	CHOCCMIX 93020010001	03 05 04
record 2	03 04 03	CHOCCMIX 93020010002	**03 04 01**
record 3	**03 05 04**	CHOCCMIX 93020010004	03 03 02
record 4	03 04 01	CHOCCMIX 93020010005	03 02 03
record 5	03 03 02	CHOCCMIX 93020010006	Ø
	Pointer to previous location	Product details: description, batch number + pallet number	Pointer to next location
	A Ø pointer indicates the head or the tail of the chain.		

Fig. 18.13

18.5 ACCESS METHODS

Serial access is used for files which are organised serially, whether they are stored on tape or disk. This means that each record is read from the tape or disk into main storage, one after the other, in the order they occur in the file.

Access to records in a *sequential file* on a serial medium like magnetic tape is also serial. Because the file is in sequence, the term sequential access is often used instead, to describe serial access to records on a sequential file. Access to a sequential file on disk is similar but since disks enable direct access there are ways of speeding up the process of accessing records such as a binary chop approach.

In this method, a block of records is copied into main memory. The key values of these records span the key value of the desired record. The middle record is inspected to find its key. If this is higher than the required value, the middle record amongst the first half of the records is inspected, and so on.

For example, to find record with key 28 :

Record key	2	5	13	17	24	28	33	41	50
Search no					1	3	2		

The required record is thus found after inspecting only two other records instead of all those preceding it.

There are three ways to access an *indexed sequential file*. If most of the records in the file need to be processed then an indexed sequential file can be treated as a sequential file, and the index is largely ignored. Each read operation transfers a block of records from the disk to main memory and all records in the block can be processed sequentially. The index is used simply to determine which block should be read next. If we are updating sequentially, then transaction files which update an indexed sequential file must first be sorted into the same key sequence as the master file.

If we use a selective sequential approach then the transaction file must also be sorted before applying changes to the indexed sequential master file, so that both files are in the same key sequence. The two files are processed together, but only those master records for which there is a corresponding entry in the transaction file are selected for updating. This method is more suitable for master files in which a relatively small number of records need to be accessed. In other words it has a low hit rate.

The file can also be updated with transactions that are not in the same sequence as the master file. In this case, the limit indexes are used to find the cylinder and track on which the required record is located. The track is then scanned sequentially to obtain the record to be processed. This is appropriate when a transaction file is not sorted into the sequence of the master file, perhaps because a transaction file is being used to update simultaneously two different master files sorted on different keys, or when the volume of transactions is low.

When accessing a *randomly organised file*, the formula that was used to store a record at a particular address must be used to retrieve that record. The transaction record keys are therefore put through the same mathematical formula or algorithm as were the keys of the master file record so that the appropriate master file record can be found and updated.

To access a record on a file with a *full index*, the index must be searched to find the required key. From this, the disk address of the record is found. To insert a record, it is necessary to create a new index entry and to insert this into the correct sequence in the index. Since this can take time it is sometimes better to use a chaining technique in the index. In this case, consecutive index entries can be physically separated but logically adjacent. The entry before the inserted one contains a pointer to the overflow area of the index where the inserted entry will reside. The inserted entry will itself have a pointer to the address of the next index entry.

There is another form of index known as the *implicit index*. In this case, an entry is held in the index for every possible key value, whether or not the record actually exists. If a record exists the index contains its address in the file. If a record does not exist, the index indicates this. It is easy to insert and delete records in such a file – the index is simply amended. Because the index is held in sequence it is not necessary to store the key values in the index. The first entry in the index corresponds to the first value of the key and so on. Searching the index is quick and from the address given in the index the required record can itself be obtained quickly.

As we have seen earlier, records in a *chained file* contain pointers to the next record in the sequence. This means that consecutive records do not need to be physically adjacent. Chaining is also used for pointing to overflow areas. For example in a sequential file on a disk, records to be inserted may be stored elsewhere on the disk. If the required record is not in its expected position in the main sequence the overflow area has to be searched serially.

18.6 FACTORS INFLUENCING FILE DESIGN

There are many factors which determine the best file organisation and access method for a particular application. The most important of these are discussed in this section.

The *purpose* of the file is likely to be the major factor in determining the most appropriate organisation and the best method of accessing records. If it is to be used for on-line enquiry, for example, direct access is needed and the file must be organised so as to permit this, perhaps using indexed sequential organisation or random organisation with key transformation techniques. These, in turn, mean that disk not tape must be used to store the file. It may be that the file has to be processed in a variety of ways. For example, a file may be accessed on-line during normal day-time running, but used to produce reports overnight as a batch process. An organisation must be chosen for both methods if it is decided not to have two or more differently organised versions. Files that are to be used for batch processing only, and where many of the records need to be processed, should be organised serially or sequentially on tape as it is cheaper than using disk.

It is important to identify any constraints imposed by the system hardware. Does the available hardware support a particular file organisation and method of access? What data storage media are available? Remember that serial devices like magnetic tape support only serial and sequential file organisation and serial access. Does the hardware configuration permit indexing of files or chaining of records, for instance?

Updating of files is usually either 'in-situ' or using the 'grandfather-father-son' method. In-situ updating involves updating, inserting or deleting records on-line, without retaining the old version of the file. There should, of course, be back-up copies of the old version, in case something goes wrong, or as archive files. In-situ updating of records in a random order requires direct access to a random or full indexed file. However an organisation which allows in-situ updating may make record accesses inefficient after several record insertions or deletions.

The father–son method involves taking a file and producing a new version, thus resulting in having two generations of the file, the father and son generations. Father–son updating is more secure than in-situ updating, because if the system fails the old version of the file should be unharmed. The updating can then be repeated when the system has been restarted. Three generations of a file 'grandfather, father and son' are often created, before the oldest version is destroyed and the storage space re-used.

A file designer needs to consider the *frequency of file accessing* for enquiry or for updating. How many records on the file will have to be accessed in one process? This is termed the *hit rate*. For example, if 100 records are processed each day in a file containing 1000 records, the hit rate is 10%. If the hit rate is high, batch updating of a sequential file may be the most efficient design. To decide whether the hit rate is 'high', the designer must calculate the 'average' time to access 100 records, then calculate the time to read through all 1000 records on the file. If it takes less time to process the whole file sequentially, that would be the more efficient choice. It is not as simple as saying that a 45% hit rate is 'low' and a 55% hit rate is 'high'.

Some files, for example library files, may rarely change, whereas others change minute by minute. Therefore, in designing a file it is important to consider how often records will need to be inserted, amended or deleted. This is termed the file *volatility*. An indexed file copes with high volatility better than sequential or indexed sequential files, since these have to be constantly reorganised to bring records added to overflow areas back into sequence. An indexed file need only have its index updated when records are added to the file. However, where volatility is low, the use of an index might be unnecessary over-engineering.

In systems which have on-line or real-time aspects, the *response time* to an external event may be critical to the performance of the overall system. Such systems include military or other safety critical systems, or process control systems supervising machinery. Fast processing of records can also be important in large batch systems because of the sheer number of records involved. Where fast response or fast processing times are required, the speed of access to records can be the dominant factor for the file design and this usually means using direct access methods with key transformation. If an indexed file is required then the system might be tailored so that the file index is stored in main memory, thus reducing access time to the index. With a magnetic tape file, processing is necessarily serial so there is no scope for improving access to the next record. But the effective data transfer rate depends on the record and block sizes being used. The longer each block is, the smaller the proportion of the tape occupied by inter-block gaps and therefore the greater the capacity of the tape and the greater its data transfer rate. There is often a trade-off between efficiency of access and efficiency of data storage. Quicker access may be achieved only at the expense of wasted space on the disk or tape.

The expected *file size* and predicted growth pattern can have a significant effect on file and system design. A file that extends over more than one disk or tape reel can cause operational problems, particularly during system recovery, when it may be time-consuming and difficult to copy the file. It is also slower to access a large file. In general, if a file has to be large it is better to use an organisation that does not use an index, because of the time taken to search the index. Random organisation using key transformation is

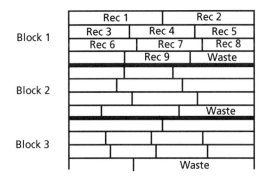

Fig. 18.14 A file comprising blocks of variable length records

likely to give quicker access to large files than, say, indexed sequential, which would require several levels of index and would be time-consuming to search. It may be better to restructure a large file, to split it into several smaller files. Small files are probably better kept on disk since they take longer to access on tape.

File records may be of either fixed or variable length. Variable length records can be dealt with straight-forwardly in sequential file organisation but can present problems for direct access files because they might be too big for the space available. In general it is quicker to access records which are of fixed length. The system application will dictate the requirement for fixed or variable length records and for the record size. The choice of record size can also affect the speed of access and the efficiency of storage. A fixed length record format must be large enough to accommodate the largest record to be stored. This may result in much wasted space in the other records that are smaller. These concepts are illustrated in figure 18.14, which shows fixed length blocks containing variable length records, and the wasted space at the end of each block.

Systems may need to be *restarted* following failure caused by hardware, software or power problems. The most efficient restart begins exactly from the point of previous failure since this means that no processing has to be redone. To achieve this situation would require a constant capture of the state of the system at every moment. This is clearly not possible. However, we do need a base point – or checkpoint as it is sometimes called – from which the system can be restarted. In a batch processing system we can begin the run again using the same transaction file and a back-up copy of the master file. For systems with very long runs this could be wasteful and we may therefore choose to set up checkpoints at stages through the processing. These effectively break down a long batch processing run into several short runs. For transaction processing systems with multiple terminals on-line to a system, restart requirements are more complex, but at least logs of the transactions can be made and snapshots made of master file records before updating, so that the system can be restarted after failure.

18.7 SPECIFYING FILES

Once the designer has decided upon the most appropriate file organisation and access method, the file design needs to be documented. There are a number of ways in which this might be done but certain information is likely to be included in the documentation, whichever method is adopted. Some examples of forms used in practice are shown in figures 18.15 to 18.19 and are described below. They provide a complete specification of a file and detailed systems specifications should contain information to this level of detail.

The file specification header page which identifies the file and acts as an index to the remaining pages of the file specification is shown in figure 18.15. A general description of the file and its attributes is shown in figure 18.16. The general description should be in language which non-technical people can understand so that all users of a file can read it and understand what the file consists of.

The attributes section documents the type of file, the medium on which it is stored and the maximum size of the file in terms of the storage medium. The size of block is given, and whether records are of fixed or variable length. The file organisation is recorded and if applicable the names of the fields comprising the key are given, and the sequence in which records are sorted, for example ascending or descending. The retention period is the length of time each version of the file is retained before being overwritten. The number of generations is the number of files in the cycle of backups, eg. great-grandfather, grandfather, father, son.

A record summary form is shown as figure 18.17. It lists all the different record types which can occur in the file. For each record type the following information is given :

- Record name : a mnemonic or short description which uniquely defines the record.
- Page ref : the page number in the file specification on which the record is specified.
- Record type : either a value in the record which uniquely identifies the type of record within the file or a notional number to be used when describing the file structure.
- Record length : the number of bytes, words or characters in a record.
- Occurrence : the minimum, maximum and normal number of records of that type on the file.

The second half of this form gives the file description and program utility. The kind of information which might go into this section includes the relationship between the various records on the file and the structure of the file. A pictorial representation of the file can also be given, probably on a new page in the file specification. This is shown as figure 18.18. The File Usage section should list all the programs which use the file, and whether they read it, write it or update it. The File Security section can be used to describe how the integrity of the data on the file is to be ensured.

Record specification forms will appear for every record type listed on the record summary form and they should be in the same sequence. Shown in figure 18.19, the record specification gives a brief description of the record and then information about the fields on the record. This includes the name of each field, its length and whether it is binary, character, packed decimal, etc. The 'format' gives the way in which the field is stored in the record. The permissible values that the field can take should be given and, finally, a brief description of the field.

If the system uses a data dictionary, some of the information in the file specification may already be defined there and so some of these forms, such as the record specifications, could be superfluous.

File Specification (a)

Reference: LCS-TRANSF
Page: 1 of 12
Version/Date: 2/March 94
Author: Danielle Sanderson

FILE NAME   TRANSACTION FILE

File Ref   TRANSF

CONTENTS   Page

1. General Description   2

2. File Attributes   2

3. Record Summary   3

4. File Description and Program Utility

    4.1 File Description   3
    4.2 File Structure Charts   4
    4.3 File Usage   5
    4.4 File Security   5

5. Record Specification   6–12

Fig. 18.15   File specification header

File Specification (b)

Reference: LCS-TRANSF
Page: 2 of 12
Version/Date: 2/March 94
Author: Danielle Sanderson

1. General Description

The Transaction File records all transactions input through the Data Capture Terminals and the Host console.

This file provides the communication from the front-end system to the batch system.

2. File Attributes

FILE TYPE				FILE ORGANISATION	
Master	☐	Transfer	☐	Sequential	
Input	☐	Other	☒		
Output	☐				
Magnetic Tape ☐		Disc ☐		KEY	
Number of Reels (Maximum)		Cyls. Trks. (Maximum)		NONE	
Punched card ☐		Other ☒		SORT SEQUENCE	
Number of Cards (Maximum)		Double density diskette		NONE	
BLOCK SIZE 521 bytes		FORMAT VARIABLE		RETENTION PERIOD 1 week	NO OF GENERATIONS 7
RECORD LENGTH 27 bytes (MAXIMUM)					

Fig. 18.16 File attributes

298 Systems Analysis and Design

File Specification (c)

Reference: LCS-TRANSF
Page: 3 of 12
Version/Date: 2/March 94
Author: Danielle Sanderson

3. Record Summary

Level	Record Name	Page Ref	Record Type	Record Length	Occurence min	max	ave
	Header 1	6	O	16	1	1	1
	Issue	7	A	26	0		
	Reserve	8	B	27	0		
	Unreserve	9	C	27	0		
	Return	10	D	18	0		
	Book Action	11	E	22	0		
	Borrower Trap	12	F	14	0		

4. File Description and Program Utility

4.1 File Description

Record type O always occupies alone block 0.

Record types A–F occupy the following blocks (blocks 1–n)

Records do not span blocks.

The first word of each block contains the number of allocated bytes in that block.

Fig. 18.17 Record summary

Systems Design: Files 299

File Specification (d)

Reference: LCS-TRANSF
Page: 4 of 12
Version/Date: 2/March 94
Author: Danielle Sanderson

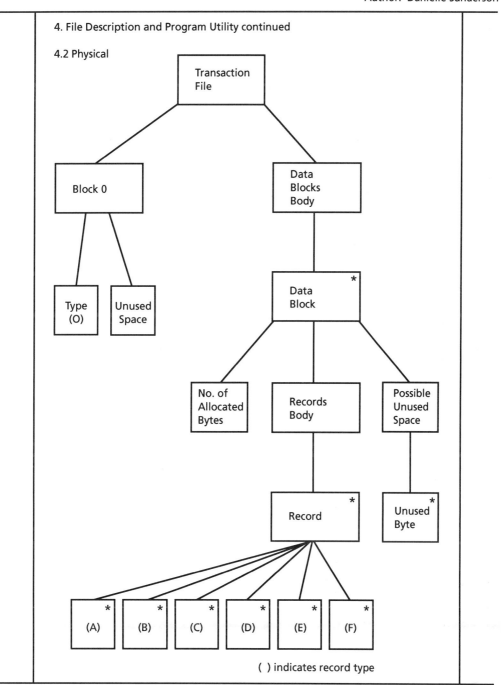

Fig. 18.18 (part 1) File description/program utility

File Specification (d)

Reference: LCS-TRANSF
Page: 5 of 12
Version/Date: 2/March 94
Author: Danielle Sanderson

4. File Description and Program Utility continued

4.3 File Usage

	READ	WRITE	UPDATE
On-line Programs		X	
RAWRIT – Write to transaction file			
RCDEQV – Process VDU message	X		
ii) Initialisation			
SYSTRT – Start-up and recovery			X
iii) Batch System	X		

4.4 File Security

A cycle of 7 files is used to maintain a weekly backup of this file.

Note: i)   the no. of files in the backup cycle which contain useful data will depend on the number of working days in the last week.

Fig. 18.18 (part 2)   File description/program utility

Record Specification

Reference: LCS-TRANSF
Version/Date: 2/March 94
Author: Danielle Sanderson

FILE NAME	TRANSACTION FILE		FILE REF	TRANSF				
RECORD NAME	BORROWER TRAP	(F)	RECORD SIZE	14				
			NOTES	This record records the setting or clearing of a borrower trap				
Field Name			Position From	Position To	Length	Mode	Format	Value Range	Description
TF-TYPE					1	C	X	'F'	Record type
TF-SRCCODE					1	C	X	'0', '1'–'6' 'A'–'F', 'Z'	Source code (values as for ISSUE)
TF-FAILFLG					1	C	X	'S' or 'F'	Success/failure flag 'S' – Success 'F' – Failure
TF-BORROWERNO					8	C	X(8)		Borrower number
TF-ACTION					1	C	X	'T' or 'C'	'T' – set borrower trap 'C' – clear borrower trap
TF-ACTCODE					2	C	X(2)	OB, UF, RR OD, WD, LC, RC FJ, AD, NM, TP IL	Trap qualifiers

Fig. 18.19 Record specification

18.8 SUMMARY

Although databases are becoming increasingly popular, there will always be a place for conventional files in computer systems. They are needed for real-time applications, where searching a database would take too long. They are also needed for many batch processing systems and for security, archive, library and audit files. In this chapter we have looked at ways of organising files and of accessing records on the files. We considered factors which help us to choose the best organisation and access method for a particular application. We have also proposed a way of specifying file design. This involves a translation from the way humans see and understand data to the way it is seen and understood by a computer system. The main storage media in use today are magnetic tapes and magnetic disks, and we explored the advantages and disadvantages of both, and when each is appropriate. Magnetic tapes are cheaper and more secure but files on them can only be accessed serially. Magnetic disks are direct access media so can be used with any of the organisations and access methods covered in this chapter. The main things to consider when specifying files are the application, the system environment, the file size, its volatility and what time is acceptable for accessing data from the file.

CHAPTER 18 CASE STUDY AND EXERCISES

Q1 What are the main issues to be considered when choosing a file organisation. What circumstances would make an indexed sequential organisation appropriate for a master file. Is there such a file in the System Telecom applications described in the case study?

CHAPTER 19

Systems Design: *Databases*

19.1 INTRODUCTION

The purpose of this chapter is to introduce the concept of databases and the various types of database currently in use, and then to describe the approach taken to designing database systems. The concept of physical data modelling is continued and is intended to pave the way for an understanding of physical data design in the following chapter.

The concept of files has already been introduced as a way of storing data. Groups of files – containing data related to each other from the business viewpoint – are often described as a database. The purists would argue that these are not true databases however, since databases have a number of characteristics that are not shared by such groups of files. It is these characteristics that will be illustrated in this chapter.

By definition, *a database is a single collection of structured data stored with the minimum of duplication of data items, to provide a consistent and controlled pool of data*. The data contained in the database is sharable by all users having the authority to access it and is independent of the programs which process the data. The main benefit of a database is that programs do not necessarily need to be changed when the data structure changes and that data structures do not need to change when programs are changed. This concept is known as *data independence*.

As we have seen elsewhere in this book, organisations have often developed systems in a relatively piecemeal fashion, typically designing systems in isolation and then transferring them to the computer: each system with its suite of programs, files, input and outputs. One disadvantage of this approach is that the systems do not truly represent the way in which organisations operate, since most organisations have business functions which are interdependent, where the exchange of information is crucial to their operation. For example, in System Telecom, customer invoicing is related to the customer details, to the charge rate details and to the number of call units used. If this information was in separate systems, then the manual effort required to collate the information for customer invoicing would be incredibly tedious.

Information obtained from a series of isolated files does not provide a complete picture of the state of the business at a point in time, since the data cannot easily be gathered or joined in an effective manner. Programs in these systems typically have to define the structure of the data in the files in order for processing to take place, which in turn creates a maintenance burden for the organisation whenever changes to the functionality or the data occur.

19.2 DATABASE CONCEPTS

A database system consists of two main components. Firstly, the physical and logical organisation of a database is controlled by a *database management system* – DBMS – as

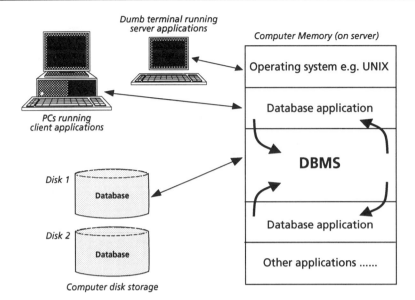

Fig. 19.1 Schematic view of a database

shown in figure 19.1. The DBMS constructs, maintains and controls access to the database. It is not normally possible to access the data except via the DBMS. Secondly, the database application code enables the data in the database to be retrieved, modified and updated. This code 'executes' a particular process or function as also shown in the diagram.

The data in a database is available for use by several users or application programs simultaneously. Each user may have a different view or picture of the data, depending on the requirements of the application. In general, an item of data is stored only once. Logically the data is a single, integrated structure which is how it is normally envisaged by the users as shown in figure 19.2. Physically, however, a database may be organised into discrete units, containing data pertinent to a local group of users. This is known as a *distributed database* which may span several machines or 'nodes', linked together using a computer network. Nevertheless, it can still be regarded as a single database. The objective is to locate the data closest to its point of use, to reduce response times and communication costs, without sacrificing the ability to process data as a single, logical database.

19.3 DATABASE MODELS

The data in a database is organised according to the data structure 'imposed' upon it by the physical data model. This physical data model is normally produced during the physical data design stage in the development life cycle and tailored specifically to the selected, or imposed, database type.

Before producing a physical database design, the designer must have a detailed knowledge of the database to be used for the implementation of the physical data model. The rules regarding this implementation and its subsequent optimisation will vary widely according to the type of DBMS. Rules are described in this chapter for a relational database.

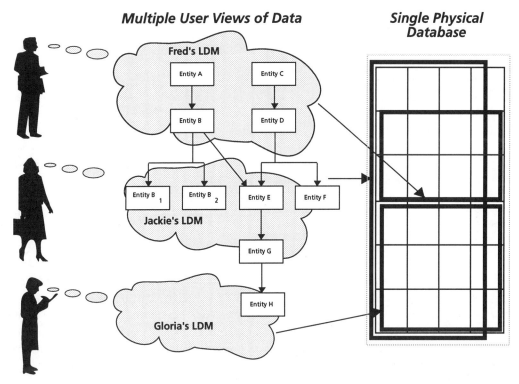

Fig. 19.2 User perspective of a database

Data independence is achieved by separating the logical structure of the data from its physical implementation. Data models are not affected by any changes in the way that data is physically stored, which gives the data model a certain degree of flexibility. The DBMS software is the vehicle which enables the data model to be mapped onto the physical storage of the data. If the storage medium or data location is changed, only the associated storage mapping needs to be modified, via the appropriate DBMS utilities.

The Database Management Systems available today can be grouped into four types. These are:

- File Management Systems (FMS)
- Hierarchical Databases (HDS)
- Network Databases (NDS)
- Relational Databases (RDBMS)

The most common databases are hierarchical, network and relational. Hierarchical Databases probably hold more gigabytes of data than any other DBMS in operation, but RDBMS implementations are rapidly outnumbering both HDS and NDS as organisations strive to produce more flexible systems with less coding and maintenance effort. Many organisations are moving away from their large-scale database systems for their business requirements, by 'downsizing' their systems to a number of smaller RDBMSs. The main enabling force behind this drive to downsize is the emergence of Open Systems standards and technology, which allow many different types of systems to intercommunicate via computer networks.

Each database model is a conceptual description – the architecture – of how the database is constructed and operates. Specifically the model describes how the data may be accessed and presented to both end-user and programmer. A database model also describes the relationships between the data items; for example, in a System Telecom database, each item of information such as customer account number is related to the particular customer that the whole record describes. That particular customer record may also be related to other items in the database such as sales area and invoice details.

With the exception of File Management Systems, the database models do not describe how the data is stored on disk. This is handled by the DBMS. However, in some circumstances, a model may indirectly place constraints on how the data is stored if the DBMS is to meet all the requirements that comprise that particular model.

In non-relational databases, the complete definition of a database is sometimes known as its *schema*. The schema relates or 'maps' the logical data structure to the actual physical organisation of the data. All data items are defined to the DBMS in the schema which is held in a data dictionary. When a user or program needs data, the DBMS will look in the schema to discover where to find the data and how to retrieve it. A subschema is a logical subset of the schema, comprising selected groups of records, record types and fields. Any number of subschemas can be defined and are coded into the program modules as data access routines. Users and programmers are usually constrained to access the database through a particular subschema. A subschema defines the particular view of the data structure that a program may access and may differ between applications which require different access views of the data. A subschema can have security controls associated with it, such as Read-Only to ensure, for example, that programmers cannot inadvertently allow certain fields in the subschema to be updated. Access to the database is provided via operators such as READ, FETCH, FIND and WRITE which operate on both RECORDs and SETs. These terms will be described in more detail in the Network Database section later.

In Relational Databases the table and column definitions are effectively the same as the schema described above, except that they are tabular in nature and are normally defined in the DBMS's internal data dictionary (see figure 19.3).

The RDBMS internal data dictionary typically contains information relating to the location of the database components, users, access requirements, individual file or table structures and application definitions. Additionally, this dictionary often holds statistics regarding data value distribution, in the form of a histogram. Statistics are collected for use by the database

Note that the DBMS internal data dictionary should not be confused with external data dictionaries, which are typically central repositories of information for systems within which the logical and physical data models may be defined and maintained in addition to descriptions of processing elements in the system. Subschemas are not defined in relational programming languages, known as 4GLs – Fourth Generation Languages, since the organisation of data and access to the data are controlled by the database and hence the application does not need to know about the physical structure, which results in less program coding for most relational systems. Relational databases are accessed using relational operators, examples of which are: RESTRICT, PROJECT, PRODUCT, JOIN and UNION. These operators are performed on relational entities, called TABLES, and are described in more detail in the RDM section in this chapter.

Relational Schema

Customer Table

Customer Code	Customer Name
M/1231	Trilbeys
M/9876	Adam Systems
M/1234	Archer
M/9944	Books Unlimited
M/4567	Arndall Electrics
M/4573	Crookshank
M/8936	Swan

Sales Order Table

Sales Order No.	Customer Code	Sales Item
S/0023A	M/9944	Mobile
S/0023B	M/9944	Exchange
S/0083X	M/1234	Portable
S/0023A	M/1234	Mobile
S/0289C	M/4567	Handset
S/0001A	M/4567	Transmit
S/0003F	M/4567	Standard

Non-Relational Schema

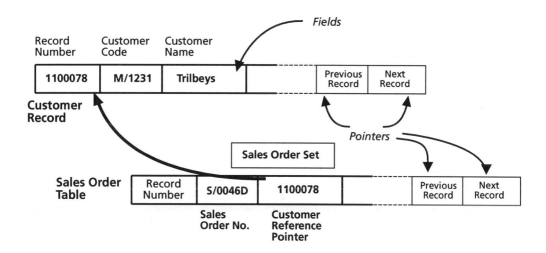

Fig. 19.3 Relational and non-relational schemas

Fig. 19.4 File management system model

19.4 FILE MANAGEMENT SYSTEMS (FMS)

File management systems are the easiest database model to understand and the only one to describe how data is physically stored. In the FMS model, each field or data item is stored sequentially on disk in one large file. In order to find a particular record, the application must start reading from the beginning of the file until a match is found. FMS was the first method used to store data in a database and simplicity is its only advantage over the other types. Figure 19.4 illustrates how the Customer File might look in an FMS database.

The disadvantages of this model become immediately apparent. Firstly, there is no indication of the relationships between the data items, so the user and the programmer have to know exactly how the data is stored in order to manipulate it. Secondly, data integrity is not enforced, requiring the programmer to build this into each module of code in a consistent manner. Thirdly, there is no way of locating a particular record quickly, since every record must be examined to find a match. It may be possible to store pointers against the file to avoid having to search from the beginning each time, but since the file is not 'organised' this may not be practical. The data could be sorted for example on customer surname, but this would have to be done after every new record insertion. A more efficient method for locating the data is to generate an index file which contains one or more of the fields from each record and a pointer to the physical record in the Customer File. Another major disadvantage of this model is that the file and associated index(es) must be completely recreated if the data structure changes, such as when a new field is added.

19.5 HIERARCHICAL DATABASE SYSTEMS (HDS)

The HDS is organised as a tree structure that originates from a root. Each level or class of data is located at different levels below that root. The data structure at each class level is known as a node and the last node in the series is known as a leaf. The tree structure of the HDS defines the parent–child – master–detail relationships between the various data items in the database. Figure 19.5 demonstrates this structure. Each main branch of the hierarchy is known as a database in its own right, effectively resulting in a set of databases comprising the complete structure.

The HDS model demonstrates the advantages over the FMS model in terms of a clearly defined 'parent–child' or 'master–detail' relationship. It also demonstrates that searching for specific data items is a more efficient and faster process; for example, if the user wants to find information at level 3 such as Customer Address, the DBMS does not have to search the entire file to locate it. The DBMS examines the request, breaks it down into its components and follows the Customer branch down to level 3, where the required data item is quickly located. As with FMS, an index on Customer Name would speed the

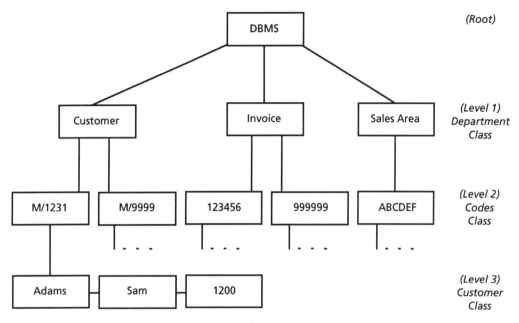

Fig. 19.5 Hierarchical database model

search further, although in HDS the index would be created on a particular class or level rather than on the entire branch or chain.

In an HDS, there is only one root node which is owned by the system or DBMS. Pointers lead to the level nodes – children of the root – where the actual database starts. Navigation through the data set is achieved by using the primary key of the parent and then by using pointers to reach the required record. Each level 1 node can have one or more level 2 children such as Customer Accounts. The children of the Customer Account would then be the actual Customer detail records at level 3. The path between the level 3 data items consists of a chain of pointers from one leaf node to the next. Each child can have pointers to numerous other children or leafs at the same level, but can only have one pointer to the parent, preserving the one-to-many relationship in a single direction.

The physical structure of the data on disk is not important in the HDS model, since the DBMS normally stores the data as linked-lists of fields with pointers from parent to child and between children, ending in a null or terminal pointer to signify the last leaf. The HDS structure enables new fields to be added at any level, since the method for achieving this consists of changing pointers to point to the new field.

One disadvantage of this model is the internal structure of the database, which is defined by the designer/programmer when the database is created. If the nature of parent–child relationships are subsequently changed, then the entire structure must be rebuilt. An example of this would be where the Customers are to be sub-divided by Sales Area. As a result of this overhead, programmers often add additional fields to a particular level in order to satisfy the new requirement by adding Sales Area to the Customer record, and in doing so, the data is often duplicated, thus leading to a maintenance overhead.

Perhaps the most significant disadvantage of the HDS model is its lack of ability to support many-to-many relationships which makes the HDS the most complicated to convert from logical to physical design. An example might be where, in a company, employees may also be managers or supervisors in which case a slightly illogical parent–child relationship might exist between a manager and themselves. This was described in detail in chapter 10. The usual solution to this would be to create a new field in the Employee file identifying the manager for each employee, but this adds to the data duplication problem and slows down searches.

Another approach to this parent–child problem might be to add second parent–child and child–child pointers to the structure, creating circular relationships. As these relationships become more complex, the architecture gradually evolves into the Network Model, where each child can have more than one parent.

19.6 NETWORK DATABASE SYSTEMS (NDS)

Network database systems are often referred to as CODASYL database systems. This is an acronym for Conference on Data System Languages. This was convened in 1959 to define and recommend standards for computing languages. Its initial task was to define a Common and Business Oriented Language, resulting in COBOL, a procedural programming language. The Database Task Group of CODASYL published its first paper in October 1969, and a revised paper in April 1971. This has formed the basis of the design and development of many subsequent CODASYL databases. The name 'Network' bears no relation to the type of architecture where the system is implemented – a computer communications network (see chapter 23). The Network model conceptually describes databases where many-to-many relationships exist.

Logical data is described in terms of data items, data aggregates, records and sets. A data item or field is the lowest addressable unit of data. It usually corresponds to a real-world attribute such as Customer Account Number. Data is stored in an NDS as occurrences within a record type. The relationships between the different data items are known as SETs. A set defines a logical relationship between two or more record types and represents a real-world relationship between two or more entities such as that between Customer and Account. One record type in a set is the owner, the others are set members. There are zero or more member records for each owner record. A set occurrence comprises one occurrence of the owner record and all its related members. Sets are implemented by means of *pointer chaining*, whereby each owner record is connected to its related members by pointers from one record to the next.

An NDS relies on either straight-line or cyclical pointers to map the relationships between the different data items. Figure 19.6 demonstrates a simple straight-line relationship between suppliers and parts. A company manager can find out who sells a particular product by searching the Parts set and then following the pointers to the Suppliers. This approach is very flexible as the NDS can also consider the combination of Suppliers and Parts as a Purchase set thereby providing two views of the same data.

There are no restrictions governing the way sets can be defined to link record types. This means that the relationships that can be supported by a CODASYL DBMS enable a full network structure to be defined and accessed. The logical data model can, if required, be mapped directly onto the physical database design without change.

The rules for converting the logical design into a CODASYL design are very simple, although there are a number of rules concerning placement of RECORDs. Initially, all

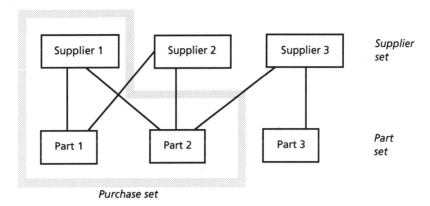

Fig. 19.6 Network database model

data groups and operational masters become RECORDs. RECORDs may be placed in one of two ways:

    CALC places the RECORD on the disk by applying a hashing algorithm – providing rapid full-key access – to the key of the RECORD to give a physical disk address at which the RECORD is stored.

    VIA places the RECORD as near as possible to its master in the specified SET.

All RECORDs at the top of the data structure, which have no masters, and RECORDs on which direct access/entry points are required, are placed by CALC. All other RECORDs must be placed by SET.

    Physical data is described in terms of AREAs and PAGEs. An area is a physical subset of the database. It is usually a constraint of the DBMS that all records of a particular record type must reside in the same area. An area is divided into pages, which is the CODASYL term for a physical block, which means the unit of retrieval from disk. With most CODASYL DBMSs, any given page may contain records of different record types. The division of the database enables areas to be taken off-line when not in use, enabling only affected areas to be recovered after a crash and enabling the designer to control placement of the data. Each area usually resides on a separate physical file.

    The flexibility of the NDS model in showing many-to-many relationships is its greatest strength, although that flexibility can be difficult to code. The interrelationships between the various sets can become very complex and difficult to map out, requiring the programmer to write the application code – usually in COBOL – to navigate the different data chains and pointers. Like HDSs, Network Databases can be very fast, especially if secondary indexes are available to be used to point directly to the physical records. Unlike HDSs, Network Databases go further in reducing the need to hold duplicated data by enabling many-to-many relationships. The Network Model also shares the main disadvantage of the HDS model when structural changes are required. Once the initial structure is created, any changes to the sets require the programmer to create an entirely new structure. Adding new data items is not such a chore as with an HDS, however, since a new set can be created and the various pointers established to generate the required relationships. The 'true' relationship model has, to date, only appeared in the Relational Database, which will be described in the next section.

19.7 RELATIONAL DATABASE SYSTEMS (RDBMS)

In 1968, Dr Ted Codd of IBM produced a paper on the mathematical concept of relational SETs, laying down 12 basic rules which he believed should be applied to such a model. A database system called 'System R' was subsequently released and resulted in an emergence of various relational databases, mainly from American universities. The Relational Database Model (RDM) has been continually refined since these early products were introduced to such an extent that in 1990 Ted Codd produced a new paper outlining 333 rules for relational databases!

19.7.1 Data structure

The RDM abandons the concept of parent–child relationships between different data items, organising the data instead into tabular structures where each field becomes a column in a table and each record becomes a row, as in Figure 19.7.

Figure 19.7 illustrates that the different sets implemented in the Network Model become different tables in the Relational Model and identify the particular data items by the column names. All the customers are grouped into one table, the sales details into another, and so on. Different relationships between the tables are defined through the use of relational set functions in the 4GL and through the definition of foreign keys as secondary indexes.

Tables often contain columns which exist in other tables like the Sales Area in the above example. The actual naming of these columns is not critical in the relational model although it is considered good practice to keep the names consistent with the design entity and

Customer Table

Customer Code	Customer Name	Sales Area	Contact Name
M/1231	Trilbeys	A16	Adams Sam
M/4567	Arndall Electrics	A16	Arndall Mike
M/9876	Adams Systems	S20	Beams Sophie
M/9944	Books Unlimited	A20	Yeates Donald

Sales Table

Salesman Code	Salesman Name	Sales Area
0001	Patter Dan	A16
0002	Spiel Ron	A20
0003	Sellem Stan	S10

Fig. 19.7 Relational database model

attribute names, provided the data in these common columns are of the same domain. Domains are essentially a method for specifying the permissible range of values for a particular column type. The datatype 'Date' could be described as one such domain, provided that all occurrences are of the same format.

There are six special properties of relational tables:

1. *Entries in columns are single-valued.*
   This property implies that columns contain no repeating groups. In other words they have been normalised.
2. *Entries in columns are of the same kind.*
   In relational terms, this means that all values in a column are drawn from the same domain.
3. *Each row is unique.*
   This ensures that no two rows are identical and that there is at least one column – usually the storage or prime key – which uniquely identifies that row.
4. *The sequence of columns is insignificant.*
   There is no hidden meaning implied by the order in which columns are stored. Each user can retrieve the columns in any order, even if this means sorting the entire table first.
5. *The sequence of rows is insignificant.*
   This property is analogous to property 4, where rows can be retrieved in any order.
6. *Each column has a unique name.*
   Columns are referenced by name and not position. Since the data model attributes are mapped to the columns, it also follows that columns should have unique names across the whole database and not just within each table, so if a customer has a Mailing Address and a Delivery Address, both columns should not be called Address.

These properties are important since they make the structure more intuitive for users, easier to validate, and flexible with respect to access requirements.

19.7.2 Data manipulation

There are a number of relational operators which can be performed on relational databases. These operators are intrinsic to the RDBMS and are invoked using part of the relational language known as SQL – Structured Query Language – which uses 12 commands for relational data definition and manipulation. The relational design helps to simplify the language, since in essence a relational database allows any data item to be retrieved from anywhere within it and relationships between data items are often specified at runtime, via the application. 4GLs combined with SQL can currently only be used to access relational databases, since they do not require a subschema in order to access the data – as is the case with most non-relational databases. Some relational products have their own 4GL operators, which are non-SQL and perform similar functions to the relational operators described here but are known by names such as Retrieve, Find, Fetch and Modify.

The product of applying the relational operators to a database results in another relational table, consisting of a set of rows and columns. This is called SET processing as opposed to row-by-row or record-at-a-time processing as required by the previous database types. This data SET is usually retrieved to the application as the result of the query. Set processing can be much more efficient than record processing, both in terms of the amount of code required and in terms of the DBMS workload. Relational operators are unaffected by the way that data is stored in the database. There are eight commonly used relational operators which are implemented in SQL:

1. RESTRICT
   This operation retrieves a subset of rows from a table based on the value(s) in a column or columns. If for example we wanted to retrieve all the Customers in Norwich, this would produce the following table:

   Customer Table

customer_name	location
Trilbeys	Norwich
Green King	Norwich
Arndall Electrics	Norwich

2. PROJECT
   This retrieves a subset of columns from a table, removing any duplicate rows from the result. If we were to project the Location from the Customer table, this would produce:

   Customer Table

location
Ambleside
Bristol
Cambridge
London
Norwich

3. PRODUCT
   This operator produces the set of all rows that are a concatenation of a row from one table with a row from the same or another table. If we were to select the product of the Customer and Sales Order tables, this would produce the following table:

   Product of Customer and Sales Order

cust_code	cust_name	location	ord_num	sales_pers	sal_area
M/1234	Archer	Bristol	S/0083X	Patter D	A16
M/1234	Archer	Bristol	S/0023A	Patter D	A16
M/9944	Books Unlimited	London	S/0023A	Spiel R	A20
M/9944	Books Unlimited	London	S/0023B	Spiel R	A20
M/4567	Arndall Electrics	Norwich	S/0001A	Sellem S	S10
M/4567	Arndall Electrics	Norwich	S/0003F	Sellem S	S10
M/4567	Arndall Electrics	Norwich	S/0009C	Sellem S	S10

   In practice, however, we would further restrict this query to remove duplicates from these columns, that is from: cust_code, cust_name, location, sales_pers and sal_area.

4. JOIN
   This is effectively a combination of PRODUCT and RESTRICT. Rows are concatenated from one or more tables according to conditions specified in the selection criteria or

WHERE clause of the SQL statement or query. For example, to obtain the total commission for each Sales Area in a company, we might use the following SQL query:

SELECT sales_area, SUM(commission)
FROM employee
WHERE sales_area IN
(SELECT sales_area FROM sales)
GROUP BY sales_area

This would result in the following table:

sales_area	commission
A16	20,550.50
A20	15,498.76
C01	5,983.00
S10	154,120.00
Y06	650.50

5. UNION

This operator vertically combines the data in rows of one table with rows in the same or another table, removing duplicates. If we considered a database which, for performance reasons, held a number of tables with the same or similar structure, such as all transactions for each year held in a separate table, then the UNION operator could concatenate all these tables together for further processing, rather than having to JOIN them all for each part of the transaction.

6. INTERSECTION

An intersection combines two or more tables where the values of the specified columns are the same. For example, if we wanted to intersect the Sales table with the Employee table on columns 'surname', 'forename', to find all sales made to employees, this would result in the following table:

Intersection of Employee and Sales Tables

surname	forename
Ainsworth	Barry
Goody	Rita
Patter	Dan
Sellem	Stan

7. DIFFERENCE

This is effectively the converse of INTERSECTION, where the result is all rows which appear in one relation which do not appear in another. The result of performing a difference on the example given above would be a list of all employees who have not purchased anything from the company.

8. DIVISION

This operator results in column values from one table for which there are other matching columns corresponding to every row in another table. For example, if we were to divide the Customer table by the Sales table, the result would be all columns except for the sales_area column, which is common to both tables.

19.8 RDBMS DESIGN

Today's relational databases adhere to the RDM and SQL standards in varying degrees. Although there is an ANSI (American National Standards Institute) standard SQL, most product vendors supply their own SQL extensions to make their product appear more attractive. An example of this is the provision of a single SQL statement which encompasses several standard SQL and 4GL statements to increase efficiency both in terms of coding and execution. Oracle's 'DECODE' function is one such example, which would take many lines of 4GL to accomplish the same result. In practice, however, these extensions can compromise the openness of a product, which may make it less attractive in open systems environments, where portability of both hardware and software is crucial.

The primary goal of the RDM is to preserve data integrity. To be considered truly relational, a DBMS must completely prevent access to the data by any means other than by the DBMS itself. The main advantages of the RDM are:

- Ease of database design from the logical design
- Developer and database administrator productivity
- Flexibility – particularly for changing data structures and code as the business changes.

RDMs enhance developer and end-user productivity in providing set-oriented access to data, independent of the underlying storage structures. The SQL language enables some consistency across products, enabling applications and designs to be ported – with some customisation – between different products. This ability is known as *database vendor independence*.

Relational products show greatest variation in the underlying mechanism used for data storage and retrieval. Although these are largely hidden from the SQL programmer and end-user, they have a great impact on the performance of the system. Some RDBMS products fail to adhere to all properties of the RDM and are often less effective in meeting the business requirements of an organisation. Designers and developers must therefore make up for product shortcomings by building customised support into the database and/or the application, which adds to development time, increases the maintenance overhead and may adversely affect performance.

It is essential to follow a design methodology in building any database system, to ensure that the design is carried out using the prescribed rules and steps for the specific database being used so that the designer constructs a robust but flexible solution. If a design methodology is not followed, then the consequences are usually that the design does not satisfy the functional and performance requirements and that the best use of the specific database is not being realised. In practice, database design is an iterative process, which begins with an initial mapping of entities to tables, attributes to columns and identification of keys. This is called the 'first-cut' design. The design is then refined to ensure that all functional and performance requirements can be met. Design methodologies specify a number of design 'rules' which describe how the data-driven approach to relational design is achieved.

Although design methodologies often specify more than a dozen rules, there are seven basic rules to achieving a relational database design, some of which constitute denormalising the database design to optimise performance. These basic rules are described briefly below.

1. Translate the logical data structure	Initially the process of converting logical to physical occurs independently from any access path requirements, transaction volumes or security requirements. This means that no attempt is made to optimise performance for the first-cut design. The steps entail identification of TABLES, COLUMNS, PRIMARY KEYS and SECONDARY INDEXES, followed by choosing appropriate storage structures.
2. Translate the logical data integrity	This process entails the enforcement of business rules and integrity constraints. The steps entail designing for business rules about entities and relationships, and for additional business rules about attributes. This may entail designing special segments of 4GL and SQL or 'database procedures' into the database itself to enforce the business rules and integrity requirements if the particular RDBMS supports these procedures.
3. Tune for access requirements	Tuning is crucial to any database design and can make the difference between a successful and an unsuccessful database implementation. Tuning techniques vary according to the facilities provided by the RDBMS and include the enabling of table 'scans' for querying, data 'clustering' techniques for efficient access, key 'hashing' for optimising record access and adding indexes to tables to optimise retrievals. In particular, a novice to relational design should place emphasis upon 'prototyping' different access techniques using a particular RDBMS to gain knowledge of the product's strengths and weaknesses.
4. Tune by adding secondary indexes	Secondary indexes are optional access structures that can complement scanning, clustering and hashing to significantly increase the ratio of rows returned by a query to the number of rows searched. They achieve this by enabling direct access to individual rows, by reducing the number of rows searched, by avoiding sorts on the rows searched and by eliminating table scans altogether.
5. Tune by introducing controlled redundancy	This entails altering the database structure to accommodate functional and performance requirements. The caveats of such changes are that they deviate from the logical design and often add complexity or detract from flexibility. Controlled redundancy is often implemented by adding columns from other tables to facilitate data retrieval by avoiding table joins. Adding duplicate data does offer distinct advantages to the business however, especially where response time requirements are short and where the speed of retrieving all the relevant data items may be critical to the success of the business.
6. Tune by redefining the database structure	This process involves redefining columns and tables to optimise the performance for queries and updates. For example, long text columns which are seldom referenced may be split off from the main table into a lookup table. This has the effect of reducing the size of the main table,

	enabling many more rows to be retrieved in a physical I/O operation. In some cases, tables may be split horizontally such as by date, vertically such as by all regularly-used columns and all seldom-used columns, or they may be combined with other tables such as Customer and Account to become Customer Account.
7. Tune for special circumstances	This last stage deals with special features rather than design steps, but is considered equally important in database design. This stage includes the provision for end-user ad-hoc requirements, implementation of security features and tuning for very large databases.

19.9 FUTURES

Despite the shortcomings of the database types covered in this chapter, the Relational Model is the most technically advanced and flexible database theory known today, even though it may be unable to meet *all* data management requirements. Many organisations currently consider the most important of these drawbacks to be support for on-line transaction processing, where the response times achieved by the hierarchical and network models are still favoured in very large systems. Even though RDBMSs are capable of storing and managing many gigabytes of data, the online transaction performance is still relatively slow. With the advent of more powerful machines, however, this performance dividing line between the network model and the relational model is likely to disappear.

Database technology seems to have lost the clear direction it had in the 1980s, as new influences such as multimedia and object orientation have emerged. Many newer RDBMSs contain significant extensions to the RDM, including features such as:

- Support for large unstructured data types like voice and graphics
- Support for triggers and database procedures to implement business rules and referential integrity in the database
- Facilities for handling real-time events, text processing extensions and advanced array handling.

RDBMS vendors are currently involved in new initiatives to increase the 'openness' of their products such that the database server can be accessed by any third-party tool including:

- MIS – Management Information Systems
- EIS – Executive Information Systems
- GIS – Graphical Information Systems
- Statistical analysis tools
- Spreadsheets
- Multimedia
- Other database vendors' application development tools.

These features make the RDM approach more attractive for complex applications and increase the range for which RDBMSs are suitable. The extensions to the RDM, which many vendors are now implementing, are crucial to the applicability of RDBMSs in large on-line client-server database applications. With the arrival of fully functional, distributed

RDBMSs, faster servers and networks, RDBMS implementations are likely to be able to solve many of the operational/geographic location problems experienced by organisations with large databases today. Many RDBMSs are being extended to incorporate object-oriented functionality. The American National Standards Institute is developing a new SQL standard (SQL/2) to include object-oriented features. It appears that object-orientation will be incorporated into RDBMSs in an evolutionary manner, although most RDBMS vendors would not consider referring to their product as an 'OODBMS' when 'Relational' has become so widely accepted and marketable.

CHAPTER 20

Systems Design: *Physical Data Design*

20.1 INTRODUCTION

There can be a big difference between a logical data design and the data model that is eventually implemented in the physical system. This is because the logical data model is constructed with no reference to physical constraints. It might therefore seem that the time and effort spent in logical modelling is largely wasted but this is not so. Logical data modelling provides the analysts and designers with a clear understanding of what data is important to an organisation and how that data is used to support the business needs. It does this by working through the first three steps in the model in figure 20.1.

The longer the analysts can keep physical constraints out of their analysis, the freer the client is to make an informed choice when the time comes and, if the system has to be ported to a different hardware or software platform, the designers don't need to rework the logical design. In this chapter we focus on the last step of figure 20.1, modelling the data for the required physical system.

Structured methods produce a high-level view of the data, in the shape of the physical data model, and show clearly how this differs from the logical data model. Throughout the chapter we assume that the physical implementation will be either a database or a number of files and only where there are different implications will we specify one or the other.

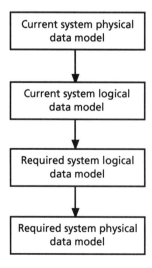

Fig. 20.1 Steps in data modelling

The first step is to explain how the initial physical data model is built and then to explore the factors that help designers tune the model to the best fit for the given hardware and the client's needs.

Before physical data modelling can begin, there are three main issues to be investigated and resolved. These are

- Quantifying the data in order to assess storage requirements
- Resolving any difficulties regarding response time or performance
- Collecting information on the hardware and software platforms to be used.

We will begin by looking at each of these in turn, before going through the process of transforming the logical data model to its physical representation.

20.2 QUANTIFYING THE DATA STORAGE REQUIREMENTS

Typically, an assessment of the volume of data to be stored and transferred will have been made before the hardware was chosen. The required system logical data model has given us a picture of how the data should be structured to serve the organisation best, but even the best data structure does not help the designer know how much storage space the data will require. Discovering the volume of data demands a different approach.

The current system is the best place to start. For example the organisation may currently hold details of 50,000 customers. In the past 3 years the customer base has grown by around 6% each year, but the organisation is now planning some major investment and expects its customer base to grow by up to 10% each year for the next 10 years. So using hard facts about past growth and the present, coupled with strategic forecasts, allows the analyst to produce initial storage volume estimates. If no such figures are available, an alternative source of data from within the current system are the existing data files and archives. They provide the same type of information, but without a forecast of future growth. It is important not to rely solely on the user's estimates without validating them in some way. This is because each user asked may volunteer a different estimate. The best way to check user information is to carry out a limited manual file check. This means duplicating effort, but the consequences of not doing so could be that memory space required by the new system is seriously underestimated.

These initial data volume estimates are at a high level because the type of question we are asking the users is 'how many customers do you have?' To decide on how much storage space is actually required we have to ask an additional question: 'How many bytes of data are held about each customer?' It would be inappropriate to ask users to answer that question partly because it's not their job to know about bytes, and partly because they can only answer based on the current system. We are interested in the required system, which will be different either because new data will be stored or because duplicated data will no longer be included, and so we'll need to refer to the data dictionary.

From the data dictionary we can get information about the size in bytes of each data item in the record. We can also get information about 'overheads', which are those bytes used by the system to manage the data, such as, for example, to record header information or the pointers that maintain data relationships. Records also usually contain unused bytes. Indeed it is recommended that between 20% and 40% of the bytes in each record are unused. This is to allow data items to be added to a record, or the size of data items to increase, with little or no impact on the data management.

Let's take 'customer' as an example. We know what data items are associated with the 'customer' entity, so we can add these together, multiply this total by the number of times the entity occurs, thus giving us a total for 'customer'. This is shown in figure 20.2.

The total in figure 20.2 is of course only for the current number of customers, so an allowance must be made for expected growth. If this is a file system the designer will then calculate how many blocks are required to store the customer data. The designer carries out this exercise for each entity in turn to build an estimate of how many bytes are required to store all the required data.

For designers using relational database systems, the system may do some of this work for you. For example, ORACLE provides a tool which, when supplied with the raw data, calculates the size of the database automatically.

20.3 ASSESSING THE REQUIRED SYSTEM PERFORMANCE

Having established the size of the system, the designers must now work with the users to define performance obectives, especially for critical functions. In the past, systems were developed and installed and only during commissioning were performance problems discovered: users had to wait too long for a system response, or the system crashed because it was unable to handle the number of users it was designed for. To some extent this still happens. A new stock trading system introduced at the London Stock Exchange

Data item name	Number of bytes
Customer name	30
Customer type	1
Customer billing address	100
Customer mailing address	100
Customer phone number	4
Contact name	30
Sales area	1
Account balance	4
Total bytes per customer	270
Add to this the data management overheads eg:	
Record header information	8
Pointers to other records	4
Empty space (for growth)	118
	400

Therefore:
$$\text{Quantity of customer data} = \text{no. of customers} \times \text{bytes/customer}$$
$$= 50{,}000 \times 400$$
$$= 20{,}000{,}000 \text{ bytes}$$

Fig. 20.2 How data is quantified

in 1987 crashed within minutes of starting up because the system could not handle the number of users who logged on. A London Ambulance Service system installed in 1993 also failed because it could not handle the number of requests for ambulances. Such mistakes are costly and, to avoid replicating these errors, analysts and designers not only have to define target system performance, but must ensure that there are few if any performance issues left unresolved when program and data specification begins. During the analysis phase, each process in the required system will have been discussed with the users, and target performance figures will have been agreed for the critical processes. Performance will be defined in terms of transaction processing time, or throughput for a volume of data, such as 100 customer orders per hour. Some of these targets will be non-negotiable requirements, but others will have a defined range within which performance is acceptable. You can reduce processing time by making the code faster using lines of assembler code embedded in a 3GL or 4GL program; the only way to make significant improvements is to make the data handling more efficient. To do this you must first investigate the likely demands on the new system.

When defining performance targets developers must always ask 'what is the consequence of not meeting this target?' If the consequence of not processing data from a sensor in a chemical plant within 0.02 seconds is that we risk a dangerous chemical reaction, this performance target is non-negotiable and indeed we would aim to err on the safe side. If a user waits 6 seconds instead of 5 for a system response the user may not notice. It is often because designers are unsure of acceptable system performance, or because unrealistic targets are set, that over-engineering occurs. So make sure that the targets are realistic and clearly understood.

20.3.1 Factors affecting system performance

There are a number of factors that will have an effect on the system's performance. A system will perform differently under different conditions or loads. System loading is influenced by

- the quantity of data stored
- the number of users logged on and
- the number of peripheral devices active.

Analysing their combined effect is probably not possible. What we are aiming for here is to ensure that we have quantified the effect of each, identified any target performance figures which risk not being met and taken action to overcome the potential problem.

Asking the users for initial estimates of data volumes is the best place to start discussing system performance because it involves the users in the process and encourages them to think more about the functionality they want. Let's assume the user says all data is to stay on the system and not be archived. The analyst can then illustrate the impact this decision will have on performance say after six months, after 2 years and after 5 years. Once the implications are understood the client is able to make an informed decision. One issue that can affect performance in terms of system loading is a disparity in estimates. Let's say we ask 'how many orders are processed per day?' The manager may give one figure which is the average number of confirmed orders per day. The sales team may give a higher figure which is the average plus the number of incomplete orders – orders started, but not confirmed. Although we don't have to be concerned about allocating storage space for these non-confirmed orders, we may need to know this information to fully assess system loading.

Another view which might be of interest is cyclic behaviour. By investigating management reports and files, the analyst might discover that the average daily number of orders triples in early December. This again has no impact on the quantity of data stored, but will have a significant impact on system performance at peak periods. Therefore, while looking for a definitive estimate for data quantity, the analyst must always record these other issues which might be mentioned in passing by users because if they are overlooked they will cause problems later, and because they each have consequences for the final system.

Response times are often affected by the number of users logged on. System users may notice that at the start of the day the system seems slower than usual because everyone has just logged on at their terminal. Or it may be slow between 9am and 5pm, and after most users have gone home performance improves significantly. It is not the fact that a user has logged on that slows the system down, it is that they are using the processor, and the scheduler has to assign CPU space and time to their tasks, and swap the tasks in and out of the CPU. Similarly, if the system has a number of sensors, all of which may be inputting data at the same time, this could slow down other response times. It is usually only event-driven input devices which have this effect; output devices such as printers usually have a lower priority and therefore less impact on performance.

20.3.2 Overheads that adversely affect system performance

The data handling mechanisms will affect the speed at which data is transferred.

Factors that are exclusive to database systems are those that depend on how the DBMS handles:

- Overheads for page management
- Overheads for pointers, indices and hashing
- Data type handling
- Data compaction.

For file systems the factors are similar:

- The time taken to locate the required block of data
- The read time which will be longer if the required block is stored in an overflow area
- Write time
- Page sizes and buffer sizes
- The time taken to swap tasks in and out of the CPU
- How much context data will be saved.

The context data of a task might simply be the values held in registers, the program counter, the program status word and the stack pointer which only amounts to a few words of data, but if the system has to carry out space management such as swapping tasks out and locating free space to assign to the incoming task, the overhead will be significant. Badly designed systems can spend more time swapping tasks in and out than actually processing the applications programs.

Other overheads result from the need to satisfy the requirements for restart and recovery of the system, for data integrity and for maintaining audit trails. For restart and recovery the system may write all, or part of, the database to another device at given time intervals, or record significant events. Data integrity may require that the system buffers a number of related data entries until the transaction is complete before committing the whole transaction to the database. This is known as *commit/backout buffering* because if at any time the user exits the transaction leaving it incomplete, the database remains unchanged. The half-complete transaction is discarded. Legal or quality registration audit trails may impose the need for transaction logging where every action carried out by a user is logged to a file which auditors can search through to check that procedures are followed correctly.

20.4 INVESTIGATING THE CHOSEN HARDWARE/SOFTWARE PLATFORM

Once the required system performance is understood, the designer will need to investigate the chosen hardware and software platform on which the new system will run. Software, in this instance, refers to the DBMS or manufacturer-supplied data handling mechanisms. Designers must accept the limitations, and exploit some of the opportunities of the chosen platform, all the time remembering that a design that is too intricate or harware-dependent may quickly become unmaintainable. The objectives of physical data design are to minimise

- Storage space
- Runtime processor usage
- Access times
- Development effort
- The need to reorganise data when modifying it... and

at the same time achieving a simple user interface! Some of these objectives may appear contradictory, for example the designer might pack data to minimise storage space but this adds to runtime processor usage because the data has to be unpacked on retrieval, and packed again before restoring.

In investigating the physical environment, the designer must ask three questions:

- How much data can the system store and in what way?
- How fast does the system transfer data?
- How is data handling affected by the programming language used?

We will consider each of these three areas.

20.4.1 Data storage

In most cases the hardware will probably be chosen after an initial estimate of the CPU space required to store programs and code has been made but for some systems the target hardware will have been identified *before* the system requirements have been analysed. This may happen where an organisation already owns a mainframe and expects new systems for all business areas within the organisation to run on the mainframe. In such a case the DP department might inform the users of the new system how much CPU space and processor runtime they can have, instead of enquiring how much they need.

Data storage is also influenced by the operating system to be used. The number of megabytes of memory is known at the time of purchase, but the designers will also need to investigate how the operating system handles memory. The features that are important here are whether the system can handle more than one block or page size, whether related data entities can be stored together, and how the system implements relationships between data entities.

The block or page size implemented in the final system depends on the operating system or DBMS used, the size of the most commonly used groups, and the CPU space the blocks or pages take up. The designer must identify which block sizes the operating system naturally handles, or which page sizes the DBMS naturally handles. Some systems are only designed to handle one size whereas others will allow a number of possible sizes, but will read or write their natural size significantly faster than any other. Often in relational database systems the DBMS defines the page size and the database administrator has to work with that as a constraint. Even if a system can be configured to use a different block or page size, this will have a cost in terms of disk management and read/write overheads. So while it may be possible to select a different block or page size, this will require more processing effort to transfer data, which will slow the system down. The designer must therefore decide whether the convenience of using a different size justifies the impact on system performance.

One way of improving performance is to group related data entities, and to store them physically close to each other. Related data could be a collection of table items or data records, and storing these close to each other maintains the relationships between them. Both DBMSs and file handlers allow related items to be grouped in the same page or block but may be restrictive, and may not allow you to store different record types together. The designer has to decide which of the available mechanisms to use. The most convenient block or page size for a developer is one which is large enough to store the largest group. If the system's natural size is smaller than this, the group must be split into two or more groups. However if two data entities have a high interdependence, such as 'customer order' and 'order line detail', we would try to keep these together. This is because separating these would result in two accesses each time we processed a customer order, which would increase the processing time.

Another reason for storing related entities together as a physical group is to improve performance by maintaining the access paths. For example if you store master and detail entities together such as 'customer order' and 'order line detail' you access the root entity directly by its key. This also supports the primary relationship between master and detail entities. All systems represent relationships using logical or physical keys. A physical key points to the physical memory area where the related data is stored. A logical key is known as a *symbolic pointer*; it doesn't give a memory address, and it is interpreted in a predefined way. It will be handled sequentially, using binary search, hashing, indexing or any other indexed sequential access method. Wherever possible we would implement logical pointers rather than physical ones because although using physical pointers is faster, it removes data independence. A system with data independence can change hardware, data handling software or applications code without affecting the data design. For this reason data independence is fundamental to system design; it allows us to isolate data from changes to the physical environment.

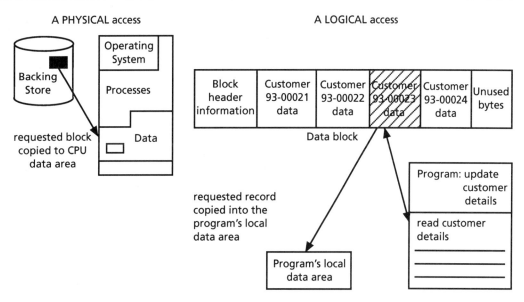

Fig. 20.3 The difference between a physical and logical read

20.4.2 Data transfer

During each read or write, the operating system's data handling mechanisms perform a physical data access. The unit of physical access is a block and this is what the hardware transfers. The unit of logical access is a record and this is the quantity of data transferred by each software access. There will usually be one or more records per block. For DBMS systems the unit of physical access is a page, and the unit of logical access is a table. Figure 20.3 shows the difference between a physical read and a logical read for file systems.

The operating system buffers any records that are resident in the block but are not yet required by the program. So if we want to change the credit limit for customer number 9300023, we first need to read the customer's details into a space in memory where we can access it. The program is unaware that a number of other customers' details have been retrieved. Customer 9300023's details are stored locally by the program while it updates the credit limit field.

The difficulty in estimating how quickly the chosen system will perform a specified task is that different manufacturers measure performance differently. They may base their timing estimates on the assumption that you are using their standard defined block or page size, or their brand of peripheral devices, or a prescribed database, or a stated communications standard. Their estimates cannot take account of the impact of any third party hardware or software or of data transfer between devices. All of this makes the task of estimating harder, and of course the reason manufacturers don't use the same benchmark is that they can then present their products in the best possible light.

20.4.3 The programming language used

The programming language has an impact on performance because different languages will take different lengths of time to perform the same instruction. It may be that one

3GL handles multiplication quickly but swaps between tasks too slowly, and a different 3GL is better at both but is not acceptable to the organisation. Let's illustrate the difference a language can make. If we want to add two numbers together and we are using a 4GL, it will access a code template. The code template may use floating point arithmetic, which is more complex than fixed point arithmetic. This involves carrying out more instructions and more complex instructions than fixed point arithmetic. If we wrote the code in assembler we could speed this up. This addition is a very simple operation, and probably not one which we would need to improve. However by identifying the operations that are inefficiently handled by the chosen programming language, we can select operations where an embedded 3GL or 2GL could be used. Although assembler languages are increasingly rarely used in commercial systems, they are still appropriate for system communications or real-time systems.

In conclusion then, designers try to keep the physical data design as close to the original logical model as possible because in this form the data is easy to understand, to access and to maintain. But the model does have to change to take account of the hardware platform and the capabilities of the operating system, database and programming language used, working within and around the constraints and exploiting opportunities.

20.5 MOVING FROM LOGICAL TO PHYSICAL DATA DESIGN

Having resolved the hardware and software issues we can now consider creating the physical design. The inputs to the process of creating a physical data design are relevant information about the required system, and about the target hardware/software platform. Earlier we looked at quantifying the volume of data to be stored and the importance of setting performance targets, and we discussed the memory capacity and performance details of the hardware. The output from this process is an optimised data design, so now let's look at what the process involves.

20.5.1 Creating a physical data design

The transformation from the required logical data design to the required physical data design begins with some simple steps. For file systems each entity on the logical model becomes a record type. Each attribute of the entity becomes a data field, and the relationships between entities are maintained by identifying keys which are handled by the applications programs. For databases each entity becomes a table, and each attribute becomes a column. The relationships between entities are maintained by using database pointers, which are always logical pointers. The DBMS itself decides on how these logical pointers will be physically implemented.

At this point the designer decides whether more than one entity – more than one record type or table – is to be stored together in the same block or page. Any entities that have a master–detail, or one-to-many relationship are candidates for being stored together. Figure 20.4 shows a possible entity group for the student enrolment example used in chapter 17.

This grouping uses the fact that 'course code' appears in the key of each of these entities. The entity 'course tutor' could alternatively be grouped together with 'tutor' because 'tutor' appears in both keys. Similarly 'enrolment' could be grouped with 'course' because 'course code' is part of the key to the 'enrolment' entity.

In figure 20.4 optional relationships have been changed to become mandatory ones. Initially all optional relationships are implemented, despite storage overheads. In this way the physical data model still supports the user's logical view of the data, which makes maintenance simpler. Later, after any difficulties in meeting storage or performance requirements have been identified, optional relationships may be reviewed. If the relationship is rarely used, the overhead in maintaining the logical pointer may not be acceptable, so the link may be removed. Users must be involved in such a decision because it will change their view of the data.

For a file system, the designer now divides the available memory area between process and data. Having identified how much memory area, both in the CPU and backing store, is allocated to data, the designer tests whether the volume of data will fit. It is difficult to estimate how much data needs to be held in the CPU at any time, and in practice, most of the data may be in backing store. For example a transaction logging file will be essentially offline, so the only CPU space required will be to hold one transaction before storing it on the file. A factor that complicates this issue is whether data is distributed across a network of CPUs or simply stored on different devices. In this case there will be overheads in maintaining the links between the distributed data areas to allow the users to access the data as if all data areas were resident on the same CPU. The designer chooses a block size large enough to accommodate the largest commonly used group. The constraint on the size chosen is that the hardware and operating system must be able to support it.

This differs for the DBMS designer who has no control over what data is retained in CPU memory, nor on the quantity. The DBMS handles this. This can make for an inflexible

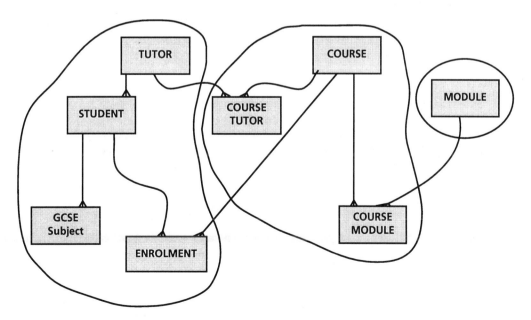

Fig. 20.4 A possible physical group of entities

system, but it means that the DBMS can ensure that if the CPU is in danger of running out of space, it can swap data out so the problem doesn't arise.

20.5.2 Testing against client requirements

It has been suggested earlier in this chapter that the users should have agreed the performance requirements, and so we have to check that the physical data model can support them. It is useful to start with a list of all of these performance targets to be met and cross-reference them, identifying those which may affect other performance targets. The key issue is to ensure that if a user or process requests data, or enters data, the user or process does not have to wait an unacceptably long time before it can continue. To work out how long the data model will take to access the data, the developer uses *logical access maps* (LAMs) in conjunction with the relevant program specification and the hardware manual.

The program specifications are used to identify which path through the code this input or output will follow, how many instructions this represents, and how long each instruction is likely to take. This is a laborious process but becomes faster after a few attempts.

Operating systems behave differently when it comes to data accesses. Some take the same time to perform each access, others can perform the first three or four accesses quickly, but any subsequent access is painfully slow. Whilst hardware manuals will give timings for accessing data, and for subsequent accesses while holding the first data set open, designers should remember that these estimates can be optimistic. To estimate file handling times, the developers may have to estimate the time to carry out the software instructions, as well as the read, write and transfer times. The problem is made worse if this processing is not yet written. It might be possible to use estimates from previous projects, bearing in mind that if the previous project was developed on a different hardware, performance times will differ. Each LAM is taken in turn, and if the number of accesses is within the acceptable limit, say four, then the data model is unchanged. However, if the LAM accesses five entities on the logical model as shown in figure 20.5, the designer has to decide which two entities to merge. Entities are merged to reduce the number of accesses and thereby meet the performance requirements. This introduces redundant data into the system, but it is done to improve speed. We call this *controlled redundancy*.

The data designer has to decide which two entities to merge and the choice is not always obvious. The designer will consider each possible option, and the decision will be based on two factors. Firstly, the designer must identify entities which have a 'natural' link. Secondly, assess how many bytes of redundant data the system is now carrying due to this merging together. If this number can be reduced by merging a different pair of entities, the designer will consider that option. Figure 20.6 shows the stages designers go through when converting the logical model into a physical data structure.

For a system with a large number of inputs and outputs, this practice will be too time consuming. We may decide it is sufficient to carry out this process for only the most important functions in the system – those functions which are used most often. If such a decision is made it should be logged, so that if performance difficulties do arise later, we understand why. It is important to understand the significance of merging data entities.

- It de-normalises the data, which means it's more difficult to see the relationships between entities.

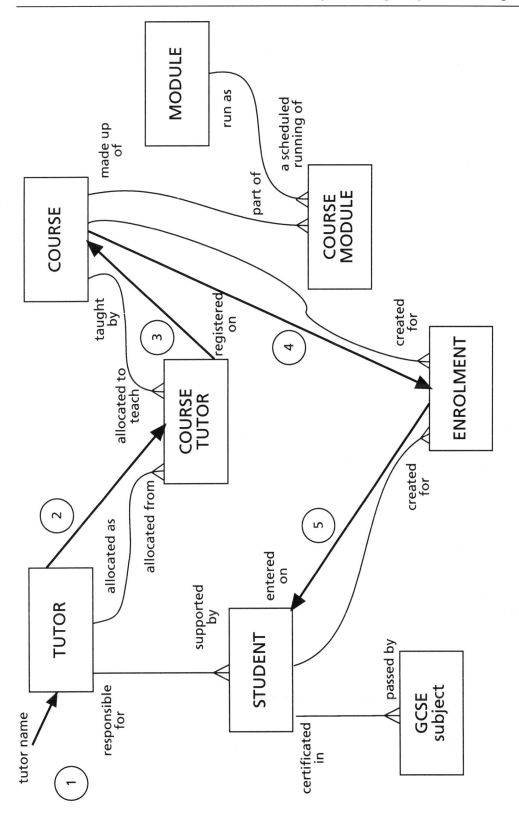

Fig. 20.5 LAM for producing a class list for a given tutor

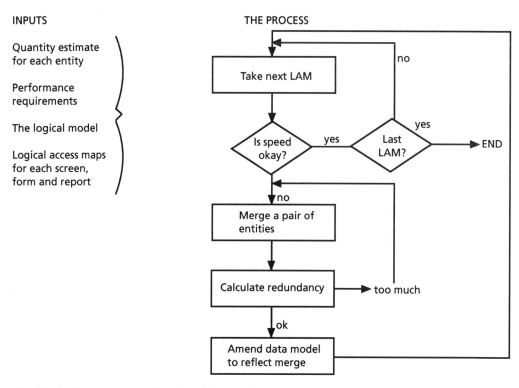

Fig. 20.6 Steps in amending the data model

- Redundant data is stored using more memory area, and perhaps reducing the system's response times.
- Read times are reduced because there is one read instead of two.
- Update times will be increased since there is more data to search through, and a larger quantity of data must be rewritten.

20.5.3 Refining the physical data design

Although optimising one function might reduce processing time for related functions, the main problem in developing a physical data design is that optimising one area can negatively affect another. If the frequently used critical processes are speeded up, and routine batch processes are slowed down, this might not adversely affect the users. However it would not be acceptable to halve the speed of an overnight batch program, causing a significant increase in online response times.

If merging entities still doesn't allow the data to meet its requirements, there are three other possibilities. The first is packing the data, although this may not be possible on DBMS systems. This means storing character strings, which would usually be held one character per byte, as numerics. This means that each program which accesses the character string has to decode the data to read it, and pack it again before restoring it. The second possibility is to create two versions of a file. If an organisation's staff file held 50 fields per employee, yet most programs accessing that file only wanted to read or update 4 fields, by creating a short version and a full version we reduce space usage and access time. Both of these entail risk. What the designer and users must do is consider whether this risk is manageable and cost effective. After all in real terms hardware gets cheaper as time goes on, so a better and safer solution might be to buy more memory. The third option is to change the accessing method; for example if using indices is too slow, try using hashing. Using a different accessing method may reduce transfer times but this is hardware and software dependent.

Prototyping may be a useful technique in checking that the estimates designers are coming up with on paper are close to how the system will actually perform. The best way to do this is to work with the users to identify critical processing. This might be processes that have to handle large quantities of data within tight performance targets, or it could be processes that access several data areas. Performance tests are designed using the maximum volume of live data that these functions should be able to process. It would also be helpful to run such a performance test on a fully loaded system, so that you not only test this function to its limit, but do so when the system is simultaneously coping with the maximum number of users logged on, the maximum number of processes active and the maximum number of peripherals active. If the targets can be met under these conditions, there won't be any performance problems when the system goes live. However, it takes a lot of time and effort to design, code and test the prototypes and simulator programs required by such thorough tests, so prototyping to this degree is still the exception. 4GL environments do allow prototyping of single functions relatively easily, and while this isn't testing the function to its limits, it will help to identify if the function *definitely cannot* meet its performance targets.

Whatever the decisions made the designers must remember to update relevant documentation, such as the system requirement specification or program specifications, to show what decision has been made and why. This is especially true if the designers had to find processing ways around the problem or if performance targets have been reduced.

20.6 SUMMARY

While the physical data model is a progression from the logical data model, it does not replace it. Developers must maintain a logical view of relationships between data because the logical model still explains the client's business functions. If they lose this, they risk building back into the system the uncontrolled redundancy and duplication that they spent so much effort stripping out.

The further the physical model is from the logical model at development time and the more modifications that occur thereafter, the greater the likelihood of reduced or unpredictable system performance. To ensure that the system meets the objectives of physical data design, the designer must focus on

1. Identifying the mandatory and high priority performance requirements
2. Resolving performance problems rather than storage space, if both cannot be met, because hardware gets cheaper
3. Meeting each program's target run times
4. Meeting the system's specified response times

5. Maintaining the user's view of data.

Since one of the purposes of DBMSs is to simplify the task of data design, it removes a lot of the control, and therefore much of the risk of error, from the data designer. Designers of file systems may not take full advantage of the operating system's data handling mechanisms because to do so might not allow this design to be portable to another system. This must be an explicit decision, not left to chance.

CHAPTER 20   CASE STUDY AND EXERCISES

Q1 Convert the System Telecom LDS shown here into an initial physical data design. Consider which primary and secondary key accesses may be required.

Q2 Consider which secondary key accesses may be required in this system.

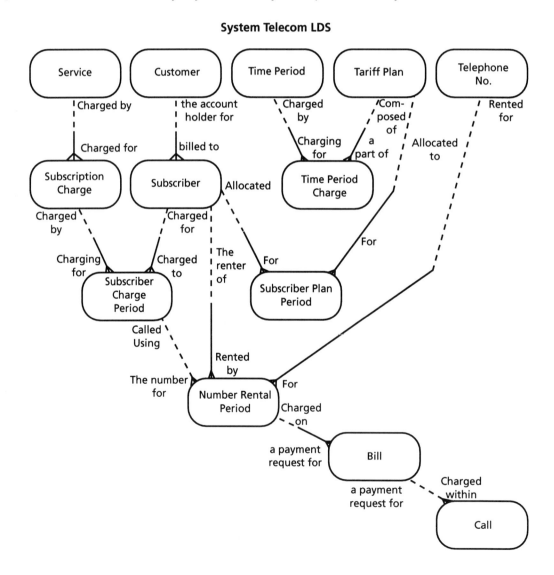

System Telecom LDS

CHAPTER 21

Systems Design: *Program Design*

21.1 INTRODUCTION

This chapter explains how to continue the structured approach to systems development into the detailed program design phase by using stand-alone specification methods, some of which have existed for over thirty years. Before that time there were no formal software systems development methodologies and every specifier worked in their own way. This caused two problems: the specification reviewer had to tune into the way the specification was written, and the progress monitor could not easily assess how close a specification was to completion. Stand-alone specification methodologies guaranteed a consistent look to the final product, the program specifications.

Since most of the errors in software systems are caused by incomplete analysis and design some methodologies end at the point where the system definition becomes detailed file or program specifications. The extent of some popular analysis and design methods is shown in figure 21.1.

	Analysis	Design	Specification	Code
SSADM	X	X	X	
Yourdon		X		
Information Engineering	X	X	X	X
Jackson Systems Design		X	X	

Fig. 21.1 Systems development phases covered by methodologies

The danger is that developers spend a great deal of time and effort stripping out errors and ambiguities up to the point of program specification, only to have them built back into the system by specifiers who have not been trained in any method. Some methodologies such as SSADM extend as far as program and data specification but others, like Information Engineering, go beyond that and provide code generators to turn the specifications into the final system code.

There are many generic approaches to program design which can be used regardless of application. In this chapter we explore the three methods we have found most helpful. These are

- Data action diagrams which are used to document general programs
- Interaction relationship charts to document human computer interaction
- State transition diagrams to express real-time processing.

21.2 WHAT IS MEANT BY PROGRAM DESIGN?

First of all we have to be clear about what is meant by 'program'. There are a number of synonyms for program: process, module, unit, routine, subroutine, procedure, function, segment, fragment and macro. All of these are lines of code with a beginning and an end, they serve one purpose, such as validate customer number, and have a well-defined interface with the rest of the system. Furthermore programs should be of a size that motivates the programmer; too small can feel belittling, too big can be daunting. The size and complexity of the program are what provide the programmer with a challenge and a sense of achievement. A well designed program is one that exhibits high internal cohesion and low dependence on other programs, low 'coupling' as it is known.

So, the following things are important:

- *One function*. Systems are partitioned by function into a hierarchy of programming modules. Each program should carry out only one function. Modularity makes the system easier to maintain.
- *Size*. Right up until the early eighties people developed systems as one big monolithic program. The developer's prime interest might be in exploring some engineering thesis and as their knowledge grew and developed so too did their programs until the challenge became not the engineering problem but how to fit all of the code into the CPU. Some systems written in this way went on to be sold to commercial organisations. However modifying the system to keep pace with changing business needs required the system to be rewritten in a more modularised, and therefore maintainable, form.
- *Cohesion* measures the strength of relations within a program. This means that what happens in one subroutine affects the other subroutines of that program. A program that performs more than one function will have low cohesion. A high degree of cohesion within programs results in lower coupling between programs. This in turn results in a more maintainable system, and therefore a higher quality solution.
- *Coupling* is a measure of the strength of the bonds between programs. Ideally programs should have little dependence on other programs in the system. This is so that any amendments to the program have little or no impact on the other programs in the system.

Systems Design: Program Design

As mentioned in these four attributes, maintainability is an important consideration when designing programs. As the client's business changes, the programs that support that business will be modified. Maintainability in isolation can be interpreted differently by each pogrammer, so it is simpler to see it as an essential by-product of adhering to the principles listed above.

21.3 WHY BREAK THE SYSTEM DOWN INTO PROGRAMS?

Analysts break the system down into functional areas, then further into programs for many reasons. The main reasons are:

- One program should be written by one programmer, so if the system was just one program, it would take one programmer months or years to write.
- Smaller programs are easier to specify, test and modify because errors or the impacts of a change are contained within fewer lines of code.
- Programmers are more motivated to code in the areas of the system that interest them.
- A large number of small programs makes rescheduling the work easier: if one programmer is taking longer than estimated on a program, one of their later programs can be assigned to someone else to write.

Of course breaking the system down brings problems of its own. The more programs there are, the more interfaces there are to be specified and tested. There is always a trade-off between the number of programs and interfaces and the program size and interface complexity. The aim is to find an optimal number of each.

Figure 21.2 shows two diagrams of the same system. The first contains several small programs and therefore several interfaces. Having smaller programs makes each one

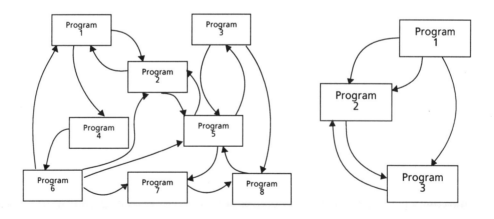

Fig. 21.2 Trade-off between number of programs and interfaces, to program size and interface complexity.

easier to design. However the interface between the programs must be thoroughly thought out otherwise the flow of data and control may be difficult to manage. The second diagram has fewer programs and consequently fewer interfaces. This time the interfaces and the programs will be more complex because each program has more to do.

Decomposing the system into its component parts also helps to identify any commonality which can then be exploited. The reason for doing this before program design even begins is so that the function is designed, coded and tested only once. If for instance every programmer coded their own error reporting, differences would be inevitable and these arbitrary differences would irritate the user.

For a function to become a possible common routine it must occur in several places within the confines of one functional area, or be used several times within one system or be common to more than one system. Alternatively, high-level design decisions may dictate a common approach to activities such as

- error handling
- error reporting
- data area updates
- communications with peripheral devices
- message passing between programs.

Using common routines serves two purposes. Firstly, it ensures consistency. If, for example, one program is called by all other programs to report errors to the user, then all errors are reported in the same way. This in turn gives the user a sense of this being one system and not a system written by several autonomous programmers. Secondly, it provides a framework for programmers to work in; they know what common processing is available so they are free to concentrate on the specific problem of their program and to use the common routines where appropriate. As programmers progress from one program to another, they are often best placed to say which common routines work well and which are too general. They may also propose and develop new common routines.

21.4 WHAT'S IN A GOOD PROGRAM SPECIFICATION?

A program specification is like any other document in that it should be accurate, brief and clear. Accuracy in this sense includes making sure that the document is complete, with each normal case and exception fully covered. Brevity is important since the more words there are, the more likelihood there is of ambiguity and confusion creeping in. Clarity ensures that each aspect of the program is explained in the simplest way.

The specifier also needs to know who the specification is written for. There was a school of thought that said program specifications should be written with the programmer in mind. It suggested that if the programmer had already been identified and was experienced and competent, then some of the detail should be left out of the specification because the programmer wouldn't need it. This idea came about as a result of the age-old argument of creativity versus consistency. Programmers insist that a too-detailed specification stifles their ability to find creative solutions to a problem. The main reason why this view has not prevailed is that specifications are also used by system maintainers and quality auditors. Both of these groups welcome the move towards a consistent approach to documenting program specifications. It also allows the programming team leader to reschedule work more easily.

Processing outlined in any program design belongs to one of three categories: processing sequence, decision or iteration. A processing sequence is a number of instructions that must be programmed in the order defined in the sequence. A decision indicates that there is more than one possible path through a program and that the path taken will depend on whether the decision criteria are met. Iteration indicates that a section within the program will be executed zero, one or more times until the exit condition is met.

21.4.1 Suggested program specification contents

Each program design must be a stand-alone document in so far as it provides the programmer with everything required to write the program. This doesn't mean, for example, that the program specification will contain copies of data dictionary entries for each data item handled by the program. This would not only make the document significantly larger than it need be, but would also introduce the possibility of different versions of data item definitions existing. To avoid duplication, the program design refers to any supporting documentation. Such cross-referencing of program specifications helps analysts and designers to check that all conditions identified in higher-level documentation, like the requirements specification or acceptance test condition list, are covered at the lowest level. It also pin-points exactly where every condition will be coded and tested. It is crucial for the success of implementation and maintenance that these cross-references are kept up to date.

The following is a suggested contents list for a program specification. In some cases it may be appropriate to leave out one or more sections. Screen and print layouts, for example, will only be required by screen and report processing programs.

- **Document details** This consists of header information which uniquely identifies this specification: title, author, reviewer, circulation list, release date, system identifier, program identifier and version number.
- **Introduction** One of the most frequently discussed aspects of program design is the level of explanation it should contain. Here again the argument of stifling creativity arises: the specifier defining the process at too low a level is doing the programmer's job. We suggest using three levels of description within the document. The introduction is the first of these and gives a brief summary of what the program does in business terms. Data referred to in the program overview should be at the data file level, not lower. This is because the overview will be read by users, project managers, team leaders, and maintenance programmers who need a short summary of what the program does.
- **Assumptions and restrictions** This section lists any constraints on the program such as 'normal path processing must complete within 0.2 seconds'. It will also add information which affects the logic of the program. This could be 'the program will be cloned to run on each user's terminal' or 'a maximum of 10 messages will be queued to this program'.
- **Attributes** This section outlines the program's environment and will cover aspects like the hardware it will run on, the operating system used and the programming language to be used. If this information is the same throughout the project, this sec-

tion might simply refer to a higher-level design document containing this information.

- **Data** The input to the program and any output it generates is detailed here. It provides a data-driven view of what the program does. The section also refers to the data dictionary to cover the format and validation rules for all shared data used in the program.
- **Functional description** This intermediate level of process description may not be necessary for small, simple programs but the default is to include it. It provides a 'big picture' view of the program's inputs, outputs and processing before going on to explain the detail. Data in this description is referred to at the record level.
- **Detailed processing** This section expands on the previous description to give a low-level detailed view of the processing paths through the program but continues to focus on *what* has to be done. Data is referred to at the data field level.
- **Errors and exception conditions** This deals with anything other than the normal case, such as events occurring out of sequence. It describes how the program will cope with these, lists any error messages and identifies the destination of each message.
- **Operational considerations** Information from this section may be incorporated into the user manual later. It describes how the operator interacts with this program in the normal case and how the operator can recover to a safe state or restart the program if anything should go wrong.
- **Subroutines** Common routines used by the program are identified as are their input parameters. Input parameters could again simply be referred to if they are defined in a higher-level document. Calls to the system executive (i.e. manufacturer–supplied functions/procedures) are also listed here. The reason for having a separate section on subroutines is so that if a subroutine changes – let's say an extra parameter is added – the modifier need only check this section rather than read through the whole document to find out whether this program must be amended.
- **Messages** This section identifies messages sent and received by the program. It lists the message identifier and its purpose.
- **Print layouts** Detailed print layouts or screen layouts are included here. Sometimes the document only includes a pro forma layout and the details are added by the programmer.

21.5 SOFTWARE CONTROLS

It is not enough to use common routines in the hope of ensuring a robust system. However useful they are, they do not address all the specific needs of each program. This is where the system development policy on building controls into software must be applied. The development project's policy defines how individual errors are to be handled and covers topics such as

- error detection
- error correction
- validation
- use of default values
- fault avoidance

- fault tolerance
- fault masking.

The system's data dictionary, explained in chapter 17, provides the program specifier and the programmer with information such as the validation rules for each data item, and default data values if appropriate. So, the data dictionary can help in error detection and validation but it cannot tell the specifier whether validation of user input should occur after each data field is entered, after a few related data fields are entered, or after a screen full of data has been entered. This is the role of the software controls policy. Validation may also cover error correction where two data items are out of step with each other. The error correction processing would have to identify which of the two fields should be accepted as correct, then set the 'incorrect' field to the expected value. As the processing involved in error correction depends on which two data items are to be checked, this is too specific a task for a common routine and must therefore be defined within the program specification. The policy also outlines the degree of fault tolerance that the system will have. This will specify

- the number of retries to read or write data
- which errors are serious enough to halt the system
- errors which will be reported to the user where the system must wait for user correction to continue
- errors which will be reported to the user but which will not stop the system.

Only by addressing these issues before program design begins in earnest will the developers ensure a robust and reliable system. The consequence of not drawing up such a policy will be a system which handles errors erratically.

21.6 DATA ACTION DIAGRAMS (DADs)

Data action diagrams are based on structured English, so we must begin by defining structured English. Structured English is a program specification method which uses a limited number of reserved English words. The vocabulary contains nouns and simple verbs but no adjectives or adverbs. The reasons for preferring this dramatically pruned version of English is to avoid using English prose which can be ambiguous, inconsistent, incomplete and verbose. The following comparison shows the benefits of using structured English.

Figure 21.3 shows that although in this example the English version of the process is brief and clear, structured English achieves more clarity by using statements written as short positive instructions.

Processing sequences are dealt with by using line after line of short instructions:

    INITIALISE customer number
    CALL validate customer

Decisions are shown as following different paths through the program by using IF, THEN, ELSE, ENDIF. An alternative to this is to use a CASE statement where there are many possible paths and where the decision is not simply true or false, as in the following example:

ENGLISH	STRUCTURED ENGLISH

For INSERTS check that the record does not exist (otherwise fail with message C) and insert in the file.
 	
For AMENDS and DELETES check first that the record does exist (otherwise fail with message D) and that the record is not locked (if it is, output message B and wait before trying again). Then lock the record and for an AMEND, update the file and unlock the record. For a DELETE, delete the record.
	CASENTRY
	  CASE insertion
	    IF record doesn't exist
	    THEN
	      insert
	    ELSE
	      fail – message C
	    ENDIF
	  CASE amendment
	    IF record exists
	    THEN
	      IF record not locked
	      THEN
	        lock
	        update
	        unlock
	      ELSE
	        wait – message B
	      ENDIF
	    ELSE
	      fail – message D
	    ENDIF
	  CASE deletion
	    IF record exists
	    THEN
	
	      etc

Fig. 21.3 English versus Structured English

```
CASENTRY day of the week
   CASE = Monday
      do the washing
   CASE = Tuesday
      do the ironing
   CASE = Wednesday
      do the shopping
   CASE = Thursday
      clean the house
   CASE = Friday
      change the beds
   CASE = Saturday
      do more shopping
   CASE = Sunday
      relax
ENDCASE
```

Iteration is dealt with in structured English by using FOR statements or DO loops such as DOWHILE, DOUNTIL, and are terminated by ENDDO. Once a program enters a loop it will not stop until the exit condition is met. Here are two simple examples:

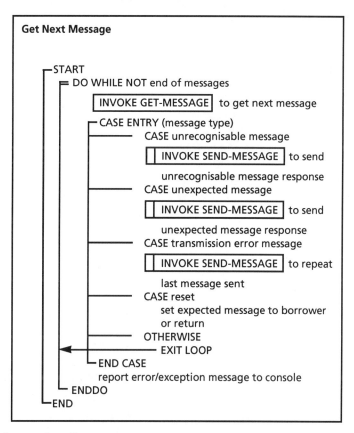

Fig. 21.4 A data action diagram

```
FOR ALL staff
   calculate monthly salary
   print payslip
   set transfer-funds flag
ENDFOR

SET day = Monday
DOWHILE day is not Saturday
   go to work
   SET day = next day
ENDDO
```

In the first example the loop executes until all staff on the staff file have been processed. In the second example the program will go round the loop for each working day, Monday to Friday, but will not execute 'go to work' or increment the day pointer on Saturday.

Structured English is not more tightly defined than this because its aim is to communicate simply with the programmer. Data action diagramming formalises this and identifies clearly the processing loops and all exit points from the program.

In figure 21.4 exits are shown as arrows pointing to the level of processing the program must continue from. In this case the program would continue from the line of code following the ENDO.

The method uses brackets to enclose a processing sequence:

```
┌─ Make tea
│     fill kettle with water
│     power on kettle
│     put teabag in cup
│     wait for kettle to boil
│     pour boiling water into cup
│     stir water and teabag
│     remove teabag
└─    add milk and sugar
```

Where there are a number of paths through the processing, at least one decision has to be made. The condition which has to be met to enter that path is highlighted as follows:

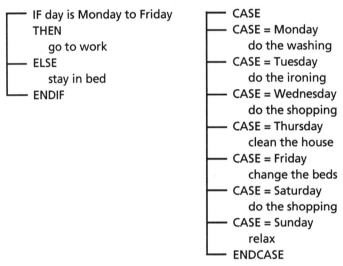

```
┌─ IF day is Monday to Friday
│  THEN
│      go to work
├─ ELSE
│      stay in bed
└─ ENDIF
```

```
┌─ CASE
├─ CASE = Monday
│      do the washing
├─ CASE = Tuesday
│      do the ironing
├─ CASE = Wednesday
│      do the shopping
├─ CASE = Thursday
│      clean the house
├─ CASE = Friday
│      change the beds
├─ CASE = Saturday
│      do the shopping
├─ CASE = Sunday
│      relax
└─ ENDCASE
```

Data action diagrams differentiate between processing sequences and iterative processing loops by clearly marking the exit point of each loop with a double bar at the top or bottom of the bracket:

```
╒═ FOR ALL staff
│    calculate monthly salary
│    print payslip
│    set transfer funds flag
└─ ENDFOR

╒═ DOWHILE day is not Saturday
│    go to work
│    increment day
└─ END DO WHILE

┌─ DOUNTIL
│    go to work
│    increment day
╘═ UNTIL day = Friday
```

The double bar on the bracket clearly shows that in the first two examples, if the condition is true at the start, the processing inside the FOR and DOWHILE loops will not be executed. However even if the condition is true at the start of the DOUNTIL loop, the processing will be carried out because the condition is not tested until the end of the loop.

21.6.1 Levels of data action diagrams

Although a low level of detail was used in the examples above to help explain how data action diagrams works, their real value is that they can be produced at any level of detail whether it is at system context diagram level, at the pre-coding level, or any level in between. To show what we mean by that, let's expand on one of the above examples.

```
FOR ALL staff
    calculate monthly salary
    print payslip
    set transfer funds flag
ENDFOR
```

Each instruction in this processing sequence begins with a verb which tells the programmer what the program has to do next. The instructions are themselves at different levels: there is a lot more involved in 'calculate monthly salary' than in 'set transfer-funds flag'. It may be that 'calculate monthly salary' is a separate program, or could be separated out into a subroutine either within this program or for common use throughout the system. DADs call these separate programs 'procedures' and identify them by enclosing the procedure name in a soft box:

```
FOR ALL staff
    ( calculate monthly salary )
    print payslip
    set transfer funds flag
ENDFOR
```

Very high-level DADs might consist only of several procedures enclosed by the bracket structure:

```
FOR ALL staff
    ( calculate monthly salary )
    ( update payroll file )
    ( update company accounts )
    ( transfer funds to staff bank accounts )
ENDFOR
```

Common procedures which are called more than once in the same DAD are shown as being repeated by a vertical line drawn inside the lefthand side of the box:

```
| report error to user |    | validate user entry |
```

Because DADs can describe system behaviour at different levels of detail they can be used to document systems functions at the analysis and design levels as well as the detailed program design. Having one approach that remains the same all the way through from analysis to implementation helps ensure consistency and completeness. At the lower level, DADs can be written in such a way that they can be used as input to code generators although this would require some minor customisation of the syntax described here to correspond to the syntax defined for the code generator.

21.7 INTERACTION RELATIONSHIP CHARTS (IRCs)

These model the behaviour of the system responding to a user. They are a graphical representation of the dialogue between the user and the system. For this reason it is our preferred method for specifying on-line processes.

There are only two constructs in an interaction relationship chart. These are a rectangular box displaying the system's response to a user action, and lines connecting the boxes such that there is only ever one entry line into a box, but there are as many exit lines from the box as there are valid user responses. To show how these might help to specify a program, we'll take the everyday example of a cash machine, an on-line process we use regularly. So that anyone reading an IR chart understands that this is a dialogue, the boxes should be labelled in conversational style. In figure 21.5 we have used quotation marks around the actual words that will be displayed on the screen. Another possibility is to end the label with a question mark to indicate that a response is required, for example "enter PIN", or simply "PIN?" The key things are to make sure that all IR charts in the system use the same labelling style, and that it reads like a conversation.

The cash machine is a fairly simple example and yet the diagram is quite complex. The label inside each box is the response to the user, or a question or an instruction from the system. The labels on the lines are the response the user has selected. The system defines the valid responses and the IR chart shows how these are handled. Invalid responses are not shown on an IR chart for a number of reasons. Firstly it is assumed that errors are handled consistently throughout the system so the IR chart for error handling will appear in the program specification for the common routine that is used to validate or report errors. Secondly the diagram would become needlessly complex, and thirdly what the programmer is most interested in is normal-case. If errors were to be handled differently in this program than in others then it would be appropriate to show it on the diagram but in on-line processes errors usually return the user to the start of the loop.

IR charts should be no bigger than an A4 page. Where there is a large number of possible user responses, layering or levelling can be used. Layering is where one A4 page models the behaviour of one route through the dialogue, so there are as many A4 pages as there are routes through the dialogue. In figure 21.5, for example, only one path is modelled through the processing. A, B and C represent the other three layers: one page showing Cash No Reciept, one page modelling Statement and another modelling Cheque Book. Levelling would show a high-level view on one A4 page, and model the detailed processing on later pages.

Each box on the chart represents one logical screen. One physical screen may be made up of several logical screens. So the screen identifier may not change but the information

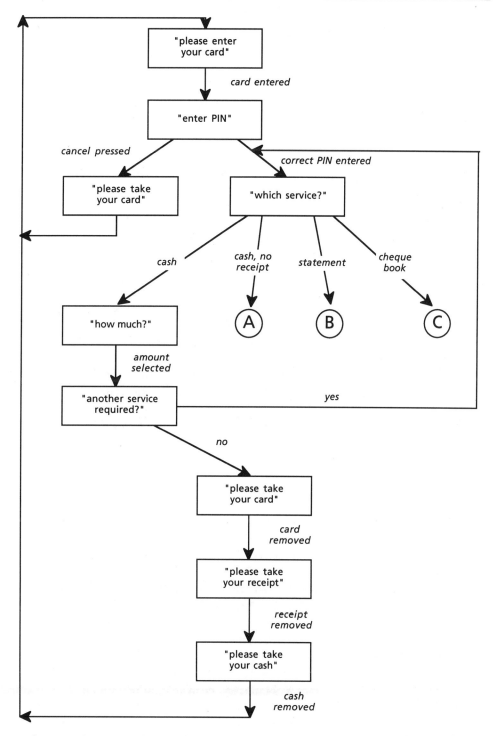

The annotations A, B and C identify other processing paths, the IR charts for which would be drawn on another page.

Fig. 21.5 An interaction relationship chart for a cash machine

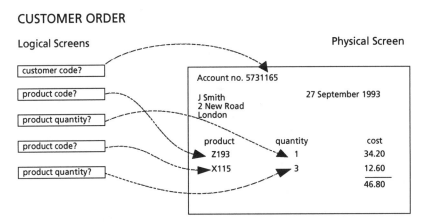

Fig. 21.6 A physical screen consisting of a number of logical screens

on the screen is built up as the system prompts the user for data. This is what typically happens with data entry screens. If a telesales person is entering a customer order the screen does not change, but field by field the sales person enters the details of goods ordered. The physical screen may be called Customer Order Details, which could consist of a number of logical screens such as Product code, Size, Quantity, etc. From this we can see that a physical screen can be one or more records, and a logical screen is one field (figure 21.6).

Interaction relationship charts are our preferred method for expressing the design of human computer interaction processes because unlike flowcharts the designer can use a level of detail which is appropriate. A programmer working from a specification which covers page after page with low-level interaction detail may lose sight of the aims of the program as a whole.

21.8 STATE TRANSITION DIAGRAMS (STDs)

These diagrams are used to model the behaviour of a finite state machine. A finite state machine is any system or machine that can be shown to be in one of a number of discrete, predictable states. Let's expand on the example of a video recorder we looked at in chapter 16: it may be ON, OFF, ON TIMER CONTROL, STANDING BY, RECORDING, PLAYING, REWINDING, FAST FORWARDING or IN ERROR. These are its known states, They each exist for a period of time. The VCR cannot be in any other state, it cannot for example be EDITING. As well as predicting what state the machine is currently in, we can list all the events that will have an effect on the machine and what state the system will move into if that event occurs.

Typically any 'machine' that can be mass produced is by definition a finite state machine. If it can be replicated, its behaviour must be predictable. A human being is a good example of a non-finite state machine. A person may be in an infinite number of states during their lifetime, or even during a day. The events that may happen to them are not predictable, neither can we predict how a person will react to a specific event, and we cannot assume that two people will react in the same way to the same event.

There are only two constructs to state transition diagrams. These are rectangular boxes labelled with the name of the state, and lines connecting the states to each other labelled

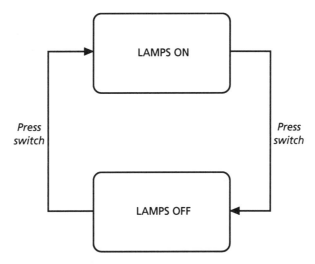

Fig. 21.7 State transition diagram for a simple finite state machine

with the name of the event. A lightbulb is a simple finite state machine as it has only two states – ON and OFF – and only one event affects the system. Pressing a switch changes the state. If the state was ON pressing the switch changes the state to OFF. If the state was OFF pressing the switch changes the state to ON.

Of course for most finite state machines the state transition diagram becomes much more complex. Let's take a closer look at the behaviour of the video cassette recorder as shown in figure 21.8.

Since the aim of any diagramming tool is to communicate as clearly and concisely as possible, it is essential to strip out any unnecessary complexity. To give a simple illustration of how STDs can become over-complicated let's expand on the lightbulb example.

A warehouse consists of three storage aisles. To save electricity the lights in the aisles are usually OFF, unless movement is detected in the aisle which switches the light ON and also sets a 'timeout'. If during that timeout period movement is again detected, the timeout period is reset. So for as long as a person is working in an aisle the light stays ON because the timeout is continually being reset. It is possible to model the behaviour of the warehouse lighting as shown in figure 21.9.

While this is a true model of the possible behaviour of the warehouse lighting system, it is overly complex since it suggests interdependencies between the behaviour of the 3 lamps which do not exist in practice. It makes no difference to the state of the light in aisle 3 whether the lights in aisles 1 and 2 are ON or OFF. A more helpful STD for this example would be as shown in figure 21.10.

This provides a far simpler model of the lights' behaviour. Another way to reduce the complexity of STDs is to show levels of detail in a similar way to DADs and IR charts. To show how this works let's look again at the video recorder.

Levelling the situation in figure 21.8 to hide complexity results in two STDs. One gives an overview of the machine's behaviour and one explores the detail for the state PLAYING (figure 21.11). Very high-level states or overview are called 'modes'. An example of a mode or high-level state for the VCR is ON. When the VCR is ON it could be in one of a

350 Systems Analysis and Design

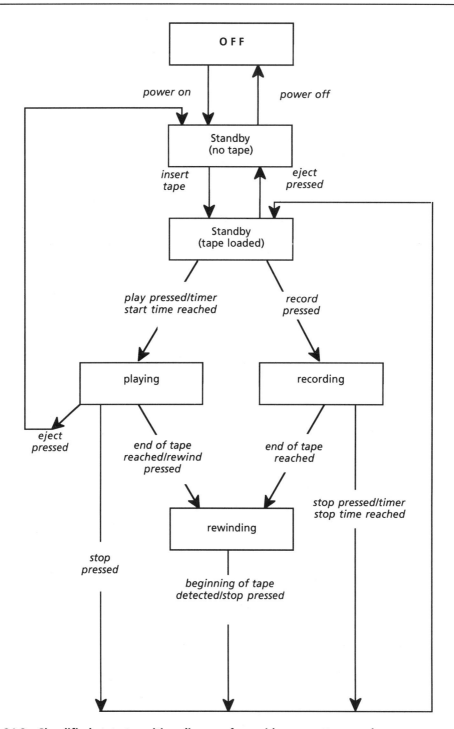

Fig. 21.8 Simplified state transition diagram for a video cassette recorder

Systems Design: Program Design 351

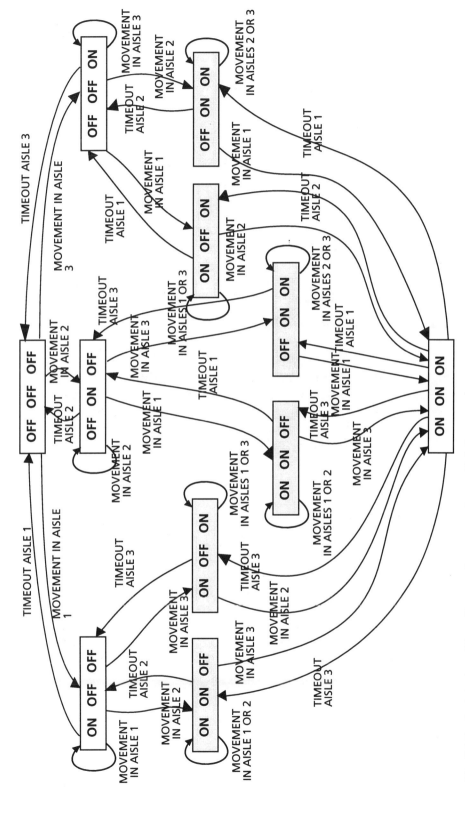

Fig. 21.9 A state transition diagram for a warehouse lighting system

352 Systems Analysis and Design

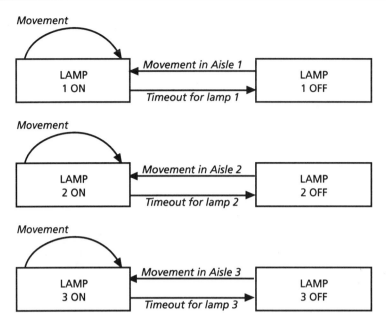

Fig. 21.10 STD for warehouse lighting

number of valid states: PLAYING, REWINDING, CUEING. So for complex systems, or to get a high-level view of the system, designers might draw MODAL state transition diagrams.

State transition diagrams are our preferred specification method for real-time applications. They are useful in design because they enforce order and predictability by providing a sequential event-driven view of the system. State transition diagrams are not appropriate to model the behaviour of programs carrying out substantial data manipulation.

21.9 SUMMARY

Whether code production will be automated or carried out by trained programmers, program design is a key phase in software development. Oversights and inaccuracies in the earlier phases will now be translated into hard code. Therefore this is the designer's last opportunity to check and clarify any ambiguities that still remain.

The program designer must also keep in mind that while a clever or elegant design may exploit the hardware capabilities to the full, it will probably prove difficult to maintain. It is sometimes said that old software never dies, it is simply modified beyond recognition. With this in mind the program designer must aim for a simple, modular, well documented system which can be easily maintained throughout the projected lifespan of the system.

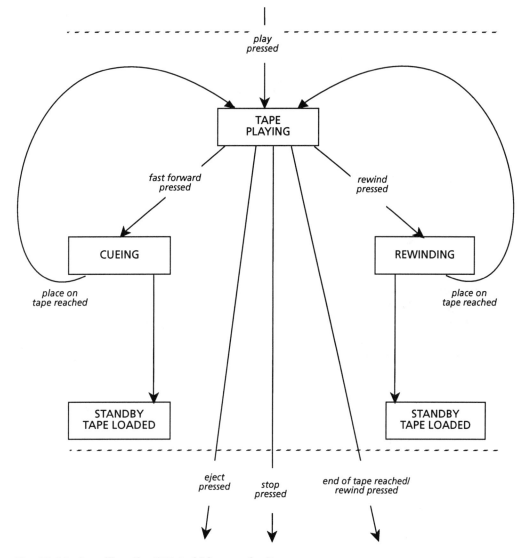

Fig. 21.11 Levelling the STD to hide complexity

CHAPTER 22

Systems Design: *Choosing Hardware*

22.1 INTRODUCTION

As a systems analyst you may find that you become involved in selecting computer hardware. This sounds a high risk activity as the cost of getting it wrong will be high even bearing in mind the continuing fall in hardware prices, so let's get it into perspective.

At the time of writing there are about one hundred million computers in the world and many more microprocessors, which are widely used in cars, washing machines, laser printers and similar devices although we don't recognise them as general-purpose computers. Of the hundred million, about 70 million are IBM PC and compatibles, about 10 million are Apple Macintoshes, about 10 million are UNIX machines, mostly workstations, and the rest are other things, including minis and mainframes. As more computers are sold the totals rise, but these proportions stay roughly the same. For most people then, choosing a computer means nothing more than choosing a 'PC'. Getting value means finding the best deal in terms of processor power, memory and disk size that's on offer anywhere that week. Don't read books to help you decide this, read magazines. The computing section of magazine stalls has swelled immensely in recent years, and you will find a dozen or more titles on any news-stand which tell you what's happening in the world of PCs. However, analysts involved in software engineering or system design need to develop a wider view. For them, choosing a computer means choosing between different competing solutions, with different operating environments and associated hardware. The first thing to understand is how computing systems have evolved.

22.2 THE EVOLUTION OF COMPUTER SYSTEMS

When you choose computer hardware you are making a *design decision* and the kind of computer system you are designing dictates the depth of understanding you need. Computer systems have evolved in the following way:

Mini and Mainframe systems: Historically, computers were first built as specialised electronic machines which would only work when pampered in air-conditioned rooms. All such computers had a *console* which the operators used to control and administer the machine. Users of the machine normally accessed it via VDU terminals, so the machine ran a 'time-sharing system' to service the needs of many users simultaneously. Until the mid-1980s most computer systems were built in this way. There are plenty of computer systems like this still in use in the world, and all are ripe for modernisation. Hence systems designers need to understand how these 'legacy systems' work, in order to better design their replacements.

Personal Computers: In the early 1980s, miniaturisation of electronics had reached the point where it was feasible to make stand-alone, desktop computers for individual people to use. The expensive, powerful versions generally ran the UNIX operating system and were called 'workstations'. The simpler versions, used for word-processing and similar tasks, were called 'personal computers'. Out of a myriad competing standards, IBM's design for the PC became widely adopted and now dominates the computer marketplace. Apple's design for the Macintosh computer became the alternative standard, and was the first computer to bring a graphical user interface to the mass-market.

There has been huge growth in recent years in the use of portable computers. These range in size from pocket-sized notepad organisers, through fold-up 'notebook' PCs the size of an A4 page, to full-blown UNIX workstations. At heart they are all personal computers.

Networks of Workstations: During the 1980s, the technology was developed for linking computers via local area networks. To begin with, LANs were used to de-couple users on terminals from needing dedicated wiring to particular minis and mainframes. From their terminal, the user could connect over the LAN to one of a variety of host computers. In the late 1980s, workstations and PCs too were integrated into the LANs, so that workstation users could also make use of central hosts, and the central facilities they supported such as printers, disks and databases.

Client/Server systems: So much computing power can now be bought in small hardware systems, that systems have developed to exploit this power. Applications are now split so that half the application runs as a 'client' in a PC or workstation on someone's desk, where local processing supports a graphical user interface which makes the application easier to use. The other half of the application runs as a 'server' on some other hardware which is specifically suited to its purpose.

There are different kinds of servers. Print-servers queue up files needing printing, and route them to appropriate printers in an orderly way. Communications-servers provide links to other computers, other parts of a LAN, or to the outside world through telephone lines and similar connections. Compute-servers provide specialised calculation power for mathematical, scientific or statistical work. File-servers provide central storage for shared files of data which more than one user may need. Database-servers are specialised file-servers tuned to support databases.

From a client system, a user can now connect across the LAN to many servers at once for different purposes. With a graphical user interface, it is possible to see programs running on various remote hosts, each program displayed in its own window. A special case of this is the X-window technology built on top of UNIX.

These descriptions are valid for general business computers irrespective of how they may be networked together. A quite different sort of computer is the 'embedded system'. Embedded computers are computers built into specialised hardware, and often perform technical jobs in real-time. Examples are computers in engine management systems, avionics and space-borne systems, factory control systems, railway signalling systems, communications, and numerous military systems. The design of these systems and the selection of hardware for them is a specialised subject outside the scope of this book, though many general principles are equally applicable to business and real-time computing.

22.3 MICROPROCESSORS

Modern computer hardware is made up of microprocessors. This wasn't always the case but their adoption as standard components has meant that computers can be built to standard designs and to use ready-made software.

Early general-purpose computers of the 1940s and early 1950s used radio valves which were inclined to overheat and be unreliable. In the 1950s individual transistors replaced valves and then the technique of making more than one transistor on a single piece of semiconducting material allowed the development of integrated circuits. With time, the number of components squeezed on to each chip of silicon rose from tens or hundreds (1960s) to thousands (1970s) to hundreds of thousands (1980s) and then to millions (mid 1990s). Then by the late 1970s integrated circuits had developed to the point where an entire central processor with its registers, arithmetic and logic unit could be squeezed onto a single chip and packaged as a multi-pin electronic component. The 'microprocessor' had arrived and it changed the rules for selecting hardware. The development of microprocessors made possible new kinds of computer system:

Embedded systems: Microprocessors were small and needed little power. They could be fitted in other machines, like cars, aircraft, factory machine tools and domestic appliances.

Personal computers: Previously, the smallest computers were the minicomputers, and were wardrobe-sized. With microprocessors, manufacturers could now build 'microcomputers' small enough for the desk-top.

Standard systems: Previously, all computer software was hand-written for a particular purpose. Apart from expensive application packages, ready-written software packages were almost unknown. Using microprocessors, manufacturers could produce computers which were cheap and built to a standard specification. The obvious example is the IBM PC. As sales of the new small computers blossomed, so software changed and packaged programs rather than bespoke programs became the norm.

By 1977, two manufacturers of microprocessors had emerged to dominate the micro world, and gave birth to the two main families of microprocessor, the Intel and Motorola microprocessors.

22.3.1 Intel

The processor which brought Intel fame was the 8080, a simple general-purpose device operating on byte values. By 1980 it had been expanded to a 16-bit version, the 8086, with a variant called the 8088. When IBM was choosing a device to power its range of Personal Computers it picked the 8088, and set the seal on Intel's fortunes. Throughout the 1980s, the Intel family developed in scope and processing power through the 80186, 80286 and 80386 variants. The 80486 device of the 90s brought more than a doubling of processing power, followed by the 'Pentium' (renamed from '586' for commercial reasons), which combines two '486s with associated control circuits on a single chip.

The family is now complicated. There are low-powered versions for use in portable computers, versions with narrow or wide data-buses, and versions with built-in floating-point units and memory management units. Intel has striven hard to ensure *backwards compatibility*. Every member of the family, even the Pentium, can run code originally written for the 8088, and even earlier designs. This approach has helped to establish the IBM PC architecture as a world standard for desktop computers for more than ten years. Unfortunately, preserving backwards compatibility has meant that the Pentium is more complex than it would be without this need, and it came to market in mid-1993 nearly two years late. Intel have indicated that the successors to Pentium will break with this tradition of backwards compatibility. The rules for choosing PCs in the future will therefore change.

22.3.2 Motorola

Motorola's rival to the 8080 was called the 6800 and a simpler derivative found widespread use in non-IBM personal computers like the BBC microcomputer of the early '80s. At this time Motorola took a brave decision to move straight to 32-bit architecture and sacrifice backwards compatibility to create a new family of microprocessors. The 68000 family was the result, and became the standard for powering all microcomputers which were not IBM-PC compatibles. 68000s were used in Commodore Amiga, Atari ST and TT, and Apple Macintosh. Indeed, so powerful were the chips that an entire new class of machine, the UNIX workstation, was developed to exploit them. The Motorola 68000 family has evolved too from the 680008 to the 680010, '020, '030 and now the '040 and with good compatibility between family members. The next new member of the family is expected to be the 68060 and, because of the clean and simple architecture, Motorola expect to be able to develop the family as long as there is demand.

22.3.3 RISC

A related issue to what microprocessor to choose is that of its capability to handle particular instructions. During the 1980s, researchers observed that central processors in computers spent most of their time doing only simple operations. The more obscure or convoluted instructions were seldom or never used. The researchers reasoned that a RISC machine (Reduced Instruction Set Computer) made with only simple but efficient instructions would out-perform a standard processor with complex instructions (CISC).

They were right. For general-purpose computing, RISC machines have been found to perform three or more times better than their CISC counterparts. CISC processors have been extensively re-engineered to compensate, but still the gap exists. RISC machines have therefore taken over from Motorola 680x0 for fast UNIX workstations, used for scientific and graphical work, and some RISC processors have been specially developed such as:

- ARM: Acorn computers – who developed the BBC micro – produced an early RISC chip called the Acorn RISC machine. Versions are used in the fast Archimedes computers popular in some schools, the Apple Newton hand-held computers, and in embedded control systems.
- SPARC: This was invented by Sun Microsystems in the mid-80s to supersede their use of Motorola 68000 series in UNIX workstations. At the byte and integer level, SPARC is designed to be compatible with the 68000.
- PowerPC: A consortium of Apple, IBM and Motorola developed this from an original IBM design. The consortium intends to use it in desktop machines such as new Apple Macintoshes and IBM PCs, UNIX workstations (IBM), and portable computers. Binary compatible versions are being developed for the embedded control market.
- DEC Alpha: Digital Equipment Company developed this as a 64-bit successor to the VAX range of minicomputers. Main uses are in workstations and servers, with possible future use in embedded systems.
- MIPS: used extensively in UNIX workstations, but not much known outside the USA.

The move from CISC to RISC now looks unstoppable. By the end of the 1990s we can expect RISC to be the computing norm. It's difficult to predict which will become the dominant architectures, but SPARC and PowerPC both have strong followings.

22.4 PROCESSING SPEED

There seems to be an insatiable demand for increased processing power. This is because

- Both system and applications software grow with time. Extra functions need extra processing power.
- Users expect to do more things at once. As system software is developed to allow multi-tasking, users expect to run many applications at once rather than just one; and expect the system to deal with background activities like printing and file-support as well.
- Computers are more widely connected by networks. They may have to run several network protocols in the background while still supporting the users' foreground programs.
- New applications are devised which demand higher processing power. The demands of the next few years will include:

    Sound and speech processing;
    Desktop video, to allow video conferencing;
    Display PostScript, and similar standards, to allow documents from a variety of sources to be seen and edited on any computer, regardless of which computer or program was used to create them originally.

Clock-speed apart, the processing power of the central processor of a personal computer depends on:

- whether it has a floating point unit (FPU)
- how wide its data paths are (8-bit, 16-bit, 32-bit, 64-bit)
- whether it uses special techniques, like cacheing.

If other things are sympathetically designed, the processing power of a computer rises linearly with increasing external clock-speed; but raising clock-speed has side-effects which increase engineering costs. It may force use of faster (more expensive) RAM; it may cause problems of electro-magnetic compatibility as the circuit frequencies move higher up the radio frequency spectrum.

If external clock-speed is held constant, then to make a computer go faster its designer must change its technology using techniques such as :

- *Internal clock-multiplying*. Some microprocessors run an internal clock at twice the frequency of the external clock, effectively doubling their internal speed. Clock tripling and quadrupling are also known.
- *Pipelining*. Many microprocessors work like factory production lines. While one part is executing an instruction, another part is decoding the next instruction and fetching its operands, while yet a third part is fetching the instruction which follows that one, and so on.
- *Cacheing*: A cache is a small area of quick memory which can hold recently used instructions and data. The best cache is one built into the microprocessor itself, but designs can use external caches which come between a processor and its main memory. Sophisticated designs have separate caches for data and code, rather than one unified cache.
- *Internal Parallelism*. Some microprocessors combine several processors – such as an integer and a floating-point processor – on one chip, with control circuits so that they can be used at the same time.
- *External Parallelism*. Some processors are designed for use in arrays of symmetric multi-processors. Examples are the transputer, the superSPARC processor, and advanced versions of the PowerPC. However, to run such arrays successfully, the operating system must be written so that it can organise, distribute and coordinate the work.

You can't make sensible decisions about choosing processors unless you can quantify a processor's power: you might as well choose by the colour of the case! To quantify power, you need not only to know clock-speed, but to know what technology is used by a particular processor, and what improvement that makes to its processing power.

22.5 PROCESSOR TECHNOLOGY

In practice, it can be hard to reap the full benefits of fast processor technology, because of side-effects:

- *Raising clock-speed*. If you raise the external clock-speed of a processor, the entire board on which it runs must also clock faster. There can be problems with signal timing and with radio-frequency interference. The advantage of clock-multiplying, which speeds up only the processor's internals without disturbing the rest of the board, is that it keeps the board's design simpler.
- *Memory limitations*. Most memory in a computer is dynamic RAM (dRAM), which is cheap and compact but rather slow. Typically, dRAM can cycle at around 80 ns, which means that you can read each item of memory no faster than at 12.5 MHz. Even modest processors nowadays have speeds of 16 MHz, while fast processors run at 50 MHz,

100 MHz or more with the upper limit doubling every few years. There is therefore an increasing mis-match between the natural speed of RAM and the speed of the processors.

Without a cache, a processor must run with 'wait-states' when using the memory. This means that at each memory access, the processor must do nothing for one or more clock cycles until the memory has completed its action. Use of a small cache on the processor chip decouples the processor from the speed of the memory, and may allow it to run several times faster than the natural speed of the memory for most of the time. However whenever the cache needs to be flushed and refilled from main memory, the whole processor is effectively slowed to the speed of the memory. If this happens often there is a bad impact on processing speed.

The cache built into the processor itself is called a 'primary cache'. Even a small primary cache enables speed to be increased. The fastest processors are built also to use a secondary cache, outside the processor chip, made from half a megabyte or more of fast static RAM.

- *Instruction interdependence.* As each instruction executes, its outcome determines the usefulness of the other activities which the chip is doing. For example, a branch instruction may switch processing right out of the current sequence of instructions, so that the contents of the pipeline and the cache are no longer useful. They must be flushed and refilled with useful information before processing is back up to speed. If there are two levels of cache, the primary on the processor chip and the secondary between the chip and the main memory, then the coordination is complex.
- *Processor interdependence.* With multiple processors, there is a need to coordinate their execution, and especially their use of shared cache and shared memory. For most processor designs there comes a point of no return after which adding extra processors brings no speed benefits. With most current designs four processors are little more powerful than three, though some can do better. Only specialised designs like the transputer can make effective use of arrays of sixty-four or more processors.

To run effectively on a superscalar RISC environment, software must be sympathetically designed and implemented. Data structures may need tuning so that they fit within the available caches. Algorithms may need partitioning between multiple processors. Compilers may need to generate tuned sequences of machine instructions so as not to disrupt pipelined parallel architectures. At the machine level, RISC architectures are hideously messy to manage.

We can see how changes to microprocessor technology can improve performance. In 1988, a large project was run on a network of Macintosh SEs each with 1 Mb RAM. By 1992, to do equivalent work required Mac LCII machines each with 4 Mb RAM. If the demand for processor power rises in the same way, what machines will be needed to do equivalent work in 1996?

Macintoshes use Motorola 680x0 processors. By running benchmark programs on machines using different processors, we can compute the scaling factor by which each change of processor technology improves on the previous technology. The scalings shown in Table 22.1 are derived from benchmark measurements.

The Mac SE uses a 68000 processor at 8 MHz with a 16-bit data bus. Let's assign it a power of One (on an arbitrary scale). The Mac LCII uses a 68030 processor at 16 MHz, still with a 16-bit data bus. Its power is therefore:

Systems Design: Choosing Hardware

Table 22.1

Change	Performance improvement	Reason for improvement
68000 to 68020	× 1.75	'020 processor is full 32-bit; it has internal pipelining, and an instruction cache.
68020 to 68030	× 1.1	'030 processor has a data cache too; its CPU is closely linked to its built-in MMU
widen data bus, 16 to 32 bits	× 1.25	Data transferred in bigger blocks
68030 to 68040	× 2.4	'040 processor uses internal clock doubling; it has large internal caches for instructions and data; it has yet more pipelining and parallelism.

One
× 1.75 (improvement from 000 to '020 processor)
× 1.1 (improvement from '020 to '030 processor)
× 2 (doubling of clock from 8 to 16 MHz)

which is nearly four. So over four years (1988 to 1992) the demanded power has risen fourfold, as has the demanded RAM. Assuming this trend continues, by 1996 the demanded power on this scale will have risen fourfold again, to sixteen, with 16 Mb RAM. What kind of machine will deliver this power?

Combining the benefits of a wider data bus (×1.25) and a change to the '040 processor (×2.4) gives an improvement of (×3). So to get a fourfold improvement we still need to raise the clock-speed by 4/3. Clock-speeds tend to have standard 'round-number' values which follow this sequence: 8, 10, 12, 16, 20, 25, 32/33, 40, 50, 66, 80, 100... Three options which give adequate processing power are:

'040 processor at 20 MHz; power is ~15.
'040 processor at 25 MHz; power is ~19.
'030 processor at 50 MHz; power is ~15.

At the time of writing, 68030 processors aren't available with a clock-speed of 50 MHz. The only option is the '040 at a clock-speed of 20 MHz or 25 MHz. So, if the need is to buy now a machine with adequate power for the work of 1996, it must have just such a processor.

It is however not just a question of processing speed, because the power of a desk-top computer depends on

- how fast it can draw on-screen, when coupled to a display, and
- how fast it can work the disk when using backing-store

In general, the circuitry that controls displays can be made arbitrarily fast to match a chosen processor speed. For extra performance, some computers have special circuitry on graphics acceleration cards to increase drawing speed. Buying this is no problem – it just costs more money. This is not so for disk speed. The fastest desk-top computers can work the disk only about five times as fast as the slowest – even though their general processing speed may be twenty-five times as fast. For work which is disk-intensive, disk speed therefore becomes the limiting factor. Examples of such work are:

- Use of large databases and library systems
- Software development with compilations and links
- The processing of large colour images
- The startup of large applications which reside on disk.

Most of these activities speed up a lot if they can be run from main memory. Adding RAM to a computer is a good way to speed up its disks, simply because the disks get used less often. For desk-top computers the operating system will usually run a disk cache in RAM, because the speed-up obtained easily outweighs the overhead of maintaining the cache.

The technique outlined above works when comparing similar computers of the same basic type, which use processors from the same family. Trying to make comparisons between different processor families is difficult and risky. Manufacturers measure processing power in different ways and commonly quote it to different standards. The best comparative benchmarks now quoted are called 'SPECmarks': SPECint92, based on general integer arithmetic, and SPECfp92, based on floating-point arithmetic.

If you have to make comparisons between different processor families, adapt the technique shown above. Find a common ground for comparison (such as SPECmark values), but normalise the results to correct for differences in quoted clock-speed. You should then be able to compute the scalings which relate one type of processor to any other when running at the same clock-speed. From there you can compute the projected power of a processor running at any clock-speed you choose.

Table 22.2 shows the approximate ratio of powers for some common processors, based on their performance with standard benchmarks. The base value of one is for a standard Intel 80386 processor clocked at 16 MHz.

Table 22.2

Processor		Clock-speed	Caches		SPEC values	Ranking
Maker	Type		Instr'n	Data		
DEC	Alpha	200	8	8	90	21
Sun	SPARC	50	20	16	65	15
Motorola	PowerPC 601	66	64		62	14.5
Intel	Pentium	66	16		64	14.5
Motorola	68060	40	16	16	60	13.8
Intel	486dx2	33	8		32	7.5
Intel	486dx	50	8		27.5	6.3
Motorola	68040	33	4	4	24	5.5
Motorola	68030	16	.25	.25	5	1.1
Intel	386dx	16			4.5	1

22.6 THE IMPACT OF OPEN SYSTEMS

Choosing hardware does mean that you need to understand the technology and recognise that this is driven by the competing technological giants who produce the microprocessors. However, there is something else. It is the evolution of UNIX and the move towards 'open systems'. UNIX's earliest beginnings can be traced to 1969, though it didn't take a settled form until the late 70s. Even ten years ago UNIX was still an academic hobby, but from the mid-80s onwards, despite its amateur pedigree, UNIX has mushroomed in commercial popularity. The main reason for its growth is that UNIX is *portable*, whereas other operating systems aren't. To port early Unix to a new platform, all that was needed was to re-write the *kernel*; the rest, written in C, would work anywhere you could find a C compiler and a linker. Porting of the operating system therefore took weeks rather than years – a dramatic and radical change for the computing world of that time. UNIX was developed at AT&T's Bell Labs, but they licensed it freely to the academic world in the USA. Though this helped make it popular, it also led to loss of standardisation as various groups at Berkeley, Stanford, and MIT added their own extensions. Later on, computer manufacturers ported the operating system to their own machines and gave it their own flavour, and UNIX is now available on IBM, DEC, Hewlet Packard, Apple and Sun systems. Apple and IBM together with Motorola have formed the PowerOpen organisation to promote an open UNIX environment on new PowerPC machines. This competes with the closed systems offered by Intel and Microsoft. These attempts to change the computer marketplace make it even more difficult to know which hardware to choose.

22.7 SUMMARY

In this chapter we've concentrated on reviewing some of the issues that affect the market for personal computers since it is most likely that analysts will first find themselves selecting hardware at this level. It's easy to appreciate the power of the forces affecting hardware design and capability. The difficult issue is judging whether today's choice will still be the correct one tomorrow.

CHAPTER 23

Systems Design: *Data Communications*

23.1 INTRODUCTION

Two hundred years ago long-distance communication could take days or even weeks. Armies might have used semaphores and signal fires but, for most people, instantaneous communication could only take place over the distance that someone could shout. Then, in 1844, the first commercial telegraph service was set up and in 1876 Alexander Graham Bell invented the telephone.

In the following 150 years the technology of communications has grown so much that today it is difficult to imagine a business without telephones. Similarly, few businesses today operate without computers. The first computers stood by themselves in splendid isolation, but now the technologies of computers and communications are merging. Computers are linked by wires and the wires themselves are managed by computers. In the not-too-distant future the technologies will be so interlocked that it will be as difficult to imagine a computer separated from a network as it is to imagine an office without a telephone.

This chapter looks at the techniques and the technology of data communications and how to use it. Whole books have been written about the theory and practice of this subject. This chapter cannot attempt to cover the whole of that ground. Rather, it will introduce the concepts and the jargon without going into too much detail.

23.2 BASIC CONCEPTS

This section introduces some of the terms used in the communications business. It's not intended to be exhaustive, but it's all you'll need at this stage.

Serial and Parallel Links. A simple piece of wire can carry only one bit at a time. To send characters down the line it is necessary to carry several bits. There are two ways to achieve this. A serial link carries the bits one at a time, perhaps on a single piece of wire. On a parallel link the bits making up a character are transmitted at the same time, each on a different piece of wire. Parallel links are widely used to connect desktop computers to printers. For longer connections the cost of providing a separate conductor for each bit is too great, so serial links are used.

Multiplexing. This is a way of allowing several independent users to make connections over the same link. Computers tend to send data in bursts with large gaps in between. This means that for much of the time the connection is not used. Sharing the connection between several users can therefore make more efficient use of the connection. There are two ways of sharing the link. These are Time Division Multiplexing and Frequency Division Multiplexing.

- **Time Division Multiplexing (TDM)** allows each user in turn access to the full capacity of the link for a short period of time. If there were ten users sharing the link, each might be allowed to transmit for one tenth of a second in each second. If the link could carry data at 1200 bits per second (bps), each user would be able to transmit at 1200 bps during this tenth of a second. In practice, from the users point of view, a speed of 120 bps appears to be available all the time.

- **Frequency Division Multiplexing (FDM)**, unlike TDM, allows users to transmit at any time but at fewer bits per second than the full link is capable of. It can be compared to the use of the radio spectrum by radio stations. They all transmit at the same time, but in different frequency bands. A filter circuit enables any particular station to be selected from out of the babble of simultaneous transmissions. Cable TV transmissions use FDM in the same way to carry many TV signals over one wire.

Data Rates. The rate at which data can be transmitted is measured in *bits per second* (bps). The term bandwidth is sometimes used, but this is a loose use of an engineering term. A high bandwidth link can carry more bits per second than a low bandwidth link. The distinction between bits per second and baud is a source of perennial confusion. Bauds are concerned with the rate at which a connection changes from one physical state to another (from one voltage to another, for example). It is quite possible for the rate at which the line changes its physical state to differ from the rate at which bits can be carried.

Reliability. It is important to be able to guarantee the reliability of data transmitted across a link. This is achieved by adding to the data a code derived from the data. This is then transmitted and, on receipt, checked. If the check is passed, then the data is accepted. If the check fails then some remedial action is taken. The particular kind of remedial action will depend on the particular link protocol being used. The data may simply be discarded or a retransmission of the data may be requested.

Checksum. The idea of a checksum is to treat a message as a sequence of numbers. These numbers are added together and the sum transmitted along with the message as a check. There is no single way to calculate a checksum and the particular method chosen may well be based on the type of data being carried.

Parity. This is the commonest form of check. It is easy to implement and in widespread use. It consists of a single 'bit' usually added to a character. The setting of the check bit is designed to make the number of 'ones' in the character either odd or even. Either can be used, so long as both ends of the communications link agree.

Cyclic Redundancy Codes. CRCs are a form of check applied to a block rather than to individual characters. The method of calculating and checking the codes is complex in software but easy in hardware. CRCs are more reliable than parity checks.

Network Topology. A computer network is a set of computers connected to each other. It is not necessary for every computer to be connected to every other. Two computers may still be able to communicate, even if not directly connected, if there are intervening computers which can act as relays. There are many different ways of connecting computers together. The commonest are (see Figure 23.1):

- **Star.** In a star, one machine forms a 'hub' to which all other machines are connected. The machines on the 'periphery' are connected only to the hub, not to each other. This

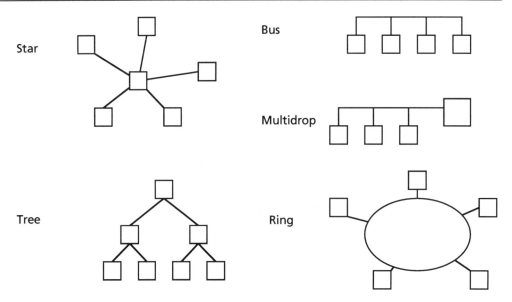

Fig. 23.1 Network topology

has the advantage that the central machine can easily control and monitor the use and performance of the network. Its weakness is that if the hub machine fails, the whole network fails. A typical example of a star network is a mainframe computer with many remote terminals connected directly to it.

- **Tree.** This is related to the star. A hub machine has a number of periphery machines connected to it. Each periphery machine may then form the hub of a further star. This may continue for several layers as the network expands.
- **Bus.** This is widely used with Ethernet Local Area Networks (LANs). All computers are connected to a single transmission medium, and all could transmit at the same time. One of the functions of a protocol in such a network is to prevent confusion arising from simultaneous transmissions.
- **Multidrop.** This approach is related to a bus. A central computer may be connected to many smaller machines or terminals, several of which share a single connection to the main computer.
- **Ring.** In a ring, each computer is connected to two others to form a circle. Rings are often used for Token Passing LANs.

Switching. In a complex network, messages may have to pass over many different links. Switching is the process by which physical links are chosen.

- **Circuit switching** is used on the telephone network. While communication is in progress, there is a physical connection between the two ends of the link. The connection is totally dedicated to that connection and cannot be used for any other purpose until the call is over. A network which uses circuit switching is ideal for carrying voice. Voice traffic is fairly continuous so it makes sense to dedicate a circuit to its use. However, computer communications tend not to be continuous. Data might be carried for a second and then nothing for the next ten seconds. For computers, circuit switching is not ideal. What is required is a way of switching the use of a circuit rather than the circuit itself.

- **Packet switching** requires data to be divided into packets of a few hundred bytes each. Packets from several different users can then be sent one after another down the same connection. In this way, several users can use the same connection at the same time.

Layering. Communications systems are often thought of as layers of software; the software in each layer being distributed across the different machines. Each layer communicates only with the layers above and below. A message is passed to the uppermost layer which passes it on, and so on. Eventually the lowest layer controls the physical link itself. The message then passes across the link and up through the corresponding layers on the receiver until it can be delivered. The parts of each layer on different machines will, in most cases, need to communicate with each other. This conversation is called a *protocol*. While the protocol seems to be carried within a layer, the only 'real' connection is at the bottom of the stack. So, the protocol in one layer is carried as messages by the next layer down. The effect is to produce a set of nested protocols each concerned with a particular aspect of the link. Layering is often used to build communications software. It enables well established 'communications' practice to be implemented in a clearly defined way. The different protocols can be separated from each other, and from processing-orientated functions. On the other hand, size and performance can suffer.

23.3 THE USE AND PROVISION OF NETWORKS

From the analyst's point of view, layering can help separate different aspects of the problem. In particular, it can be used to separate the use being made of a network from the problems of providing the network. Imagine you are making a phone call. You care about what you want to say. You don't care about whether your voice is carried by microwave or copper wire, or the number of telephone exchanges which the connection must pass through. The telephone company, on the other hand, does care about how the connection is made, but does not care about your conversation.

The same distinction, between use and provision, can be made between programs which use network connections and programs which provide them. In this case, however, the two programs could well be running alongside each other in the same computer. The programs which use the connection will be aware only of the end-to-end connection. Intermediate points where, perhaps, the packets are switched would be invisible.

In figure 23.3, the arrows represent protocols. The lower-layer protocols are concerned with the provision of the network and with the distance over which information has to be carried; something considered in more detail in sections 23.4 and 23.5. The upper layer protocols are directly concerned with the needs of the application. If, for example, the application concerned banking, this protocol might be concerned with checking that credit cards are valid, or that a withdrawal will not exceed the overdraft limit.

If these 'user' and 'provider' layers are not separated, then a change to the way the information is carried could require a change to the application itself. It is as if, in a telephone network, a change to the local telephone exchange required you to speak in a different language!

Both the upper and lower layers can be complex and are often divided into several smaller layers. As we shall see in section 23.5, the OSI Reference Model uses seven layers in addition to the application itself.

368 Systems Analysis and Design

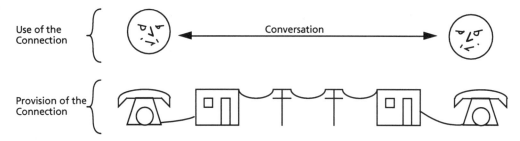

Fig. 23.2 The use and provision of a connection

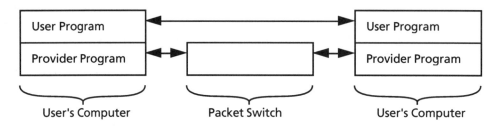

Fig. 23.3 User and provider layers

23.4 CARRYING INFORMATION ACROSS NETWORKS

Networks can be categorised, according to their size, into Wide Area Networks (WANs) and Local Area Networks (LANs). The differences lie not only in the physical distances involved, but in the speed and number of the channels provided. A typical Ethernet LAN would provide a speed of up to 16 megabits per second for a maximum distance of about two and a half kilometres. While there are, in practice, ways of linking Ethernets over much longer distances, there are still limits. WANs are big. The size of a WAN is limited only by the ability to provide a physical medium on which to transmit. A WAN might provide only a few kilobits per second but could provide it from one side of the Earth to the other. Unlike a LAN, the rules for using a WAN do not set a limit on the overall size of the network. So, you might use a LAN within an office, and a WAN to link offices together.

23.4.1 Local Area Networks

You can think of Local Area Networks (LANs) in terms of the protocols used or in terms of the way in which the physical link is used. There are two ways in which the physical link can be used, known as broadband and baseband. A *Broadband Network* has a very great capacity and uses frequency division multiplexing to provide a number of separate logical channels. The channels may be of different capacities and may be used for very different purposes. You might use one for carrying a video signal, another for analogue voice and use a third treated as a baseband channel in its own right. A *Baseband Network* is one in which all signals are transmitted at their 'base' frequency. That is they are not combined with any 'carrier frequency' to provide separate logical channels on the physical link.

The simplest way to coordinate a small network is polling. One computer is designated the Master and the others the Slaves. No Slave may transmit until given permission to do so by the Master. The Master will ask each Slave in turn whether it has anything to transmit. The Slave may then transmit, or relinquish the right until polled again. This simple procedure prevents contention, but may lead to a backlog of messages if the Master fails to poll for some time. For more complex LANs, Token Passing and Ethernet are the available options.

In a *Token Passing Network* the connections to the network have a predefined sequence. The ends of this sequence are joined to form a logical circle round which the information flows. In the case of a *Token Ring* the physical sequence of connections to the ring is used. While there are several different protocols used for token passing, the principle is that a bit pattern called a 'token' is passed like a relay baton round the circle. Only the user who has the token can transmit, and rules of the protocol guarantee that the token reaches each user in turn. *Ethernet* is the popular name for a protocol more correctly known as CSMA/CD. The acronym CSMA/CD stands for Carrier Sense Multiple Access with Collision Detect. The general idea is that anyone wishing to transmit must first listen and wait for a gap in the conversation. When a gap in the conversation is found, the user can start transmitting. The message might be transmitted successfully. Alternatively, another user will also have listened to the line, heard the gap in the conversation and started to transmit. In this case, the two messages will collide. Both messages will be garbled and will be unrecognisable. If messages collide, the two would-be-users of the line each wait for a random length of time and try again. If the times are selected well, the chance of them colliding again is very small.

23.4.2 Wide Area Networks

For good historical reasons the majority of communications links in place in the world are designed for voice traffic. This is unfortunate for the transmission of data because the requirements for data and voice are very different.

Let's consider a few of the differences. In the first place, telephone conversations are always two way but data transmissions may be two way or one way. When data traffic is two way it is often 'both ways at the same time'. Two-way voice traffic tends to alternate in direction; at any given time only one person is speaking. Secondly, data must be transmitted without noise. Voice traffic, however, as we know to our cost, is assumed to be able to cope with all the crackles and pops which the circuit can deliver. We can usually make sense from a noisy message, but a computer usually can't. Also computer data tends to be transmitted in bursts. Telephone traffic is fairly continuous; gaps where neither person is speaking are rare. We also prefer voice traffic to be delivered immediately. A gap of only a few seconds when no one is speaking seems like a lifetime and is often intolerable. Some data has this characteristic too, but most can be delivered later – like a letter sent by post. Finally, the way in which bits are represented on an electrical connection differs greatly from the way in which voice is represented. The effect is that a telephone line, which is designed to carry voice, distorts data to the point where, after only a short distance, it is unrecognisable.

You can, however, use voice lines to carry data. The process of making data suitable for transmission over telephone lines is called *modulation*. Interpreting the tones at the receiving end is called *demodulation*. Hence the acronym MODEM used to refer to the boxes which perform the MOdulation-DEModulation. Modems vary very much in complexity. The longer the distance over which data is to be sent, and the faster the data has to be sent, the more complex the modem. For low-distance and low-speed transmissions a simple device called a Line Driver may be adequate. This consists of little more than an amplifier. For longer distances and higher speeds the way the digital signal is 'modulated' is more complex. The form of this 'modulation' is specified in a number of internationally recognised recommendations. Effectively, the job of a modem is to 'disguise' the data as a signal which 'sounds like' voice and, thereby, to reduce the distortion.

The maximum practical data transmission rate using modems on a voice-grade telephone line is about 19.2 kbps. However, higher-capacity digital links are available. British Telecom markets a 64 kbps service as Kilostream and a 2.048 mbps service as Megastream. Circuits similar to these are available from other suppliers such as Mercury. In the USA, since the designs of the telephone networks differ, the corresponding circuits have different capacities; 56 kbps and 1.544 mbps. A 1.544 mbps circuit is known as a Bell-T1 circuit. Not only are individual links available from commercial suppliers such as British Telecom, but access to complete digital networks can be provided. These are often called X.25 networks. X.25 is the name of a standard which describes how to connect a computer to such a network.

23.5 STANDARDS AND STANDARDS-MAKING BODIES

An Open System is a system which is available to everyone. That is, it is not closed to people who don't, for example, buy from one particular hardware manufacturer. The idea of Open Systems Interconnection (OSI) is that any two Open Systems should be able to exchange data with the minimum of difficulty. This requires standards. Some standards exist because everyone decided to copy a particular approach. So, for example, the widespread use of the IBM-compatible personal computer is a consequence of other manufacturers choosing to standardise on this design, not because some international standards organisation told them to conform. In data communications, however, it is the standards organisations which reign supreme. Whenever someone uses jargon such as X.25, IEEE 802.3 or RS-232 they are using the reference numbers of particular standards documents.

There are two organisations which are particularly important in data communications. They are the International Standards Organisation (ISO) and Comité Consultatif International Téléphonique et Télégraphique (CCITT). The members of CCITT are telecommunications providers and are mainly concerned with how connections should be made to and between networks, rather than how networks should actually work. For our immediate concern, ISO is the important body.

23.5.1 The OSI Reference Model

When the ISO decided to produce standards for Open Systems Interconnection (OSI), it very soon found that the problem was too complex for a single standard. The problem had to be broken down into a number of smaller problems. The Reference Model was

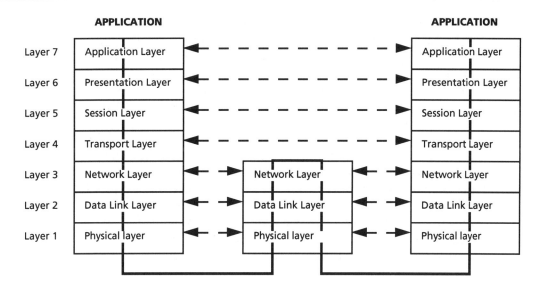

Fig. 23.4   The seven-layer model

the result. Each of the seven layers of the Reference Model has a particular function to perform. It will usually be the case that the communications network, whether OSI or not, will be supplied. Even so, the seven layers provide a useful checklist of the concerns which a network designer must address.

The upper three layers are concerned with the needs of the application; the lower three with making the connection. In the middle, the Transport Layer exists to overcome possible mismatch between the service requested by the upper layers and the service provided by the lower layers.

Figure 23.4 illustrates these seven layers of software. This form of diagram, with the layers of software shown piled on top of each other, has led to the use of the term Protocol Stack. The dotted lines show where the protocols in each layer seem to be exchanging information directly. So, for example, the application on one machine seems to be exchanging information directly with the application on the other machine. In practice, however, the path along which the real data passes is shown by the heavy line.

The three-layer stack in the middle is a switch. The lower layers are aware of and talk to the switch, the upper layers are not aware of it and talk directly from one end to the other.

23.5.2 The upper layers

Layer 7 – Application Layer   The Application Layer is the interface between the communications system and the application processes. The main services of the Application Layer are:

1. File Transfer, Access and Management which describes how to send files from one place to another.
2. Virtual Terminal which involves message exchanges in which small amounts of data are transported in both directions. Virtual Terminal concerns itself with the

characteristics of different types of display. There is no point, for example, in sending colour graphics information to a monochrome VDU.
3. Electronic Mail. The CCITT X.400 Message Handling System (MHS) recommendations describe what in popular parlance is known as electronic mail.
4. Directory Services. To complement the X.400 recommendation on Electronic Mail, there is a series of recommendations called X.500 which describe how to maintain and distribute directories of users of electronic mail.

Layer 6 – Presentation Layer An Open Network might contain many different types of computer using different, and conflicting, ways to represent data. Some computers, for example, might represent letters of the Roman Alphabet with an ASCII representation, others might use EBCDIC. The purpose of the Presentation Layer is to resolve these conflicts. It is concerned with data representation during transmission between Open Systems.

A connection between two similar machines may not need a Presentation Layer. If the machines differ greatly it may be necessary for the designer to simplify the data formats in order to make the Presentation Layer manageable.

Layer 5 – Session Layer The Session Layer controls how connections are made. Connections can be Duplex or Half Duplex. A Duplex connection is one in which both parties can transmit at the same time. A Half Duplex connection still allows communication in both directions, but not at the same time like a telephone conversation. The term Simplex is also used. This refers to a link on which communication may only take place in one direction such as in radio and television.

The structure of a conversation over a Duplex connection is usually left to the application program. A Half Duplex connection, on the other hand, may well be coordinated by the Session Layer. It is necessary for the two ends of a Half Duplex conversation to know at all times who is transmitting, and who is receiving. One way of arranging this is to use a token such as those used to coordinate token-passing networks. The token represents 'permission to transmit' and is passed between the parts of the session layer at each end of the link.

Since the computers at two ends of a link run independently, neither can predict the stage which processing has reached on the other machine. If synchronisation is required it could be provided by the Session Layer. The machines on the two ends of the link agree to a number of Synchronisation Points in the processing which they should reach at the same time. If one machine is ahead, it might wait for the other to catch up, or in some cases might 'backspace' back to the other machine.

23.5.3 The lower layers

These layers provide the connection. If the network is a LAN, then these layers implement the Ethernet or token-passing protocols. If the network is a WAN, then these layers would be concerned with the connection to the network. In this later case, a standard known as X.25 is widely used. It describes the protocols used by these three lower layers when a user's computer is to be connected to a packet switched network.

Layer 1 – Physical Layer The Physical Layer is concerned with the physical form of the transmission medium: the wire, optical fibre, microwave link or smoke signal. It is concerned with the way in which bits are represented on the physical link. We touched on this earlier when discussing modems. A modem is used to convert the signal into a form which can be carried by a telephone line.

Layer 2 – Data Link Layer   The Physical Layer transmits data but makes no guarantees about the accuracy of the transmitted data. The Data Link Layer provides that guarantee. The Data Link Layer is not concerned with the content of the data being transmitted or received; its only concern is that the data is delivered with no errors.

Layer 3 – Network Layer   It is the responsibility of the Network Layer to ensure that the correct point-to-point links are chosen. Route selection can be based on many different criteria. The ultimate destination is obviously important, but so is the current network loading. If a particular link is overloaded it may be best to send data to its destination indirectly. Not all communication links however conform to the definitions of the Physical and Data Link Layers. There are many communication systems such as the conventional telephone system that were created before the ISO–OSI Reference Model was devised, and a communications strategy that refused to use these systems for data communication would be severely limiting. One function of the Network Layer is to bridge these differences; to bring them all up to a common Network Service. This can involve joining together a number of different subnetworks to provide a single end-to-end network. In theory the Transport Layer need not know what actual communications service is being used. In practice different Transport protocols (called Classes) are used to bridge networks of different reliability.

The extent to which the actual communications services must be enhanced to produce the required Network Service varies. The telephone system for example provides little more than a physical link whereas the X.25 protocol for connecting computers to public packet switched networks comes very close to providing the required service. Very little additional enhancement is needed in this case.

23.5.4 The Transport Layer

The job of the Transport Layer is to bridge the gap between the services provided by the Network, that is the service provided across the interface between the Network and Transport Layers, and the services required by the Application, that is the service required across the interface between the Session and Transport Layers.

Depending on how big the gap between the requirements of the Session Layer and the service provided by the Network Layer, several different Transport protocols, called Classes, can be used. The commonest are Class 2 which assumes a reliable network, and Class 4 which allows multiplexing over an unreliable network.

23.5.5 The X and V series recommendations

The X series recommendations are produced by the CCITT to describe the use of digital networks. The V series are concerned with analogue communication including modems. X.25 is a protocol for connecting computers to public packet switched networks. It corresponds to the lower three layers of the OSI model. Its importance lies in the fact that

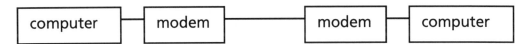

Fig. 23.5 Computers connected by modems

if you want to use a public packet-switched network, such as PSS from British Telecom, you will have to use this protocol. The X.25 recommendation assumes a user's computer is capable of supporting the protocol. Not all computers are this powerful. In these cases some kind of interface device is required. Such a device is called a PAD (Packet Assembler/Disassembler). A PAD can receive data over an asynchronous connection, convert the data into packets and transmit them over the network. Similarly it can receive packets, strip off the protocol and transmit the data to the user on an asynchronous connection. RS-232 and V.24 are, respectively, the IEEE and CCITT definitions of how to connect devices together using D connectors. RS-232, the American terminology, tends to dominate over V.24 which is often seen as a more European approach. Each D connector has many different circuits. Some of these are used to carry data while others carry control and timing information. Both standards envisage a situation in which devices such as a computers, terminals or printers are connected via modems, as in figure 23.5 for example.

In this case the modems are given the name DCE (Data Circuit-terminating Equipment) and the computer and printer the name DTE (Data Terminating Equipment). RS-232 and V.24, therefore, describe how to connect a DTE to a DCE. Unfortunately, the most common use of RS-232 and V.24 is to connect a DTE (such as a computer) directly to another DTE (such as a printer). The process of re-interpreting the standards in order to achieve a DTE to DTE link is seen by many as a 'magic art'.

23.5.6 TCP/IP

TCP/IP (Transmission Control Protocol/Internet Protocol) is a standard which was developed in the USA before the OSI standards became available. It corresponds roughly to the OSI Network and Transport Layers. It is important because it is both widely used and widely supported, particularly in the USA. TCP/IP was originally devised by the US Department of Defense to be used over a WAN linking many sites across the United States. The network still exists, now known as ARPANET, linking universities. Today, with transatlantic links, many European universities are also connected. As indicated earlier, the upper layers of a communications stack are not concerned with distance, but with the needs of the application. It is not surprising, therefore, to find TCP/IP also used over LANs.

23.6 DESIGNING A NETWORK

The main point to remember about designing a network is that, in the vast majority of cases, the network is not being designed in isolation. The problem which the introduction of a network is designed to solve must already be being solved to some extent otherwise the organisation would not be capable of operating. In other words, there is already some kind of network even if it consists only of a few separate modem links, or someone walking down the corridor once a day carrying a diskette.

Much is made, in many textbooks, of the analysis of network topology. In practice however there are few options. So, for example, a manufacturing company which has a head office in London and a factory in Manchester may well already have a point-to-point data link, Kilostream perhaps, linking the two sites. When other sites are to be linked in, this existing link can be expected to become the backbone of the new network.

23.6.1 Wide Area Networks

The first step in designing a wide area network, as with all systems design, is analysis. The first thing we want to know is how much data is passing from where to where and when? Having said that a rudimentary network is likely already to be in place you should find that the peaks of the existing traffic are being supported by the existing leased lines.

The question then is to decide whether the remaining traffic should be supported, and if so, how? The first of these questions may come as a surprise. After all aren't we installing a network in order to carry traffic? The point is that the installation of any piece of office equipment, from a PC to a leased line, has a cost. If the traffic from some outlying office consists of a just couple of kilobytes once a day, a modem over an ordinary dial-up telephone line could be the most cost effective option.

Having decided where to install lines, the next decision concerns the type of line. The criteria are the cost, the volume of data and the reliability required. High capacity, high reliability lines are not cheap. Indeed, it may be that the ideal link is not available. Kilostream for example can only be supplied from some of British Telecom's exchanges. If your local exchange doesn't support it, you can't have it. A good way of supplying multi-megabit links is through microwave links but they require that the aerials can see each other. If you can't see the local microwave tower from the roof of your building, you can't have microwave.

So, having decided on the links, what goes on the end of them? This will depend on the network protocols. If you have chosen, for example, IBM's SNA (System Network Architecture) then the links will go straight to the IBM machines on each site. These machines will then supply all the required switching as well as running application software. If, on the other hand, you have chosen X.25, then there are many suppliers who will supply dedicated switches. These machines do nothing but switch data traffic. Machines running application programs are each connected to their local switch. This connection could be a direct cable, or it could be a LAN.

23.6.2 Local Area Networks

The first point to make about a local area network is that no matter how much you analyse the requirement today, in a year's time a large number of those who need to use the network will have moved from one part of the office to another. The key difference, then, between designing a WAN and designing a LAN is the rate at which the configuration can be expected to change. Flexibility is the keyword. No-one will thank you if, when the Chairman has decided to move from the fifth floor to the sixth floor, you have to tell him that he can't be connected to the network for six months.

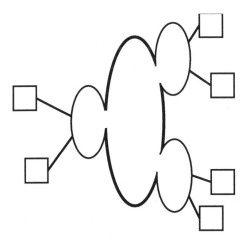

Fig. 23.6 Token ring made of a spine and feeders

Structured wiring Many new office blocks are pre-wired. Structured wiring consists of permanently laid cable which runs between plug panels. By plugging these permanent cables together in different ways many different network configurations can be built. In each different configuration the use to be made of some particular cable might differ. In one case it could be part of a point-to-point link; in another, part of a token ring.

Spine and feeder A common structure for a LAN is a central spine which could link buildings on a site, or floors in a building. From this central link smaller feeder networks are connected.

A token ring network might look like figure 23.6. Here a main loop connects smaller loops, offices perhaps, to which users are connected. The Ethernet equivalent of this is based on a bus structure as in figure 23.7. The central spine might run through a service shaft in a tall building. The feeders connected to it could service different floors in the building.

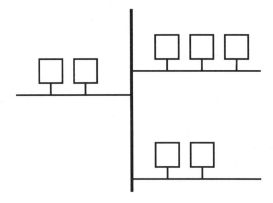

Fig. 23.7 Ethernet made of a spine and feeders

23.7 SUMMARY

There are a few key points which you need to remember. Firstly, in most cases, a potential user of a communications network will already be solving the networking problems in some way. There can be many different ways in which this can happen:

- Uncoordinated leased lines
- Dial-up connection
- The postal service
- Motor cycle courier
- Fax.

These existing information flows indicate places where the user has found a need to provide ad hoc solutions to parts of the networking problem. They are places where the analyst must pay particular attention and get answers to the following questions :

- How much data has to be carried?
- How often?
- How far?

Secondly, having decided where the information must flow, start the network design by separating the layers of the network. The lower layers will normally be provided by the computer manufacturer. The upper layers, however, will be concerned with what the user wants to do. The upper layers are where you will spend most effort. This is where the application problems must be solved. The first stage in solving these problems is to recognise that the analysis of a communications problem is no different from the analysis of any other problem. The user has a problem. You have to understand it. Having understood the problem as a whole, you will be able to see where there are requirements to carry data from one place to another. At this point you need to start to consider the technology to be employed to carry the data. That is, you need to start to consider the lower layers:

- Does the volume of data merit a dedicated line?
- Do the distances involved suggest a LAN, a WAN or both?
- What communications products are offered by the user's existing suppliers?

The answers to these questions will give some idea of the structure of the network and the products which might be used to build it.

Having obtained some idea of the requirement and of the structure of a network which could satisfy it, return to the design of the upper layers. Use the OSI Reference model as a checklist:

- Application Layer:
  Is the problem one involving simply file transfer, or is a more complex interaction required?
  Is there a need to provide an electronic mail service which may need to store messages within the network?
- Presentation Layer:
  Are there several different types of computer involved?
  If so, are their data formats compatible?

- Session Layer:
  How will the connection be managed?
  Who will make the connections, and who will break them?
  What needs to be done if a connection fails?

Finally, don't forget the costs. It is difficult to give any guidance in this area as prices are continually changing. However, if you are building a complex WAN it is usually cheaper to buy your own switches, and to rent the links between them. Also office workers tend to move desks often, so, ideally, a LAN should be easily reconfigured. The trade-off is between the initial high cost of structured wiring, and the higher continuing cost of reconfiguring an inflexible installation.

This last point leads to one final consideration which must always be borne in mind. The objective is not to install a communications system, but to use a communications system to solve a problem. The cost of the communications system must always be seen in the light of the benefits to the user in terms of solving the problem. A sophisticated network is no use if it doesn't solve the original problem or if the user can't afford it

CHAPTER 23   CASE STUDY AND EXERCISES

Q1 Describe three different local area network protocols and identify the circumstances for their likely use.

Q2 LANS and WANS are used in different circumstances What are the different circumstances for using them? When is one more appropriate than another?

Q3 One of the regional offices of System Telecom operates in South Western France as an 'Agence'. It sells System Telecom products and services as well as other telecommunication products such as pagers, answer phones etc. It now wishes to install a computer system for office administration. It could have a centralised machine or a LAN-connected PC network. What are some of the factors to be considered in deciding which to install?

CHAPTER 24

Systems Design: *Systems Implementation*

24.1 INTRODUCTION

Buying a new bespoke software system is different from any other purchase an organisation may make. Premises, furnishings or company cars may take weeks or months to purchase but often they are instantly available. The client knows exactly what is wanted before buying and can return things that are unsatisfactory. Bespoke software systems can take months or even years to analyse and design and years to develop. So the client expects the system to be in place for a number of years and to be able to cope with changing business needs over its lifetime. For these reasons the implementation phase must result in a system that meets the client's needs at the time of system delivery, and yet is flexible enough to adapt to unforeseen future requirements.

24.2 CODING AND UNIT TEST

This is the first step in the implementation stage after low-level design. There are two ways of producing the software code. There is, and probably always will be, the traditional method of employing a programmer to code, test and debug each software module. The greatest cost to most projects is during this activity because there are more people to pay and with that comes a higher probability of confusion, inconsistency and slippage and the need for more management. The alternative to using programmers is to use a code generator. A *code generator* is a suite of programs that matches the input to an appropriate code template and from these produces modules of code. To prevent errors, the input is necessarily tightly defined. Some code generators can take diagrams such as structure

Assets	Investments
Decision to buy is simple.	Decision to buy takes time and deliberation.
Short purchase time.	Long delay between purchase and delivery.
Immediate business improvement. 'No quibble' guarantee.	Return on investment is long term. Customer may be as responsible as the developer for problems.

Fig. 24.1 The difference between assets and investments

charts as their input, whilst others require a formal structured English. Of course both methods – programmers or code generators – have advantages and disadvantages so it is essential to return to the system requirements before making the choice.

24.2.1 Employing programmers to write code

Programmers are trained to write code in second, third or fourth generation programming languages. First generation programming languages are the languages the processors understand: pre-defined binary patterns. We still refer to 'machine code' when talking about the collection of 0s and 1s that are indecipherable to us, but essential to the processor. No-one programs in machine code today because of the high probability of error (it's easy to inadvertently swap a 0 for a 1), and the low probability of tracing such an error (there are no machine language level debugging aids).

Second generation programming languages are usually called assemblers or macro code. A second generation language is specific to a particular hardware and may not even be portable across different hardware from the same manufacturer. For example macro 10 ran on a DEC-10, macro 20 ran on a DEC-20 and although there were similiarities, macro 10 wasn't portable to the DEC-20.

Third generation programming languages came into existence in the late 1950s and 1960s when the need for software portability and transferable skills grew. In the 60s it was estimated that around 70% of systems development projects were never completed, and of the 30% that were delivered few, if any, were error free. Nowadays the customer takes it for granted not only that the system will be delivered, but that it will work.

Fourth generation programming languages came into being in the late 1970s and began to be widely used in the 1980s. There are two main reasons for their success. Businesses were experiencing rapid growth and needed systems delivered and updated quickly to support that growth. Advances in technology led users to expect user-friendly systems that they could develop or modify themselves. Anyone with experience of using a computer can learn enough about a 4GL in one day to write some basic report and screen input programs. 4GLs take, as their input, one line of near English and pick up the appropriate code templates to produce the required results. Because each template may contain redundant code, a program developed in a 4GL is not as tuned to the user's problem as a 2GL or 3GL program. The final system will therefore have a percentage of unused code which takes up memory space and slows down the system performance.

One line of a 4GL is equivalent to about 10 lines of a 3GL, each line of a 3GL in its turn is equivalent to about 10 lines of a 2GL, each of which is equivalent to about 10 lines of machine code. So each generation gets closer to English, and is exponentially more powerful than the preceding generation (figure 24.3).

Taking account of the non-coding activities of program design, test and documentation, it is estimated that the average programmer produces ten lines of working code per day. The more powerful the language therefore, the more productive the programmer. The short learning curve and high output associated with 4GLs should mean that there is less need for specialist programmers, and that users can write their own systems. The reality however is that instead of 4GLs making programmers redundant, programmers with 4GL experience are in more demand than 3GL programmers.

The main benefits of employing programmers are:

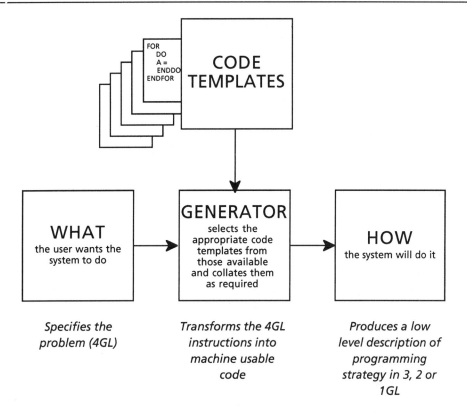

Fig. 24.2 How a 4GL works

- The programmer is a trained expert in the programming language, and perhaps in the hardware and operating system, to be used.
- Programmers think and they make connections. They might see how something in one program conflicts with something in another program and by raising the issue they can save time, trouble and money.

Obviously there are problems with employing programmers – they are human! They may not have the skills to do the job which could mean training them; they may not want to do it which could mean supporting and motivating them; they may leave, which could generate recruitment and familiarisation costs. The biggest problem in employing programmers is that, just like everyone else, their human behaviour is unpredictable!

24.2.2 Using code generators

The alternative to using programmers to write code is to buy a software package that generates the code automatically. These are tools that take as their input, program structure diagrams or structured English, and convert these into lines of third or fourth generation code. Some gurus such as James Martin predict that eventually the programmer will become obsolete. People are expensive, hardware gets cheaper, so they argue that it is more efficient to buy a larger, faster machine and use code generators. To date, experience

4GL	3GL	2GL	1GL
Fourth Generation	Third Generation (High Level Languages)	Second Generation (Macro Code or Assembly Language)	First Generation (Machine Code)

JOIN PRODUCT	A = B + C + D	MOVE B, R1
		ADD C, R1
		ADD D, R1
		MOVE R1, A
			0001 001000 000001
			0110 001100 000001
			0110 001110 000001
			0001 000001 001111
			1010 001110 010101
			1110 110100 100110

Fig. 24.3 Increasing power of programming generations

of using code generators suggests that programmers are still required to tune the generated code to meet performance requirements. This is likely to be the case for the foreseeable future.

The main benefits of using a code generator are

- It is relatively cheap. Once bought the generator can be used to generate several systems.
- Changes are incorporated by updating the design and regenerating the system.

Using code generators is only feasible where the cost of the generator is less than the estimated cost of employing programmers. Another point to take account of is what happens to embedded code, such as code written by a programmer to improve system performance, when the system is regenerated? Such code should be stored in separate subroutines to ensure that it remains unaffected by any system regeneration.

24.3 TESTING: ENSURING THE QUALITY

An error or anomaly in program code can remain undetected indefinitely. To prevent this from happening the code is tested at each of the levels shown in figure 24.4. To successfully test a system, each condition, and combination of conditions, has to be tested. Just as the system was decomposed by the analysts into more manageable subsystems or functional areas and then further into programs and subroutines, now these are used like building blocks. Each program is tested and linked to other programs. This unit of programs is tested, and linked to other units and so on until the full system has been tested.

The first level of test is *unit testing*. The purpose of unit testing is to ensure that each program is fully tested. To do this the programmer writes a test plan. The plan consists of a number of test runs such as the valid paths through the code, and the exception and error handling paths. For each test run there is a list of conditions tested, the test data used and results expected. A designer or team leader then reviews the plan to check that each path through the code will be tested correctly. The programmer is responsible for creating, or selecting from a ready prepared pool, test data which will produce the required test conditions. An example from a test plan is shown in figure 24.5.

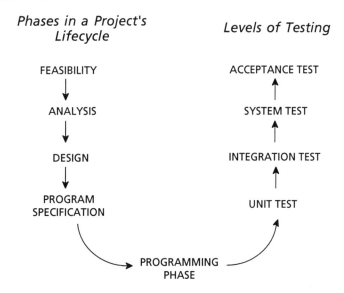

Fig. 24.4 Levels of decomposition and testing

Following testing, all errors are investigated by the programmer with the support of a reviewer. The errors may be in the program code, the test data used or even in the expected results if these were incorrectly specified. The deliverable from this phase is a working program, its design documentation and the related test plan including expected and actual results proving that the program works as specified.

The next step is *integration testing*. Sometimes this is called link, subsystem or level one testing, and it is an intermediate step between testing each program in isolation and testing the whole system. The purpose of integration testing is to test the interfaces between programs in the same functional area. The output from unit testing becomes the input to integration testing. The tests are defined by designers and are carried out under their supervision. Each program is linked to all the other programs with which it interacts. What is being tested is not only that the data is correct and in the correct format, but that it happens in the specified sequence and within the specified response time. The deliverable from this phase is a number of integrated subsystems again accompanied by test plans, expected and actual results. Figure 24.6 shows how integration testing works for the VCR example (page 350) where all the interfaces between the 'respond to operator command' program and all programs it interacts with are tested.

All of the applications programs are now linked together for *system testing*. The purpose of system testing is to test the whole system exhaustively including any additional housekeeping functions like file archiving. This is the developers' last opportunity to check that the system works before asking the client to accept it. For large systems, system test is run by teams of programmers supervised by analysts or designers who may have to resolve, with client representatives, any issues that arise. Because the test plan must be followed without any deviation non-technical staff such as data entry clerks are sometimes used as system test runners. They run the tests, collate the results and identify mismatches between actual and expected results. In reality no system is ever completely

384 Systems Analysis and Design

Video Cassette Recorder Conditions list	
CONDITION	COMMENT

1 key pressed
  1a record
  1b play
  1c rewind
  1d fast forward
  1e eject
  1f stop
  1g pause
	The sequence and timing of key presses cannot be predicted

2 current state of VCR
  2a recording
  2b playing
  2c rewinding
  2d fast forwarding
  2e stopped
  2f paused
	The system may be left in this 'current state' by a previous test run or the tester may have to set the VCR into this state before running their test

3 VCR tape status
  3a non-write-protected tape inserted in VCR
  3b write-protected tape inserted in VCR
  3c no tape in VCR
	Check whether tape is write-protected before insertion

4	
5	

	Condition tested	Expected response	Actual response
1	2e, 1f	VCR remains stopped	√ worked as expected 25.7.93
2	2e, 3a, 1b	tape plays, picture appears on TV screen	√ worked 26.7.93
3	2e, 3c, 1b	VCR remains 'stopped', indicates 'tape not inserted'	√ worked 26.7.93
4	2e. 3a, 1a	VCR remains 'stopped', indicates 'write protected'	X VCR started recording

Fig. 24.5 A sample test plan

tested, usually because of time constraints but also because it is impossible to predict and accurately simulate every combination and every sequence of events that may happen in the live environment.

Since system testing is more complex than unit or integration testing, it is often split between test running and bug fixing. This helps ensure adequate control over the way corrections are applied to the system and system test version control documents are kept

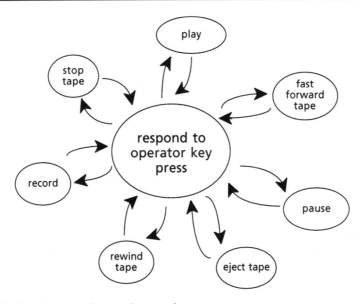

Fig. 24.6 How integration testing works

as in figure 24.7. The system testers document the errors, each bug is assigned a priority, the bug fixers amend the code, and a new version of the system is released. This doesn't mean that all top-priority bugs will be fixed before the next system release. Some may not be completed in time. Also some lower-priority bugs may be included in the next release because they can be fixed quickly and easily, or as a by-product of fixing a high-priority bug.

System testing incorporates a number of other classes of testing: performance testing, volume/soak testing and regression testing. Each of these three requires that the whole system has been tested before they can begin. The purpose of *performance testing* is to validate that all response times or transaction periods specified in the functional specification can be met by the system, especially when it is fully loaded. This will include timing how long the system takes to respond to a user request, timing normal case paths through processing, and exception cases. *Volume or soak testing* ensures that the system can handle the expected number of users or transactions. Let's say a system is to have 10 keyboard operators logged on, processing up to 5000 transactions per day, with a growth factor of 20%. Then a volume test of this system would be 12 users logged on, processing 6000 transactions. Since this would be labour intensive, it helps to write a program that simulates system use by the required number of people over a specified number of hours. The final stages of system test usually involve a *regression test*. This may comprise a

	Error number & description	Test document reference	Priority	To be released in system version
1	VCR records onto write-protected tape	REC/04	1	3	Oct 94
52	'write-protected' indicator does not flash	INS/17	3	2	Aug 94

Fig. 24.7 System test version control

selection of system test runs, or may be specifically written to test only key functions. The purpose of regression testing is to ensure that corrections during system test have not introduced other bugs. This test may be kept for later use by the maintenance team to check that any enhancements they carry out do not introduce any errors into the system. The deliverable from the system test phase is a fully operational system, accompanied by the system test planning documents, test scripts, expected and actual results.

The user formally accepts the system when it has sucessfully passed the *acceptance test*. The purpose of an acceptance test is to prove to the client that the system meets the business requirements agreed in the functional specification. Simulator programs may still be used in place of third party systems, but where possible these are replaced by the real systems. Any remaining test data is replaced with live data provided by the client. The acceptance tests are run by client staff or under client supervision to ensure that the developers are using the system as the eventual users will use it.

During the course of acceptance testing, the client records all errors, discrepancies and aspects of the system with which they are unhappy. These are then discussed with the project manager to identify why they happened and who bears the cost of correction, the client or the developer. Errors are usually corrected at the expense of the software developers, whilst changes are incorporated at the expense of the client, and the tests are rerun. When all problems have been resolved, the client signs for acceptance of the system. The deliverable from this phase is a system that works to the satisfaction of the client as defined in the requirements specification document and any related change request documents.

24.4 DATA TAKEON AND CONVERSION

In the transition from the old system to the new it is essential that the data the organisation already has is safely transferred to the new system. It is unlikely that the format of the data on both systems will be the same as the organisation may, for example, be moving from file-based data to a database system.

One of the number of ways of getting the data onto the new system in its new format is to employ data entry clerks who can quickly enter quantities of data. This was the traditional method when moving from a manual system to an automated one, and it is still used in some cases. However because of the repetitive nature of the task, errors do occur and can be difficult to trace. Data can also be entered onto the system by the users. This is a useful way of building users' skills with the new system and is often done as part of their familiarisation and training. If this is done early enough in the project lifecycle it can save programmers, designers and analysts time defining test data and provide them with a plentiful supply of normal-case data. Error and exception case data would still have to be defined. Helpful as it is, there are two major drawbacks with this method. Firstly it is the most error-prone method because it is being carried out by user staff whose priority is to learn how to use the system. They may not be checking the accuracy of their input, as they are just getting used to and appraising the user interface. Secondly, if live data is taken on early in the project development to be used as test data, and the system goes live months or years later, some of that data may be out of date.

An approach to data takeon which is less prone to error is *data conversion*. Data in the old format is run through a program, or series of programs, to convert it into the new format. The conversion program is not part of the applications code and has nothing to do with the application under development. It is written to enable the transition from one system to another and once that transition is successful it will never be needed again. The conversion programs have to be analysed, designed, specified, coded, tested and reviewed like any other small system and may therefore be costed separately to the cost of developing the application system.

Conversion can also be from one hardware medium to another. An independent bureau is usually employed to do this. These bureaux are companies whose business is to take input data in one form and return it to their customer stored on another medium in the appropriate format. To illustrate this, imagine someone buys a new hi-fi comprising a CD player, cassette player and radio. What do they do with their old records? A hardware conversion solution would be to hook up their existing record player to the cassette player and record the LPs onto cassette tapes. Some organisations use bureau services regularly, for example to archive data from magnetic tape onto microfiche. Other organisations will only require these services at times of changing from one hardware platform to another.

At the same time it is also important to consider whether archived data should be converted to the new format, or whether it is better to maintain a method of accessing archived data in the old format. In our example that would mean keeping the record player and using it whenever the owner wanted to listen to that music. If the client decides that some or all of their archived data should be accessible to the new system, it is not necessary to wait until after the system is installed. Conversion can be done beforehand but not if it diverts essential manpower at a critical time, for example if it would postpone the 'going live' date. Each situation is different. The key is to make sure nothing is overlooked. If data on the old system is changing right up to the time when the old system is switched off and the new one switched on, it's important to know how those changes will be recognised and added to the new system in the correct format. The logistics of this would be difficult if data conversion relied solely on bureau services.

These are the main points to consider for a sucessful data takeon.

- How much memory storage area will the data require, including any archived data that may be required on backing store?
- Will the client require archived data on 'going live'?
- Is it cheaper to provide a system prototype and ask users to enter a quantity of live data, or allow time for designers to create test data?
- Which approach to data conversion suits the client's and the developers' needs best? Using bureau services? Using specifically written software?

Errors made during conversion are serious. The new system cannot go live with corrupt or partially complete data areas and if system changeover fails due to incorrect data there must be a contingency plan. This might involve running a minimum service on the new system, or reverting to the old system.

24.5 USER TRAINING

To be successful, user training requires a learning environment that includes competent trainers, and enough time to train properly, based on well defined training objectives.

The training should include not only training in the day-to-day functions but in all the functions the system offers. Training takes place over a long enough period of time to allow users to feel competent with each skill area before developing the next. Where possible users should be trained in the live environment or at least in a simulated one. This is especially helpful when learning how to use fallback procedures because if the users ever have to do this in practice they will be under pressure and not feel confident in their skills.

When deciding who trains the users there are three possibilities: the system developers, more experienced client staff or a professional training company. Systems developers are sometimes employed to train the users because they understand how the system works, but they often have a technical bias and train the users by constantly focusing on the technical operations of the system, rather than how to use the system to serve their business needs. It is tempting to think that experienced staff, perhaps supervisors or managers, make the best trainers, but this is often not true either. Although experienced client staff will have a greater knowledge of the business they may be so far removed from the users abilities and difficulties that they don't see them, and if the user doesn't feel free to admit to difficulties because the boss is the trainer the training will fail. Often professional training companies sell the fact that they don't know the system as their major strength. Because they have no pre-knowledge, they make no assumptions and they can see things from the users' point of view. They don't use the technical jargon of systems developers and once they have learnt about the system they can communicate that knowledge to the users in a way that is helpful to them.

This last option is usually the most expensive, but to ensure that a system is fully used and is well recieved by the users it could prove to be money well spent. A painful introduction to a new system will prejudice users against it for a long time afterwards.

There remains the question of whether to train all system users, or to train a few key users who will then train staff in their area. The first option allows for all users to make mistakes in a safe environment but if there are a lot of users to be trained it can take a long time. The second option is cheaper but makes it critical that those who are trained develop a sound knowledge of all the functions of the system.

Training should be scheduled to take place during the transition from the current system to the new one. Learning from a training programme that takes place weeks or months before work with the live system will fade. If the training takes place after the user has already worked with the system they may be able to iron out some difficulties they have experienced. Conversely the users may have already acquired some bad habits.

Improving the system's user interface will also ease the transition from the old system to the new. The aim is to allow the user to sit at a terminal and within a short time to be able to use the system. This means discarding the traditional command languages which can be cryptic and difficult to remember in favour of WIMP interfaces: windows, icons, mice and pull-down menus. The pros and cons of all of these are covered in detail in chapter 15. Any difficulties the user encounters are dealt with by the system's context-sensitive help functions. These identify the problem and suggest steps to resolve it. In this way the user learns about the system as and when they need it. A good user interface means less formal training is required. However providing a user-friendly interface but no training, or a poor interface and thorough training, is not helpful. Both areas need to be user-centred to provide the best results. These issues are explored further in the next chapter.

24.6 GOING LIVE

Thorough preparation for going live is essential to ensure the success of the system. The steps in this process are:

- Installation on site
- Site commissioning
- System changeover

The task of installing a new system has degrees of difficulty depending on whether the hardware is already in place and installation is simply a matter of loading new software, or whether the hardware and software both have to be installed. If the hardware is already running then only the new software system has to be installed. This type of installation might occur where the hardware was previously installed for earlier applications or if maintenance programmers are releasing a new version of the system to implement a change. If hardware has to be installed then there are issues of site preparation to be considered. The main ones are:

- Is the area big enough?
- Is the layout appropriate?
- Is the environment appropriate? air conditioned? dust free?
- What communications lines need to be installed?
- What other hardware has to be linked to the new system?
- Does it require a separate, or backup, electricity supply?
- Is the site secure? from people? from natural disasters?
- How will the hardware be transported?
- When will it be delivered to site?

Installation must be planned like a project to ensure that all tasks are carried out in the appropriate sequence and within the timescale and may well be managed by an estates department or by a clerk of works. The risk is that the site may not be ready on time.

Site commissioning occurs after the system has been accepted as complete by the client and has been installed on site and connected up to any third party components. For example a railway station control system could consist of plant management software that can be connected to other subsystems as in figure 24.8.

The system may have been accepted by the client when the control system ran using simulator programs which behaved in the way the signal, train location and points control subsystems were expected to behave. However the way the subsystems are expected to behave, and how they actually behave, may be quite different. This is why site commissioning tests are run, to identify all discrepancies in the interfaces between systems and to correct them.

The first activity involved in commissioning a system is monitoring the interfaces. This entails connecting the station control system to the signal subsystem, attaching a data line monitor to the communications link between the two systems. Having set the hardware up in this way, the installers then check that for each message sent the specified reply is recieved within the defined time period. The interfaces between all subsystems are checked in the same way. All errors are corrected before the acceptance tests are run again. This time the real subsystems take the place of simulators. Commissioning can also be done by running the systems live, but it must be remembered that work produced in this way

390 Systems Analysis and Design

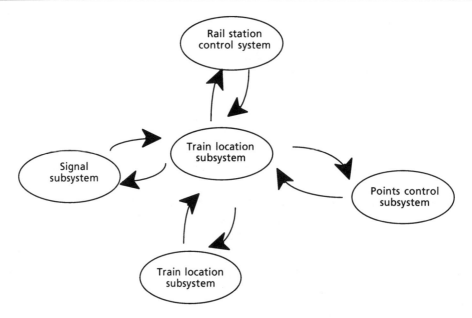

- Are "hand shakes" between the subsystems and the control system the same as between the simulator and the control system?
- Are message formats the same?
- Is the sequence of messages received by the control system as tested using simulators in place of the subsystems?
- Do the subsystems handle error/exception conditions as expected?

Fig. 24.8 Commissioning a system

may be wrong, or will at least take longer to complete than true live running. This is because the main reason for this live running is to test the interfaces, not to run the organisation's day-to-day business. Until the system is proved to work, that is a secondary consideration.

System changeover. Some decisions about development, user training and data takeon will be affected by the method of system changeover the user has selected. The choices are phased installation, direct changeover or parallel running and this will need to have been agreed before the analysis phase is complete because it will affect the system design and will impact on project planning.

Let's assume the latest possible date for a system to go live is December 1st 1995. Figure 24.9 shows what the three possible types of changeover might look like.

Phased installation is where one part of a system is installed and run live for a period of time. Then a second part of the system is added and both are run live for a period. Any remaining parts of the system are dealt with in the same way. Systems can be phased by functionality and by geographical location.

If changeover is phased by functionality, the core functions of the system are developed and installed, and as phase 1 development nears completion, phase 2 development begins. As phase 2 nears completion, phase 3 development begins. This allows the user to become familiar with the basic system, then resolve the integration problems between the part

Systems Design: Systems Implementation 391

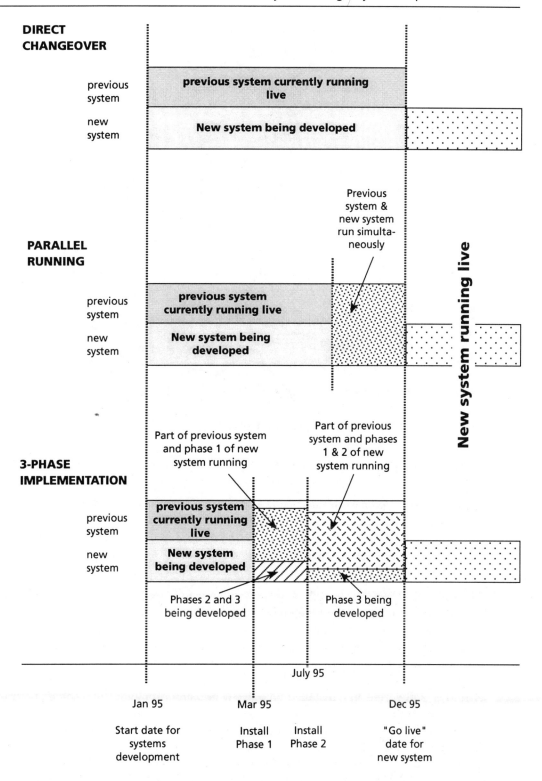

Fig. 24.9 The three types of system changeover

Fig. 24.10   Installation phased by location

of the system already installed and the newly installed phase. The user might have a number of reasons for choosing this type of changeover:

- They get the core part of the system faster than if they had to wait for the complete system, so the system may start to pay for itself earlier.
- User learning is spread over a longer period of time so it is less stressful.
- In spreading the development over a longer period, the client organisation will also be able to spread the cost.

This is also the type of approach people might have to buying personal computers. First they buy something that meets their immediate needs: a monitor, keyboard and disk drive for running a word processing package. Then they buy a printer, then a mouse, then a windows interface package, then they might decide to upgrade the disk drive and so on. The pace is determined by their needs, their speed of learning and their available funds.

It is also possible to phase installation by geographical location, as shown in figure 24.10. The software is first installed on one site and any problems are resolved before it is installed on the further sites one after the other. Some clients do this by identifying a test site and select one of their branches to receive and test new software before it is released to all other sites. Sometimes systems are phased by location so that the installers learn from the first experience and apply that at subsequent locations. In figure 24.10 the sites are ranked from lowest risk to the business if problems occur, to the most damaging risk. In this case the Bristol operation is the smallest so that is installed first, and the London operation is the head office so that is installed after all the errors have been found at the other sites. Alternatively, the client may have decided to install the system on the test site, test it thoroughly and then install it at their main office to get the most out of the new system quickly. The client must identify the objectives when installing by location to ensure that the sequence and timing of installations is the best possible.

Direct changeover occurs when at a given time on a given date one system must end and its replacement must start. This type of changeover often results from changes in legislation. Most new financial legislation comes into effect at the beginning of a new financial year; changes that may require significant alterations to computer systems to support the new policy must be ready on time.

There are advantages to this type of changeover. It is the cheapest option and it provides a clean break between the old and the new. But when it goes wrong it can leave the client unable to carry out normal business. This was what happened on the day of the 'big bang' when the London stock market was deregulated. Trading started, the systems could not cope with the number of people logged on and the systems crashed. Because this kind of failure during a direct changeover is so widely publicised, the system and its developers get a bad reputation, and the users are instantly wary about using it. Developers need to have a contingency plan in place so that they know what to do if the system fails irretrievably on changeover.

Parallel running is the most expensive changeover option, and as such may be beyond the financial resources of the client. It requires that both the old system and the new run side by side over a period of time. This time period should be chosen to ensure that cyclic variations such as end of month processing are covered.

The aim of parallel running is to validate the new system by checking the results it produces against the results produced by the old system. Obviously if the results are not directly comparable this option isn't appropriate. To validate the new system in this way it is not necessary to run all of the daily transaction through the new system. It is possible to replicate a percentage, say every tenth or every hundredth transaction, and check the results of these. Alternatively a number of each type of transaction could be selected and followed through the system. Of course to be totally certain, every transaction would have to be carried out twice: first on the existing system and then on the new system.

Parallel running means that both systems must be staffed and this issue alone may well influence the decision about whether every transaction or only a selection of them are run twice. The client will have to employ temporary staff and decide where to use them. Is it best to run the old system using temporary staff, which could be less efficient than using the permanent staff and may introduce errors, or to use them to run the new system, and deprive the permanent staff of a valuable learning opportunity. Whichever option is chosen, the challenge is to manage the staffing to best effect.

24.7 THE MAINTENANCE CYCLE

However long the development period, the maintenance cycle will be several times longer. This is because the system must first of all pay for itself, and then it must provide a return on the investment. Development is complete and the system is now in daily use, serving

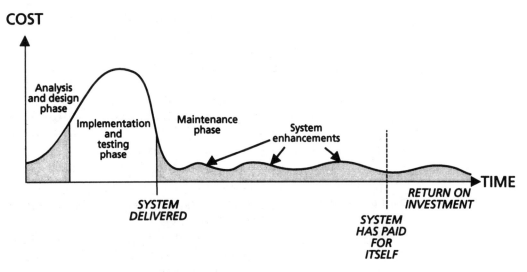

Fig. 24.11 Lifetime system costs and return

the business needs of the client organisation. Initially the client and developers may negotiate a warranty contract. This is similar to guarantees that accompany hardware with the important difference that no warranty work is done free of charge. The client estimates how much support they are likely to need, and a monthly or quarterly schedule of payment is agreed. If the client uses less than they contracted for, they are still liable to pay for the full contract, although this is often negotiable. If the system is relatively stable the contract is reduced and if the client wants some major enhancements the contract is increased. So, even after the system has been delivered it still costs the client money.

Operation is the normal everyday use of the system by the users. As they operate the system under business conditions, they recognise what they like and don't like about it. They evaluate it. This is especially true if they encounter errors or difficulties. Occasionally they will suggest improvements, additions or amendments that would make their job easier. This evaluation phase often provides more work for the software developers, and so the maintenance cycle continues into a new inception phase, and so on round the loop once again as shown in the b-model in chapter 7.

24.8 SUMMARY

Implementation is the longest phase on the project and the most labour intensive. The deliverables of this may seem less significant to the developers, and the tasks more repetitive. Each developer may have a limited view of the system, understanding only the area they are working on. For all of these reasons the major difficulty during system implementation is managing the software developers. Analysts and designers have already defined the system, the working methods, procedures, and standards to be followed. Implementation can only succeed if these are followed. This means that much of the implementation effort is spent in reviewing and testing.

CHAPTER 25
Change Management

25.1 INTRODUCTION

Up to now the focus of this book has been the technical issues surrounding software development projects. The preceding chapters have attempted to provide a thorough grounding in the concepts underlying Systems Analysis and Design, as well as plenty of practice in using structured methods. However, there are a number of non-technical issues that organisations must address in order to make the transition from old system to new as smooth as possible. Indeed, there is mounting evidence that unless these non-technical issues are handled carefully, the system will not deliver the benefits it was designed to achieve.

In the 1960s and 1970s, when the automation of back-office tasks was the primary objective of many IT systems, IT delivered tangible gains in productivity. Today, the picture is less bright. Many organisations are reporting disappointment in the payoffs from their IT investment. In 1988, the Organisation for Economic Co-operation and Development (OECD) stated: 'IT is not linked to overall productivity increases'. This conclusion was echoed in the same year by the Kobler Unit, whose researchers found no correlation between the overall IT spend and business efficiency. A major survey in 1990 of senior IT executives from businesses across Europe found that only 27% felt that their organisations were very successful at exploiting IT. And in 1991 the *Financial Times* ran a story about the failure of banks to get the desired benefit out of their new systems. The 1980s were littered with investments in technology that failed to reduce headcounts.

All this is in stark contrast to the optimistic claims made for IT by theorists and practitioners. According to some, IT should not just be helping an organisation reduce headcounts, it should be supporting improvements in service, linking customers and suppliers, providing high-quality information to managers, and so on. But very few organisations even get close to realising these wider benefits of IT. What is happening here, why are so many IT systems seemingly unable to deliver the goods? Is there any way of reversing this trend? And what is the role of the systems analyst or designer in all this? In this chapter we discuss some of these non-technical reasons and suggest that for every technical project there needs to be a parallel people project. This people project will be concerned with the management of change and so this chapter introduces the theory of change management and describes some of the activities that must be carried out and the role that analysts and designers can play in the management of change.

25.2 INFORMATION TECHNOLOGY AND PEOPLE

The crisis outlined above has arisen largely because IT professionals and managers in organisations seeking to benefit from investments in technology have overlooked a critical success factor: the people who actually use IT systems. Without adequate consideration of their needs the system being introduced is likely to fail no matter how well-designed it is. For example, we need to be very clear about

- Who they are
- How they are motivated
- What they know about the system they're getting
- The aspects of the change to a new system they are likely to resist, or embrace
- The skills and guidance they will need so as to get maximum benefit out of the new system

The best way to make sure that these issues are addressed is to develop a people project in parallel with the technical IT project. This people project would use the theory of change management as its framework. It would aim to make the impact of new IT systems on an organisation as positive as possible, by ensuring that users and managers

- understand the objectives of the change to a new system, and are committed to achieving them
- have realistic expectations of what life with the new system will be like
- know exactly what will be required of them before, during and after implementation
- get the right level of support throughout the change.

To highlight the importance of the people project, let's look at a typical organisation that's just implemented a new computer system. Two weeks after implementation the office is in disarray. Clerical staff are floundering in a sea of paper because an enormous backlog of work has built up, and whenever a customer rings in with a query, they're told: 'Sorry, we've got a new computer, I'll have to call you back.' Disagreements about who is responsible for updating a key piece of data have led to disagreements between the customer service and marketing departments. One group is actually staging an informal strike. The system doesn't work the way it was supposed to; their representative says 'We were promised things eighteen months ago but now they won't be delivered till Phase 2. When is Phase 2?' Meanwhile, the supervisors have disappeared into their offices, dazed and confused. All the tasks they used to do – checking their subordinates' work, giving expert advice, authorising payments – are now done automatically by the computer. As far as they can tell, supervisors no longer have any role to play. Upstairs in the conference room, senior managers are demanding an explanation from the IT director. Why doesn't the computer system work? It works perfectly, replies the IT director, and it does work perfectly but the people who need to use it are not ready, willing or able to use it.

Now let's imagine the alternative. Two weeks after implementation, the office is running smoothly though it's a very different place from what it was before the new system was installed. Clerical staff have organised a telephone section to deal with enquiries from the public, leaving the computer users to work without interruption, at their own pace. Issues of data ownership have all been resolved by a task force of users drawn from the different departments affected by computerisation. Key users are on call to help with any difficulties that staff experience; as the first line of support, they're able to filter out ninety percent of the minor problems that would otherwise be flooding in to the computer department help desk. Staff who won't be affected by the changeover until Phase 2 have had a few months to get used to the idea, and understand why things are being done in this order and are taking the opportunity to learn from their colleagues' experience. Meanwhile the supervisors are monitoring workflows and making a note of best practices for a seminar with their colleagues from another site. Upstairs, the IT director is receiving the congratulations of senior managers. The system works well, the IT director agrees. But it's the people that are really making the difference. These scenarios are not exaggerations or caricatures: they are drawn from life. A people project really can make the difference between a successful IT system implementation and a disastrous one.

The foundation for a sound people project is effective communication. A good communication plan is therefore an important part of any change management programme. In chapter 3 we discussed some of the skills that people need in order to communicate effectively with one another but in the people project effective communication has the specific goals of:

- Raising awareness of the objectives and potential benefits of a change
- Giving users information about what to expect from a new system, and what their role will be
- Ensuring that people know what support is available, and how to access it.

Remember too that communicating means influencing people as much as informing them, helping to shape their attitudes, building their commitment and gaining their co-operation.

25.2.1 The role of analysts and designers

The primary role of systems analysts and designers is, of course, to produce a computer system solution to a problem that meets the customer's requirements. This task can easily be so absorbing in itself that there is seemingly no time left over for thinking about the non-technical issues surrounding the introduction of a new IT system, much less for setting up a people project to address them.

So even if the people project is not driven by analysts, designers, or even IT managers, it needs their active support. Many of the tasks carried out by analysts in the early stages of an IT development project have outputs which the people project will need to draw on. For example, the process of creating data models and data flow diagrams may raise questions of data ownership, which need to be fed to the people project to resolve, perhaps through a redefinition of roles and responsibilities or the introduction of a new procedure. Likewise, if systems analysts have done a detailed assessment of costs and benefits, this will give the people project some idea of the messages they can use to sell the new IT system to users and managers.

Analysts can also draw on the people project for valuable help in areas such as human–computer interface design discussed in chapter 15. The look and feel of the HCI can be one of the most significant factors in determining a user's response to a system. The people project can help create the conditions in which HCI design can be done collaboratively, thus ensuring that both sides get what they need from this all-important aspect of the system.

Very often, analysts and designers do more communicating about an IT project than any other group. Analysts have extensive contact with the people who will ultimately use the system, as well as with the budget-holders who are actually paying for and steering the development project. They probably know more about the user community's expectations and desires than line managers. They are often more up-to-date on decisions about the final size and shape of the system, when it will be implemented, even what it will be called, than anyone else outside the project steering group. In short, by virtue of their position and responsibilities on the project, they hold a great deal of information, and can play a major role in influencing the user community. For this reason analysts must consider carefully their own communications to users, helping to manage user expectations of the system throughout its development. In practice this means being aware of key messages about the project that its sponsors hope to deliver and reinforcing these messages whenever possible, by word as well as deed, and in any case never contradicting them.

In some organisations, the distinction between IT and the business has blurred sufficiently for a new breed of change agents or internal consultants to emerge. These people have backgrounds in either IT or the business, or both, and can therefore provide a range of services such as advising the Board of Directors on issues and opportunities in IT, writing specifications, designing new business procedures and organising training. Clearly, where such people exist, they have a large part to play in managing the change driven by a new IT system.

25.3 CHANGE MANAGEMENT

The theory of change management draws on a body of research from areas such as group dynamics and organisational development, as well as a vast pool of ideas based on the practical experience of managers. It is not concerned with the rights and wrongs of any particular change. It looks at the process of change itself. We need to begin with two ideas that are particularly useful in understanding the process of change, and defining change management. These are the S-curve and the concept of unfreezing.

For many years the consultants Nolan, Norton have been using a very simple model of what happens in a change process. In this model which is called the S-curve, the x axis is time and the y axis is performance, profit and happiness (figure 25.1).

After the implementation of a new system, performance dips at first because the change has disrupted things. Gradually, the organisation begins to get benefits from the change, but after a while there tends to be a levelling-off, as people get used to things. There might even be a slight decline in performance. At this point the organisation tends to start its next change. Change management is about optimising the curve. It's not concerned with whether the change is the right one, it aims to:

- minimise the depth of the dip, A on the diagram
- optimise the angle of ascent, B on the diagram, so as to
- prevent or minimise the second dip, C on the diagram.

This S-curve is a rather mathematical view of change. The psychologist Kurt Lewin, who worked in the 1930s and 1940s on the behaviour of groups of people at work, developed a different approach.

Lewin's theory says that there are three stages to helping people change. First, you need to *unfreeze* them, make them feel restless and dissatisfied with the present situation. Next, you *move* them and show them the way forward. Finally, you need to *refreeze* them

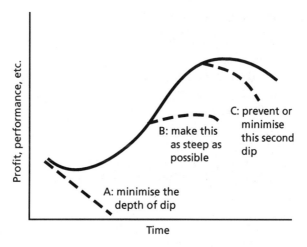

Fig. 25.1 The S-curve

into the new circumstances. Usually, organisations concentrate on the middle step and put lots of time and money into trying to move people to accept new ideas or behaviours. They often fail to convince people of the need to change, and also do not stabilise them once the change has been accomplished. Consequently, the change does not work and people revert to their old ways of working because they feel no motivation to change. To some extent this model maps on to the Nolan, Norton S-curve. The initial dip may be the inevitable cost of unfreezing, the climb is where the real moving happens, and if refreezing does not take place the climb is held back.

Both models have one assumption in common: it is that people resist change. In common with other observations that are so true as to hardly deserve mentioning, this one is typically overlooked. The sponsors of a change often assume that everyone will share their enthusiasm for it, or at the very least accept that it is needed. This is rarely the case. Among the reasons that people give for resisting a change to a new computer system are:

- Scepticism about the ability of any computer system to understand their job
- Reluctance to trust the machine with information that might get lost
- Anxiety about making a mistake that will break the system
- Anxiety about making too many mistakes, and being exposed as ignorant or inefficient
- Anger at becoming deskilled and a slave to the machine
- Fear that the new system will cost them a job
- Fear that the new system or rather the host of computer-related disorders it may bring, such as headaches, repetitive strain injury, lower back pain, eyestrain, and radiation poisoning, will severely damage their quality of life.

These concerns need to be handled sensitively, but in many cases what underlies them all is simply fear of change itself. Organisational change, whether or not it is driven by IT, can threaten a person's status and job security. It brings about new roles and responsibilities, and forces new modes of behaviour on people. The literature of change management furnishes many examples of the old saying that people prefer a known evil to an unknown one; in fact, they will fight to retain it.

25.3.1 Unfreezing, moving and refreezing

But is it really true that people are really so resistant to change? If it were, life would be a much more static and predictable affair than it is. People do undertake to change things about themselves and their world all the time. The key difference is that people don't mind change, so long as it's not imposed on them. If it is, they will resist. If, on the other hand, they feel as though they own the change, they will be more comfortable with it, and be less likely to offer resistance. That's why the first step in Kurt Lewins model, 'unfreezing', is about giving people a reason to change and enabling them to see the necessity for it themselves. The reasons that organisations in the 1990s have for changing – world recession, demographic trends, the growing power of consumers, a new-found interest in the environment – by no means dictate the action that organisations should take. That would be too easy. In fact, once a business, a corporation or a government department has accepted the need to change, there are any number of paths it could follow, each one hedged with uncertainties. Some of the questions asked are:

- Why have we chosen to build a new computer system, rather than buy one?
- Why now?
- How will it help us achieve our business objectives?
- Why are we computerising some functions and not others?
- What will my job be like?

These questions aren't new or surprising, but what is often surprising is the urgency and passion behind them. It is the job of senior managers to deal with these questions and to communicate the way forward as clearly as possible. Without this communication, staff will very quickly become bewildered and anxious, and won't be able to contribute as much as they should to bringing the new world into existence. The dip in the S-curve will be steep and, indeed, there may be no climbing out of it.

The second step in Lewin's model, 'moving', is about telling people why one particular route has been chosen over all others, and preparing them to go down it. People don't change overnight. However much we consciously applaud and embrace the new world, deep down we may still be clinging to the old one. Psychologists talk about the tendency of people under pressure to revert to forms of behaviour that they believe themselves to have outgrown. Observers of the change process make the same point: it's human nature to drop back into familiar patterns of living and working, no matter how obvious the need to change. It is this tendency to revert to old ways after a change has supposedly been accomplished that causes the S-curve to level out early, before any real benefits have been achieved. Users of a new computer system may revert to old ways for any number of reasons. Problems with using the system may give them an excuse to mistrust it. Difficulties getting through to the help desk may leave them feeling stranded or out on a limb. Lack of recognition or positive feedback about the benefits achieved so far may cause them to wonder if this change is worth the trouble. And so on, until one day someone discovers that staff throughout the organisation are running their old, paper-based system in parallel with, or instead of, the computer-based system.

The final stage of the change process is 'refreezing' people, stabilising them in relation to the change. This can be done by catching and solving problems fast, giving people information that reinforces their training, measuring success and sharing information about best practices. In short, by offering plenty of support. This is not to say that refreezing should prevent further change. If it did the new regime would become as repressive as the old.

Although we said earlier that the first step in helping people change, unfreezing, was about overcoming resistance, in truth all the steps are about overcoming resistance. Resistance to change can come from any quarter, at any time. People who once embraced the change become willing to say anything for the sake of appearances but fundamentally remain uncommitted. Perhaps they back out at the last minute, or their commitment dwindles due to lack of positive reinforcement, and they gradually slip back into the old ways. On a computer project, resistance may take various forms. Sometimes, users and managers may:

- spread negative rumours about the new system
- stay away from the walk-throughs and project reviews
- leave data conversion until it's almost too late
- even go on strike.

But careful change management in the form of an effective people project can minimise the risk of these things occurring.

25.4 THE PEOPLE PROJECT

The people project has four stages:

- Creating involvement
- Building commitment
- Providing skills
- Managing the benefits.

Let's look in turn at each of these stages.

25.4.1 Creating involvement

The main activities during this first stage of the people project include:

- Appointing a sponsor
- Setting up steering groups, focus groups, and so on
- Selecting change agents
- Examining the implications of the change to a new computer system for customers both internal and external
- Interpreting user requirements
- Setting high-level business objectives for the project
- Designing new procedures.

A *sponsor* is a senior manager from the business side who acts as change manager and leader of the entire project. Sponsors are a powerful influence on the project, and often become identified with its success or failure. They must be willing to work closely with their IT colleagues, sell the system to other senior managers and generate enthusiasm among users and managers. It is an extremely challenging job, but an essential one. In a recent survey the Amdahl Institute found that all the stories of outstanding achievement with IT always had a specific individual associated with them who had the vision and drive that made it happen.

Focus groups are small groups of consumers whose role is to give feedback to the suppliers of a service or product. The term originated in market research, but is now applied to groups within organisations that have some or all of these characteristics:

- A medium to long-term lifespan (six months or more)
- An advisory rather than a decision-making role
- A membership drawn from the ranks of middle management, to give managers some influence over a change that they might otherwise feel victimised by
- The potential for commissioning work to tackle problems.

A focus group on a computer project might give feedback on priorities for development, evaluate proposals for training users, or help plan the timing of the roll-out to district offices.

No matter how inspiring the sponsor, no matter how helpful the focus group, large-scale change cannot simply be created from above and cascaded through an organisation. The process works best when there are supporters at every level and every site. Ideally, they should be people who are proven influencers, known and respected by their colleagues. In addition, organisations are increasingly recognising that change works best when those most affected participate in its design. *Change agents* may be champions whose role is to influence their peers or they may be full-time members of the project team whose role is to create the change. On an IT project, for example, they may define new business processes, specify requirements, or design an awareness programme for staff.

The other activities typically carried out during this first stage examining the impact on customers, interpreting user requirements, setting objectives, and establishing new business procedures may all draw on work done by systems analysts in the early days of the project. It is important for analysts to bear in mind the aims of the people project to win commitment from everyone affected by the system, to make sure people are adequately prepared for the change, to follow through on the change with the right level of support while carrying out their own research into customer and user requirements, or preparing a cost benefit case, or identifying data flows. The following checklist of questions may be useful:

- Who are the customers for the system, both internal and external? How will they be affected by the introduction of the system? Will they need to provide new or different data? Will the format of reports change?
- Is the user's understanding of the statement of requirements/functional specification the same as yours? Are the users really prepared to put time into this project, or are they showing signs of resisting?
- Are the objectives for this project clear, measurable, and specific? Do they mean anything to users? Is there more than one objective? If so, are some objectives likely to be more important to one group than another? How do the objectives translate into benefits that can be used to sell the system?
- Are there any questions of data ownership? Is there a process in place to resolve them? Is the resolution likely to entail procedural changes, or just a clarification of roles and responsibilities?

25.4.2 Building commitment

The main activities during this stage of the people project include:

- Drawing up a communication plan
- Educating branch/unit managers on the implications of the change
- Marketing the change to everyone affected
- Collaboratively designing the user interface and office layouts.

A *communication plan* sets out the project's approach to communicating the introduction of the system to everyone involved. Earlier we said that effective communication is a key element of any change management programme, and this simple truth cannot be overemphasised. A communication plan specifies the key messages to be delivered to each group of people affected by the change, and the methods to be used for delivering them. It is informed by a detailed analysis of the audiences for communication about the project, the actions required of them, the barriers that may have to be overcome to persuade them to carry out these actions, and the benefits that will result if they do. Clearly, the preparation of a communication plan is another area to which analysts and designers, with their detailed knowledge of who will be affected by the system, can contribute.

A good communication plan will specify who is responsible for communicating to whom. Analysts are often in a position to communicate project news to users about changes in the scope of a system, or revisions to the development timetable, but it's important that users hear this type of news from the right source. This is best done by their own manager, or someone identified with the business rather than with IT. Unless this happens, users may feel that IT is imposing decisions on them and when the news is bad, perhaps about the project being descoped or falling behind, it can damage the IT department's image and relationship with users. Analysts and designers should therefore be wary of falling into the trap of delivering bad news to users. Good news and bad should be delivered by user management.

Educating branch or unit managers about the implications of the new system is particularly important as it is often this group that offers most resistance to change. It is not difficult to see why. IT has often had the effect of empowering frontline staff, of breaking down departmental barriers, or of reorganising a business away from the old functional departments. All these impacts may threaten middle managers and, if they are not fully involved, the project may fail.

Marketing the change can involve face-to-face selling by the sponsor to individuals affected. This might include a corporate video that paints a picture of the brave new world that the change will create, or the design of a unique identity – a logo or an image – that helps give the project a special profile and so on. Very often it will involve all these things and many more in a co-ordinated programme of carefully designed and targeted communication. Marketing a new computer-based system may be different from marketing a car but the objective is the same: to persuade people to buy. They show that they have bought by investing time and effort in learning what the system can do and how to use it.

The design of the *user interface* is typically the responsibility of system designers, but a moments thought will make clear that human-computer interaction has an extremely important non-technical dimension. Involving users in HCI design can help ensure that the system looks inviting and works in a way that makes sense. Even if the design cannot be truly collaborative, users will welcome the chance to be involved, and this will increase their commitment and satisfaction with the system once it's delivered. And the same lessons apply to the planning of *office layouts*: if these can be designed collaboratively, user commitment and satisfaction will rise.

25.4.3 Providing skills

The third stage of the people project includes such activities as:

- Carrying out a task analysis
- Writing learning objectives
- Designing training materials, including tutorial guides
- Designing user guidance materials
- Training those affected, and planning continuing education.

Task analysis is an essential step in producing practical user guides and training material. The aim is to understand exactly who uses the system, and how they use it. It is then possible to tailor the user guidance and training to suit the needs of different groups of users. For the purposes of designing effective user guidance and training, task analysis should have the following outputs:

- List of user groups
- Brief description of each job
- List of all tasks carried out on the system
- Details of what triggers each task and how regularly this task happens
- Flowchart which shows where the computer tasks fit into clerical procedures
- List of everything that will change or be difficult for each user group once the new system is introduced
- Matrix showing which user groups perform which tasks
- Matrix showing which system modules or menu options they use to perform each task.

Systems analysts may have already carried out some task analysis before the people project is underway. Usually it falls to the people project to examine the impact of the system but it may be that lists of user groups and tasks can be made available by systems analysts.

Learning objectives define what a person will know or be able to do at the end of a training session. For systems training, learning objectives usually have a practical value for students in their work context. For example, the learning objectives of an introductory session in using a new office computer might be:

- to be able to log on or off the system
- to know which application to use and under what circumstances
- to be able to move through menus to get to the screen needed.

Once defined, learning objectives feed into the design of *training materials* such as tutorial guides and workbooks. These should relate very closely to *user guidance material* such as reference guides, problem-solving guides, keyboard templates, and other desktop reminders. In general, users work better if the documentation they receive is task-based rather than system-based, as this reflects the users reality. Moreover, they only need guidance on the tasks they themselves will carry out, which cuts down the size of the manuals they need, and makes them more friendly.

Training of users has been discussed already in chapter 24, but it is worth considering how different approaches to training may map onto the model of change management that we have been using. Training people away from their workplace, for example in learning centres, can help the process of unfreezing. Workshops and walk-throughs can help move people in a common direction. Finally, workplace training is good for meshing learning into day-to-day work, in other words for bringing about refreezing. Effective refreezing is also the goal of continuing education, which may take the form of refresher training, best practice seminars, or advanced courses that encourage and enable people to extend their knowledge of the system and get more benefit from it.

25.4.4 Managing the benefits

The final stage in the people project lifecycle involves:

- Running a help desk
- Post-implementation reviews
- Education or consultancy to help branch/unit managers to manage the benefits
- Collecting learning points from this project for the next one.

By this stage, the system has been running for some time, and the objective is to ensure that any problems are caught quickly so that people don't become demotivated and that knowledge about benefits and best practices is spread as widely as possible. The *help desk* clearly has a major role to play in both these areas. As the focus for feedback about the system, the help desk can identify bugs in the system as well as remedial training needs. It can also capture ideas for using the system effectively, and shares these throughout the organisation.

Education of managers is essential to ensuring that the benefits achieved are actual, rather than notional. Finally, reviews and the collection of *learning points* at specially-designed seminars or workshops can help ensure that future changes, whether they involve the implementation of the next system, or a move to a new building, or the reorganisation of departments, are handled as smoothly as this one was.

25.5 THE CHANGE MANAGEMENT PAYOFF

The four stages in the people project lifecycle may, of course, overlap and the list of activities just given for each stage is by no means exhaustive. However, the scope of these activities, and their significance to the success of a new system, will now we hope be evident. Two things in particular should be clear: that the people project is of equal importance to the ultimate success of the system as the development project, and that the two projects complement each other. The people project, and the numerous strands of activity associated with it, imply that a great deal of time and effort must be focussed outside software development in order for a new IT system to succeed. The cost of all this influencing, communicating, and support is by no means negligible, although compared with the cost of developing the system itself it will probably seem very small indeed. Nevertheless, it must be justified in some way. There are two difficulties. Firstly, in most cases, you have only one chance to implement a system. It's difficult to compare how things went with how things would have gone if a people project had been set up. Secondly, there are very few hard ways of measuring success. To some extent, you have to rely on subjective evaluations of success.

To overcome the first difficulty, some organisations have taken to using standard questionnaires on IT implementations, that enable one implementation to be judged against another. The questionnaires attempt to track progress against the original business objectives, as well as gathering perceptions from users and customers about the quality of the system, the fit between expectations and reality, the smoothness of the implementation, the level of training and communication, and the responsiveness and efficiency of post-implementation support functions. Over time, it becomes possible to state critical success factors for any IT investment. Many of these, as we have already stated, are non-technical. The best system ever designed may fail to make any difference to an organisation's bottom line if users are not committed, organised, trained, or supported.

Overcoming the second difficulty requires us to acknowledge that hard measures that can be translated more or less directly into money are simply not appropriate here. IT systems are successful because the people who use them are motivated and confident. It would be meaningless to try and put a value in pounds on motivation and confidence. In some cases a few minutes at the start of every team meeting spent discussing the imminent arrival of a new computer system may be enough to keep users feeling enthusiastic and involved. In others, outside professionals may be required to co-ordinate and help deliver a marketing campaign complete with videos, newsletters, training packs, and management seminars.

The key is to focus on the areas that will make the most difference. You will never be able to eliminate resistance to change altogether. Try to identify the main sources of resistance – middle managers anxious about their new role or users disillusioned by past experience, or anyone else – and concentrate on winning them over. The Pareto principle states that if there are 100 levers you could pull, there are almost certainly 20 or so that will achieve 80% of what you want.

The truth is, organisations that do not attempt to manage change when developing a new IT system are not avoiding the non-technical issues, they are simply addressing them in a bad way, and that is where the real money goes. Unhappy users make mistakes, boycott the system, spread negative rumours, go on strike, hold the entire organisation back. The incremental cost of addressing their issues in a sensitive and thorough-going way is trivial. A good change management programme will help ensure that the system is implemented smoothly, and yields the benefits for which it was designed.

25.6 SUMMARY

To sum up this chapter on the role of change management in an IT project, we've seen that there is a crisis in IT. Computer systems are not delivering the benefits for which they were designed. One important reason for this is that *people* have been overlooked.

All technical projects should therefore be accompanied by a people project, designed to help users and managers make the most of new computer systems. Change management theory gives us insights into how such a people project should be set up and run. Effective communication is a key part of any change management programme. Systems analysts and designers can feed into the people project in a number of areas, and need to support the all-important communication effort, helping to ensure a smooth transition to the new system.

The discipline of change management studies the process of change to see if there are ways of helping it succeed. Two models, the S-curve, and the concept of change as a three-stage process of unfreezing, moving, and refreezing, are particularly helpful. On an IT project, success may be partially equated with overcoming resistance among users and managers to the changes wrought by a new computer system. Resistance can be overcome by giving people ownership of the change, providing them with a vision of the new world and the knowledge and skills to master it, and supporting them to ensure they do not revert to old ways.

The people project can be divided into four main stages. Creating involvement requires appointing a sponsor, setting up focus groups, and selecting change agents, as well as checking the user requirements and setting business objectives that everyone can understand. Building commitment can be done through co-ordinated communication and marketing efforts and collaborative work on interface design and office layouts; the focus of much activity in this stage is on middle management. Providing skills includes training people and producing effective user guidance materials. Finally, managing the benefits involves catching and solving problems fast through an adequately resourced and well-briefed help desk, and spreading good practices throughout the organisation.

The benefits of all this change management activity are hard to quantify. However, unless the people issues are addressed in a sensitive and thorough way, those IT projects that ignore them will continue to fail to meet their objectives.

CHAPTER 26
What Next?

26.1 INTRODUCTION

So far in this book we've addressed three aspects of the time dimension of analysis and design. Firstly we've looked at the unchanging aspects of analysis and design and, for example, considered fact-finding interviews and communication. Secondly we've examined some of the aspects of the analyst's job that have changed as a greater understanding has been gained about methods and techniques. The chapters dealing with structured methods and the management of change are examples of this aspect. Finally there are chapters where the content is driven by technological change and the opportunities it offers to improve the solutions we offer to system problems. This final chapter deals with a different time dimension, it deals with the future; not the immediate technological future but with the broad sweeps of change that will affect business and information systems until the next century. It's important to recognise that all of the ideas, forecasts, suggestions and predictions do not fit neatly together. It is a turbulent world that lies ahead. You need to decide for yourself, in your situation, the parts of this chapter that offer you an opportunity to improve what you do. Judgement is required; there are different views of the future. Although you should expect this chapter to offer a sound view of likely trends, it won't all be right.

26.2 HOW DID WE GET HERE?

Information processing is concerned with processing data to generate information. Information systems receive input, store it, process it and output information; nowadays they also transmit or communicate locally or remotely. These functions have existed for a long time and been in business use for at least a hundred years since the US Census in 1890. Developments in the 1960s and 1970s concentrated on making these functions go faster and cost less. Many people suggest that this 'more speed, less cost' phenomenon began with the 1939/45 World War. A key difference however between IT and other office technologies – the typewriter or the adding machine for example – is that whilst other office technologies generated cost savings when introduced, only IT has consistently generated year on year cost savings through technological developments and progressively increased functionality. Since we so often use Japanese industry as a model for the effective implementation of new technologies let's take two Japanese examples – one of reduced cost and one of increased functionality – and see how information technology has dramatically affected two older technologies that, like the typewriter and the adding machine, would not have been changed otherwise.

The first example concerns the watch industry, effectively controlled by the Swiss with their intricate and beautifully crafted clockwork mechanisms until the late '70s/early

'80s when Japanese digital watches using chip technology flooded the market and, costing less than 10% of Swiss products, decimated the Swiss watch market. The other example is the fax, the facsimile transmitter which was revolutionised by the Japanese for whom, with their image-based language, the older telegraph technology with its use of Morse code was unsuited. Nowadays of course with the ever-falling cost of hardware it is possible to plug a portable computer into a hotel telephone socket and transmit text and drawings to a hard copy printer in another continent.

Over time, information technology has been a driver for change in isolation and when linked with other technologies. The development and use of IT has been revolutionary.

26.3 WHAT'S HAPPENING TO WORK?

IT is also revolutionising the nature of work. Until the 1960s and 1970s, the organisation of work was all about planning and control, and Adam Smith and Frederick Winslow Taylor relived their lives in factories and offices throughout the world. Taylor's view was that there was a best way of accomplishing every task. This best way was worked out by managers and technical experts who then instructed the workforce to 'do it this way'. Managers knew best, the workers knew nothing, the system ruled operations, and supervision and personnel systems focused on enforcing this way of life through the use of the carrot and the stick; beat the donkey most of the time and occasionally let it reach the carrot and take a bite of it.

The key assumption of the Taylorist model is that the workforce has nothing to contribute to the production process, to the system. This view is advocated by some people on the grounds of technical efficiency but others say that it found favour – and still does – because it enables management to control the workforce by removing knowledge, and hence power, from the shop or office floor.

There are of course alternative perspectives based on quite different views about people, jobs and the role of management. Many of these views are grouped into what is often called sociotechnical systems design. Their general thrust is that productivity rises if job design, social needs and technology are considered in an holistic way. More recently as we shall see later, people have talked about 'empowerment' and 'inverting the triangle'. Whilst we can see a general evolution away from Taylorist principles towards more higher skill work we shouldn't suppose that all implementation of new IT systems is designed on this basis. Just as there are implementations that give greater freedom and more opportunity for initiative to system users, so there are those that still deskill the users' jobs.

IT systems have some importantly different characteristics from other technologies and these characteristics have significant implications for the way work is organised. IT systems break down organisational barriers and enable organisations to be restructured in quite different ways.

To take advantage of the potential of this technology smarter employees need to be able to work with data and with systems and there is a greater need for employers to train and for managers to coach employees so that their full potential is realised.

26.4 HOW WILL WE SURVIVE?

So if IT systems will impel so many changes, how will we survive? All organisations are interested in their future. Survival is the first rule of business; it comes even before profit. There is a lot of evidence about how technology is changing the way organisations operate. The great improvements in productivity in the eighties came from the application of technology. But where is technology taking us, and in particular where is information technology taking us? To answer this question, the 'Management in the 1990s' Research Programme (MIT90s) was set up to

> 'develop a better understanding of the managerial issues of the 1990s and how to deal most effectively with them, particularly as these issues revolve around anticipated advances in information technology.'

It therefore addresses the future shape of business and the development of computer-based systems, and is therefore central to the organisation and management climate that analysts could expect to find in the years leading up to the next millennium. To help in this research, the team at the Sloan School of Management at MIT (Massachusetts Institute of Technology) created the following model which they called the MIT90s Paradigm – the MIT90s example or pattern. This is shown in figure 26.1.

The limit of the organisation is shown by the dotted line. Outside it is the technical environment and the socio-economic environment, both of which cause change inside the organisation. Changes in information technology have, for example, shifted the balance of power in organisations away from technology-led central computer departments to user departments with microcomputers and local area networks. In Europe at the time of writing there is a recession and the economic climate is one of retrenchment, of

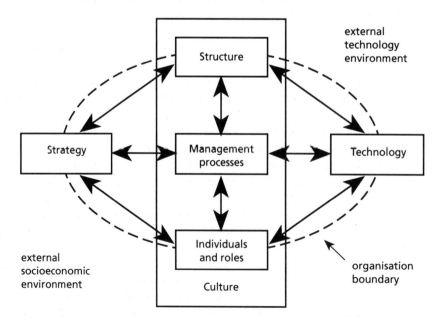

Fig. 26.1 The MIT90s framework

downsizing in organisations and reductions in investment in new computer-based systems. Central to the MIT90s paradigm are the management processes that link together corporate strategy and information technology. The paradigm therefore emphasises the need for IT to be viewed as a strategic resource. Note the importance of management – or business – processes and how they are separated from the structure of the organisation. As we shall see later in section five, this separation has important implications.

One of the UK sponsors of this research programme was ICL (International Computers Ltd), and this company identified nine important top-level issues from the research output. ICL says that it is now deriving benefits from this research. Whilst all of these findings are interesting, two are particularly relevant to this book.

Firstly, there is unlikely to be any reduction in the rate of development of technology in electronics or computing. This will lead to ever-increasing computer power becoming available in smaller and smaller boxes at reducing cost. Applications that were previously uneconomic will become possible, and computer technology will be incorporated more and more into manufacturing processes and products. The main difficulties will be in developing software quickly and cheaply to enable this capability to be used. As consumers increasingly expect products to be 'smart' we shall see increasing demand for software developments using fuzzy logic and expert systems.

Secondly, there has been no evidence that the implementation of computer-based systems has given organisations sustainable competitive advantage. This may be a surprising finding particularly for people who spend their lives developing new systems for organisations. There are two important facts that lie behind it. Most organisations implementing new computer-based systems don't evaluate the business benefits that new systems generate. Indeed there is a view that says that organisations make a very poor attempt to quantify the benefits they expect to get. Whether that's true or not, more quantification of the result and business benefits would clearly be of value. There is also a significant qualifying word in the finding; the word is 'sustainable'. The finding says that some temporary advantage may be gained, but sustainable advantage is not achieved because IT systems are easily copied. It seems to me difficult to accept that 'IT systems are easily copied'. If you've read through this book up to here then 'easily' is probably not the first word to spring to mind when thinking about the development of an IT system. Also you have to ask yourself whether systems are copied so that the business can be changed as a consequence, or whether competing organisations face similar competitive pressures which drive them to develop similar systems.

The MIT90s programme also paid attention to organisational change issues as well as issues of technological change, and the implementation of change was seen as the key challenge. We've already looked at this in the previous chapter where we emphasised the need for a people project to run alongside the system development project. The MIT90s project offers a wider, more management-oriented view in the context of the whole organisation and suggests that unless the technology and the organisation are aligned together then there cannot be an effective implementation of IT. This concept of *alignment* comes from the evidence that the requirements of the IT system must be matched by the capabilities of the organisation. So systems that need highly committed users who can engage in complex diagnostic tasks will call for highly motivated, well educated and trained staff. Similarly, *alignment* means that the requirements of the organisation are supported by the design of the IT systems. Decentralised organisational structures with local decision-making means that systems need to deliver to the managers' desks the information they need to take their local decisions. This issue of alignment is examined from a project manager's viewpoint in *Project Management for Information Systems* (Pitman 1991).

Four groups of people play an important role in aligning the technology and the organisation. Top management has to provide a clear vision of the kind of organisation it wants and the steps needed to get to it. Middle management – the group that is most at risk in today's organisations – has a crucial role. Typically they are instrumental in guiding IT projects from feasibility to implementation yet the impact of organisational change is often to eliminate management at this level. Throughout this book we've referred to the importance of user involvement in the development of new systems. The ideas – in the previous chapter – of a people project further emphasise its importance. This group – the users, or customers, or clients – of the new system are the third key group. Finally, new technology cannot be introduced successfully where unions represent users unless the unions are involved. The Thatcherite ideas that permeated business and society throughout the '80s have led to the decline of trade union power in the UK but this is not so everywhere.

So, how will we survive? Let's recognise that

- The future will not be an extrapolation of the past except in hardware. The future software requirements of excellent organisations will not be for centralised, deterministic systems. Decentralised, heuristic, personal, flexible and responsive will be the key characteristics.
- Empowerment of individuals will be the key; control will be the death knell. This chapter has already said something about the nature of work. Modern views about the nature of organisation suggest that the old triangular structures of a senior manager sitting on top of junior managers who in turn sit on top of workers will be replaced by flatter inverted triangles where empowered workers exercise greater responsibility and are supported by their manager.
- Finally there will be greater partnership between the suppliers of IT solutions and their customers. This will be true whether the supplier is the in-house department or an external organisation.

Survival then depends, even more in the future than in the past, on taking a proactive rather than a reactive view about change.

26.5 BUSINESS PROCESS REENGINEERING

One important proactive approach to business and systems development aims to completely break apart the way organisations are structured and to reorganise them in a completely new way. This approach is called 'business process reengineering' (BPR). Information technology is the key enabler in this new approach.

We need to begin by looking back over 200 years to Adam Smith's *Wealth of Nations* which was published in 1776. This book described how the then industrial revolution could be used to increase productivity by orders of magnitude. The single principle that underpinned the new approach was the division or specialisation of labour and the consequent fragmentation of work into small tasks. Smith described each as coming from an increase in the dexterity of individuals who worked more efficiently because the scope of their job was limited, from the saving in time that resulted when people didn't have to change from one task to another, and finally from the invention of machinery to automate the now simple and relatively deskilled tasks. Mass production became possible. The same principle was used by Henry Ford to establish his automobile empire and Alfred Sloan developed new management principles which applied the division of labour in the management of the huge individual enterprises. Still later, in the 1950s and 1960s organisations grew by continually expanding operations at the base of the organisational pyramid and filling the missing management layers.

The thrust of BPR is that these old ways of organising business simply don't work anymore. Business Process Reengineering, say Hammer and Champy, is 'the fundamental rethinking and radical design of business processes to achieve dramatic improvements in critical, contemporary measures of performance such as cost, quality, service and speed'. It is about

- Fundamental change. It doesn't ask the question about how to do some things better but about why do we do it at all?
- Radical change. When we know what we have to do, what is the best way of doing it irrespective of how it might be done now or how we are organised now.
- Dramatic improvement. BPR is for organisations in deep trouble, or who see trouble coming, or in good shape and well managed and who want to put their competitors in deep trouble.

One of the big UK accounting and consulting practices in the UK has a 'top ten checklist' to help identify whether or not an organisation is ripe for BPR. ' If you recognise four or more of the following signs then you should be looking at BPR ' they say:

- You don't know how your competitors do what they do
- Customer complaints are rising
- There's no common understanding of your organisations key performance measures
- You still do things the old way and haven't introduced significant new technology in the last three years
- Similarly, you've not been first in the market with a new product or service in the last three years
- No one would recognise senior managers if they walked round the sites or offices
- Individual managers have their own agendas and empires
- Individuals from different departments rarely work together
- You haven't reorganised in the last three years
- Staff development is based on improving technical skills rather than developing a broad range of experience.

Finally, BPR is about processes. This is the most difficult part of BPR to understand because most managements are focused on tasks. A process is a collection of activities that takes various inputs and then creates an output that's valuable to a customer. In a training centre for example, 'course scheduling' can be a task or a process:

- If it's a task it's a list of courses, with dates and venues that don't clash.

- If it's a process it's a list of courses, with dates and venues, and assigned trainers, with joining instructions for the student, course materials for trainers and students, a class list for the receptionist and a billing list for the accounts department so that the customer gets invoiced. In other words it is complete, it's done because it has to be done, and it has value to the customer.

BPR stands the Adam Smith, FW Taylor models on their heads. Instead of having hundreds of simple tasks linked together by complex processes, what we need are simple processes that enable the organisation to do what it needs to do. In reengineered organisations

- Jobs are combined. The person who sells you the idea and then designs and estimates your new kitchen using kitchenco products also schedules the installer, checks that the work has been done and sends you the bill. This change from job specialisation to job integration is called moving to a case worker process.
- Workers make decisions. The man who calls to install your new gas meter doesn't say 'I'll have to ask the supervisor, I'll be back in the morning', but makes his own decision about the installation as he is empowered to do through a less complex process.
- Work is done where it makes most sense to do it. This may well mean relocating work across organisational boundaries to improve the overall performance of the process.
- Checks and controls are reduced. Reengineered processes use controls only to the extent that they make economic sense. Expense claims in a company have to be signed by a division's financial controller if they are for more than £100. He's based at a different site so the claim is copied in case the original goes astray. When the financial controller gets the claim – assuming he's not away – he authorises it and sends it to the central DP site for processing. This happens at the end of the week and a few days later the individual is paid. The process takes up to three weeks, by which time the individual is on the road again. He gets fed up with funding the company's cash balance so uses the 'advance on expenses' procedure as well in order to get a cash advance. This also involves forms, and signatures and controls. It's a clear case for reengineering how employees' travel costs are covered.

Modern information technology is part of the reengineering effort but only if it is part of the revolution. Using IT to do faster what we already do now is not using IT to help reengineer the organisation. The key question is 'How can we use IT to do things that we're not doing now'.

26.6 CONCLUSION

This chapter has tried to encourage you to question why things are the way they are, and to suggest that the life of a systems analyst will be different in the future. The future will not be like the past. If you're interested enough to read more, there are three books listed in the bibliography.

Finally though there are two observable trends that will run forward through the 1990s. Firstly, the business reasons for investing in IT:

- from the 1960s onwards they were to improve productivity
- from the 1970s onwards – to improve managerial effectiveness

- from the 1980s onwards – to create competitive advantage
- from the 1990s onwards – to do business differently by reducing costs and improving quality, by improving customer service and by increasing flexibility to improve responsiveness to change.

These reasons for investing in IT will operate within a framework of organisational transformation that will see

- the orientation of business move from product to customer
- mass production will be replaced by flexible, on-demand production
- increasing value will come from intangibles like customer satisfaction and quality
- an organisation's intellectual assets will be with knowledge workers and not with management
- rewards will be given for performance and not for loyalty and seniority
- competition will become global and not national
- organisation structures will move from functional or hierarchical structures to networked and matrix structures
- economic relationships will move from take-overs and vertical integration to alliances.

It is within these new and radically different circumstances that new analysts will work.

Bibliography

Apple Computer Inc., *Macintosh Human Interface Guidelines*, Addison-Wesley 1992.
Crosby, Philip, *Quality is Free*, McGraw-Hill 1979.
Deming, W. Edwards, *Out of the Crisis*, Cambridge University Press 1982.
Juran Joseph, *Quality Control Handbook*, McGraw-Hill 1979.
Open University, *Usability Now! A guide to usability*, 1990.
Scholtes, Peter, *The Team Handbook*, Joiner Associates 1988.
Hammer, Michael, and Champy, James, *Reengineering the Corporation*, Nicholas Brealey Publishing 1993
Scott Morton, M. (Ed.), *The Corporation of the 1990s*, Oxford University Press 1991.

INDEX

Action Diagram, see DAD
Active listening, see Listening
Ada, 40, 258, 260
    'Rendezvous', 258, 260
Aesthetic integrity, 246
'Albatross system', see System
Alignment, 411
Alphabetic keyboard, see Keyboard
Analysts, attributes of, 8-10
Anatomy, see Ergonomics
Animation, 239
ANSI (American National Standards Institute), 316, 319
Apple Macintosh, 239, 245, 354, 360, 363
    human interface guidelines, 245
Area, in a database, 311
    see also DBMS
Archive file, see File
ARM (Acorn RISC Machine), 358
ASCII (American National Standard Code for Information Exchange), 263, 371
Audio-visual equipment, 57
    see also Presentations
Audit file, see File

b-model of system development, 109-110
Backwards compatability, 357
Bandswitch, 14
Bandwith, see Data rate
Bar code reader, 235
Bar chart, see Graphics
Barchart, 98
Baseband, 368
    see also Network
Batch system, see System
BCS (British Computer Society), 62, 74
Behavioural model, 29
    see also Yourdon, Edward
Bell, Alexander Graham, 23
Benefits, of proposed solution, 193
Birrell, N and Ould, M, 109
Block, 327
Body language, see Communication
Böhm & Jacopini, 10
BOMP (bill of material processing) structure, 160
BPR (Business Process Reengineering), 412-413
Brainstorming, structured, 187
Broadband, 368
    see also Network
BS 5750, see ISO 9000
Buffering, 263
Bugs, see Code defects
Bus, 366
    see also Network topology

Business analysis, 82-90, 92

C programming language, 258, 363
Cacheing, 358, 359, 360, 362
Cameron, John, 31
Card reader, 262
CASE (Computer Aided Software Engineering) tools, 29, 33, 36, 37, 41, 119, 149, 165, 177
CCITT (Comité Consultatif International Téléphonique et Télégraphique), 370, 371, 373, 374
CCTA (Central Computer and Telecommunications Agency), 34
CCTV derived dispays, 249
CD-ROM (Compact Disk Read Only Memory) drive, 262
Chain diagrams, 38
Chained file, see File
Change agent, 402
Change control, 104
Change count, 216
Channel transfer time, 282
    see also Magnetic media
Checksum, 365
Chord keyboard, see Keyboard
CISC (Complex Instruction Set Computer), 357, 358
Class, see Transport protocol
Client, see Customer
COBOL (COmmon Business Oriented Language), 101, 118, 310
CODASYL (Conference on Data System Languages) model, 38, 290, 310
Codd, Ted, 312
Code defects, 205
    see also Corruption
Code generator, 379, 381
Cognitive skills, see Ergonomics
Cohesion, 336
Colour, and screen design, 239
Colour blindness, 239
Command language, see Dialogue
Commissioning, 389-390
Communication, 108, 252, 261, 364, 367
    asynchronous, 260
    non-verbal (body language), 43-46, 55, 57, 58, 59, 134, 140, 192-195
    synchronous, 260
    verbal, 43-46, 47, 51-56, 59, 108, 134, 140
    written, 43, 44, 45, 47, 49, 50, 51, 59
Communication plan, 403
Computer Training Services, 118, 139
Concatenation (artificial speech), 227
Configuration management, 40, 105
Consistency, 245, 246
Context diagram, 155

Core system, 186
Corruption
    code, 206-207
    data, 210-212, 215
Costs
    development, 189
    'hard' and 'soft', 191
    hardware, 190
    lifetime, 191
    maintenance, 191
Coupling, 336-337
CPU (Central Processing Unit), 324-326, 329, 330, 336
CRC (Cyclic Redundancy Code), 365
Critical path, 97
Crosby, Philip, 63-65
CSA (Computer Services Association), 61
CSMA/CD (Carrier Sense Multiple Access with Collision Detection), 369
Current logical system, 111-112
Current physical system, 111-112
Cursor control key, 234
Customer, 2, 12, 27, 40, 62, 108, 150, 213, 220, 224, 251, 320, 321-322

DAD (Data Action Diagram), 33, 336, 341, 343-346
Data, 265, 266, 268-274, 277, 279, 304, 306, 323, 327, 330, 333, 338, 339-341, 364, 366, 375
Data catalogue, 164, 179
    see also Data dictionary
Data connector, 374
Data conversion, 386-387
Data dictionary, 12, 148-149, 154, 164, 165, 169, 197, 201, 233, 274-277, 295, 306, 321, 339, 341
Data encapsulation, 253
Data encryption, 216
Data entity, see Entity
Data flow, 150, 151, 154, 155, 170, 233, 253, 277
Data flow diagram, see DFD
Data independence, 303
Data integrity error, see Error processing
Data lock, 215-216
Data Navigation Diagram, see DND
Data model, 113, 154, 265, 272-274, 320
Data processing, 325
Data rate, 365
Data structure, 12, 278, 312, 321
    see also Entity relationship
Data takeon, 386
Database, 180-181, 201, 251, 279, 302, 303, 308, 320, 328, 355
    definition of, 303
    distributed, 304

Index

Database environments, 6
Database models, 304, 306
Data store, 16, 150, 151, 154, 169
    see also DFD
Data store/entity cross-reference table, 168
DBMS (Database management system), 38, 149, 201, 303-305, 306, 310, 314. 324, 325-327, 329, 330, 334
DCE (Data Circuit-terminating Equipment), 374
De Marco, Tom, 108
DEC (Digital Equipment Company), 358-363
Decision, 339
    see also Program specification
Deferred online system, see System
Deming, W. Edwards, 62, 63
Demodulation, see Modem
Department of Trade and Industry, see DTI
Designers, attributes of, 8
Development costs, 189-191
Development timescales, 189
Device driver, 263
DFD (Data Flow Diagram), 12, 16, 25, 31, 33, 36, 38, 51, 96, 113, 138, 148-154, 168, 170, 181, 187, 197, 198, 223-224, 229, 233, 241, 255, 257
DFM (Data Flow Model), 149, 154, 164-170, 171, 175, 179, 182, 183
Dialogue, 223, 224, 225, 237-239, 242-243, 245, 249
    command language, 242-243
    form filling (template), 241-242
    menu, 240-241
    natural language, 243
    question and answer, 241
Dijkstra, Professor, 10
Direct manipulation, 245
Directory services, 372
    see also Layering
Display chart, 230, 232
Display screen equipment regulations, 247
DND (Data Navigation Diagram), 198, 273
Document analysis, 145
Document flow diagram, 155, 157
DOS, see MS.DOS
DTE (Data Terminating Equipment), 374
DTI (Department of Trade and Industry), 61, 73, 74
Dump file (security file), see File
Duplex, 371
Dvorak keyboard, see Keyboard

EAP (Enquiry Access Path), 179-180
EBCDIC (Extended Binary Coded Decimal Interchange Code), 263, 371
EC (European Community), 13, 14, 39, 247
Effect
    optional, 177
    simultaneous, 177-178
ECD (Effect Correspondence Diagram), 170, 171-172, 177-179
EIS (Executive Information Systems), 318
ELH (Entity Life History), 36, 38, 170-177, 182, 184
Electronic mail, 372
Elementary process description, 16
Embedded system, 355
EN 29000, see ISO 9000
Enquiry access path, see EAP
Entity, 266, 267, 269, 277, 326, 328, 329-330
Entity-event matrix, 171-172, 176
Entity-event modelling, 171
Entity life history, see ELH
Entity model, 156, 167, 169, 197, 201, 265, 268, 269, 273
Entity relationship, 33, 148, 158-164, 176, 179, 265, 267, 272, 273, 277
    degree of, 158
    one to many, 158, 267
    recursive, 159, 161
    see also Entity model
Entity relationship matrix, 267-268, 277, 278
Environmental model, 29
    see also Yourdon, Edward
Ergonomics, 201, 246
Error processing, 175
Essential system, 29
    see also Yourdon, Edward
Ethernet, 369, 376
    see also LAN
Eurogroup consortium, 39
Eurolab, 14
Euromethod, 21, 37, 39
European community, see EC
European space agency, 40
Event
    and function, 180
    business, 170
    random, 176
    system-recognised, 171
    system, 170
    time-based, 171
Event recognition, 170
External customer, see Customer
External entity, 16, 150-151, 154, 155, 169, 171
    see also DFD

Facsimile (fax), 227, 376
Fagan inspection, 65, 74, 76-78, 103, 185, 221
    see also Structured walkthrough
Fagan, Michael, 76

Faraday shield, 220
Feasibility, 6, 7, 107
    report, 124-125, 126
    study, 108, 110, 113, 115, 116, 123-124, 126, 127
Field, 279
    see also Record
File, 150, 279, 284, 285, 293-297, 299, 300, 302, 320, 332, 355
    archive, 280, 285, 321
    audit, 281
    chained, 289-290, 292
    computer, 150
    dump (security), 220, 280, 285
    indexed (full), 288-289, 293, 333
    indexed sequential, 286-287
    library, 280, 293
    master, 279-280, 286, 288, 291
    output, 280, 285
    paper, 150
    transaction, 280, 285, 286, 288, 291
    transfer, 280
File access, 291-292
File design, 201
File generations, 292-293, 295
File organisation, 284-291
File specification, 294-302
File transfer, 371
Financial Times, 395
Fishbone diagram, 63
'5 C's test', 194
Float, 97
FPU (Floating Point Unit), 358
Floppy disk, 281, 282
Flowchart, 38, 151, 348
FMS (File Management System), 306
    see also DBMS
Focus group, 402
Forgiveness, 245
Form, input, 223, 230, 233, 237, 241, 269
FORTRAN, 258
Four Ps, of effective speaking, 56
    see also Speaking, effective
4GL (fourth generation language), 306, 313, 316, 317, 333
Fraud, 205, 209, 210, 219
Function, 180, 181, 186
Function key, 234
Functional specification, 194-196

Gibson/Nolan model, 5
GIS (Graphical Information System), 318
GOTO statement, 10
Graphics, 229, 239
Grindley, Kit, 7
GUI (Graphical User Interface), 239

Hackers, 205, 208, 209, 214, 218
Half duplex, 371
Halon gas, 204
Hammer, M and Champy, J, 413
Handshake, 219, 221, 390
Hard disk, 281
Hardware, 190, 354-364
Hashing, 287-288, 317, 326, 333

Index

HCI (Human Computer Interaction), 238, 249, 252, 397, 403
HDS (Hierarchical Database System), 305, 308-309, 311
    see also DBMS
Head movement (seek time), 282
    see also Magnetic media
Head switching, 282
    see also Magnetic media
Header, 295, 321, 339
    see also File specification
Hewlet Packard, 363
Highlighting, 238
Hit rate, 293
HOOD (Hierarchical Object Oriented Design), 40
    see also OOD
Hot standby, 219-220

IBM, 64, 76, 200, 263, 354, 356, 363, 370
ICL, 411
Icon, 239
IE (Information Engineering), 29, 32, 33, 41, 335
IEEE (Institute of Electronic and Electrical Engineers), 374
Implementation, 7, 127, 379-394
Index
    cylinder and track, 286-287
    full, 288-289, 291-292, 293
    implicit, 292
Indexed file, *see* File
Indexed sequential file, *see* File
Informate, 4
Information Engineering, *see* IE
Information System, *see* IS
Information Technology, *see* IT
Input, 224, 225, 233-234, 236
Instant Image Corporation, 115-116
Intel microprocessor, 356-357, 363
Interface, 17, 62, 205, 223, 224, 238, 245, 249, 250, 251-254, 257-258, 263, 264, 336-338, 371, 383, 389, 493
Internal clock, 359
Internal customer, *see* Customer
Interview, 128, 129-141, 145
I/O (Input/output), 154, 164, 179-180, 182, 200, 269, 272, 330, 340
IP (Internet Protocol), 374
IRC (Interaction Relationship Chart), 336, 346-349
IS (Information System), 4, 5, 6, 23, 24, 28, 31-34, 38, 82, 87-89, 91, 148
ISEB (Information Systems Examination Board), 34, 36
ISO (International Standards Organisation), 370, 373
ISO 9000, 61, 72-73, 104
    see also QMS
IT (Information Technology), 7, 23, 24, 34, 39, 61, 70, 85-89, 91, 101, 104, 117, 395
Iteration, 112-113, 339, 342

Jackson, Michael, 31, 172
    see also JSD, JSP
JMA (James Martin Associates), 32
JSD (Jackson System Development), 31, 32, 41, 335
JSP (Jackson Structured Programming), 31, 34, 36, 37
Juran, Joseph, 63

Key (key attribute), 162, 266, 270, 272, 278, 326, 329
    compound, 162
    foreign, 162
    simple, 162
Keyboard, 233, 234-235, 236, 240, 262
Kilostream link, *see* Point-to-point data link

Label, 285
LAM (Logical Access Map), 198, 273, 275, 330-331
LAN (Local Area Network), 201, 355, 366, 368, 369, 372, 374-375, 377, 378
Language, programming, 380-382
Latency, *see* Rotational delay
Layering, 346, 367, 368, 371-372
    Application layer, 371, 377
    Data link layer, 373
    Network layer, 373
    Physical layer, 372
    Presentation layer, 372, 377
    Session layer, 372, 378
    Transport layer, 373, 374
LBMS (Learmonth & Burchett Management Systems), 34
LCD (Liquid crystal display), 249
LDM (Logical Data Model), 161, 164, 166, 168, 171, 175, 179, 183, 202, 267-269, 273, 278, 320, 321, 333
LDS (Logical Data Structure), 36, 40, 96, 148, 156, 161-164, 176, 177, 183, 266
    see also Logical Data Model, Entity model
LED (Light emitting diode), 249
Levelling, 151, 346, 353
Lewin, Kurt, 398, 400
    see also Unfreezing
Library file, *see* File
Line driver, 370
Line graph, *see* Graphics
Listening, 49, 109, 128, 130, 133, 135
    see also Communication
Logical access map, *see* LAM
Logical data store, 168, 169
Logical model, 197-198, 224
Logicalisation, 168
Login prompt, 209
Logon, 219, 221

Magnetic drum, 284
Magnetic ink character recognition, *see* MICR
Magnetic media, 227
    disk, 262, 263, 281-282
    tape, 281, 283, 292, 293, 302
Mainframe system, 354
Martin, James, 273

MASCOT, 335
Master file, *see* File
Meeting, 51-52
    see also Presentations
Memory management unit, 357
Menu, 23, 240, 243, 246
    pull down, 241
Menu driven system, 180
Menu hierarchy, 180, 181, 240
Merise, 21, 37-38, 41
Meta-method, 39
Metaphor, 245
MHS (Message Handling System), 371
MICR (Magnetic Ink Character Recognition), 235
Micro Writer, 234
Microfiche, 227
Microfilm, 227
Microprocessor, 356
Microsoft, 363
Mind map, 138
Mini system, 354
MIS (Management Information System), 318
MIT90s (Management in the 1990s) research programme, 410-411
Modell, 9
Modelling, 112
Modem, 203, 208, 218, 369-370, 372, 373, 374
    demodulation, 370
    modulation, 369-370
Motorola microprocessor, 356, 357, 360, 363
Mouse, 236, 240, 262
MS.DOS, 242, 259, 260, 261
Multidrop, 366
    see also Network topology
Multipart stationery, 230
Multiplexing
    frequency division (FDM), 364-365, 368
    time division (TDM), 364-365

Natural language, *see* Dialogue
NCC (National Computing Centre), 70, 138, 145, 230
    clerical document specification, 145, 146
    interview report, 138
NCR (no carbon required) paper, 230
Network, *see* LAN, WAN
Network database system, 305, 310-311, 367
    see also DBMS
Network diagram, 97
Network topology, 365-366, 374
Normalisation of data, 268, 330
    see also RDA
Numeric keypad, *see* Keyboard

Observation, 143
Objectives, learning, 404-405
    see also Training
OCR (Optical Character Recognition), 235

Index

OECD (Organisation for Economic Co-operation and Development), 395
OMR (Optical Mark Recognition), 233, 235
Online system, *see* System
OOA (Object Oriented Analysis), 253
　see also OOD
OOD (Object Oriented Design), 39-40, 253
　see also HOOD
Open system architecture, 6
Open system standards, 305
Open systems interconnection, *see* OSI
Operating system, 363
Operations, 172
Optical disk, 227
Optical fibre, 262
Optionality, 158-159, 165
　see also Entity relationship
Options, quantification of, 189-191
ORACLE, 322
OSI (Open Systems Interconnection), 201, 367, 370, 373, 374, 377
Output file, *see* File
Output, 224, 225, 230
Overheads, 321

Packet switched network, public, 373
PAD (Packet Assembler/Disassembler), 373-374
Page, in a database, 311, 327
　see also DBMS
Palmer, I R, 32
Paper tape, 284
Parallel life, 172-173, 175, 176
　see also ELH
Parallel running, 391-393
Parallel structure, *see* Parallel life
Parallelism, 359
Pareto principle, 406
PARIS model of systems analysis, 107, 113-114
Parity, 365
Partitioning, 112, 253, 264
Pass
　by reference, 258
　by value, 258
Password, 217, 219, 221, 251
PC (Personal computer), 355, 356, 357, 392
Pentium, 357
　see also Intel microprocessor
Perceptual motor performance, *see* Ergonomics
Peters, Tom, 63
Petri net, 169
Physical data design, 328-333
Physical design, 127, 328
Physical storage media, 202
Physiology, *see* Ergonomics
Pie chart, *see* Graphics
Pipelining, 359
Plotter, 227
Pointer, 289, 290, 308-310, 321, 324, 328
　symbolic, 326

Pointing device, 236
Point-to-point data link, 375
Pope, Alexander, 10
POS (point of sale) equipment, 236
PowerPC, 358, 359, 363
PREP model, 56
　see also Speaking, effective
Presentations, 53-59, 192-193
　see also Speaking, effective
Price Waterhouse, 7, 61
PRINCE (Projects IN Controlled Environments), 34, 91, 96, 121
Print layout chart, 230-231
Printers
　as peripheral devices, 262, 263
　impact, 226-227
　non-impact, 227
Procedure call, 258
Procedure, *see* DAD
Process, 150-151, 154, 156, 168, 169, 180,
　external, 180
　internal, 180
Process decomposition diagram, 33
Process design, 201
Process-entity matrix, 333
Processing sequence, 339
Program specification, 199, 336, 338-343, 348, 356
Program testing, *see* Testing, unit
Program, 336-338
Programmer, 380-381
'Programmer-friendly', 199
Project Management for Information Systems, 412
Protocol, 367, 368, 369, 373
Protocol stack, 371
Prototyping, 188-189, 230
　evolutionary, 188
　throwaway, 188, 189
PSS, *see* Packet switched network
Psychology, *see* Ergonomics
Punched card, 284

QA (quality assurance), 69, 72
QC (quality control), 70-72, 103
QMS (Quality Management System), 69-71, 72, 73
Quality, 61-74, 94, 96, 99, 103
　definition of, 62
Quality circle, 63
Quality management, *see* QMS
Quality manager, 71
Quality manual, 71
Quarantine, 208
Questionnaire, 141-143, 145, 148, 406
Questions
　context, 130-131
　in interviews, 128-131, 135-136
　in presentations, 58-59
Quit and resume, 172, 173, 176-177
　see also ELH
QWERTY keyboard, *see* Keyboard

RAM (Random Access Memory), 359, 361, 362

RDA (Relational Data Analysis), 36, 169, 268-273
RDBMS (Relational Database Management System), 305, 306, 311, 312, 316-319
　see also DBMS
Reading, effective, 48
　see also Communication
Record, 279-285, 293-294, 301, 306, 308, 310, 322, 327
Record searching, 143
Redundancy, controlled, 330-33
Relational data analysis, *see* RDA
Relational database, *see* RDBMS
Relational operators, 306
Relational programming language, *see* 4GL
Report, written, 193-194, 223, 269
Required logical system, 111-112, 233
Required physical system, 111-112, 320, 321
Requirements analysis, 7, 21, 62, 92, 95, 104, 114-115, 128, 131, 165
Requirements catalogue, 164, 166, 169, 181
Requirements specification, 7, 21, 62, 114, 278
Reverse video, 239
Reversion, 177
Ring, 366
　see also Network topology
RISC (Reduced Instruction Set Computer), 357-358, 360
Rotational delay (latency), 282
　see also Magnetic media

Scanner, automatic, 235
S-curve, 398-399, 407
Schema, 306, 307
　see also Database
Scholtes, Peter, 67
SCOPE model, *see* Terms of reference
'Scraping burnt toast', 67
Screen, 223, 224, 233, 237-239, 346, 348
Security file, *see* File
See-and-point, 245
Seek time, *see* Head movement
Semaphore, 262
Sequential access, 291
Serial access, 291
Set, 306, 310, 313
Shared memory, 259, 261
Single system view, 3
Site commissioning, 389
Smith, Adam, 409, 412, 414
Software, 280, 354, 356, 367, 371
SPARC processor, 358
SPC (Statistical Process control), 63
Speaking, effective, 51-56
　see also Communication
SPECmark, 362
Speech recognition, 236-237
Spine and feeder, 376
Sponsor, 401
SQL (Structured Query Language), 313, 316

Index

SSADM (Structured Systems Analysis and Design Method), 7, 12, 21, 34-37, 38, 41, 83, 102, 113, 149, 150, 154, 156, 158, 167, 169, 171, 173, 179-183, 200, 335, 336
Stand-alone specification methods, 335, 339
Star, 365
 see also Network topology
State, 55-256
State indicator, 172, 173, 175
 see also ELH
STD (State Transition Diagram), 169, 255-256, 336, 348-353
Stationery, preprinted, 230
Structured analysis, 10-12
Structured English, 341-342
Structured programming, 10
 see also JSP
Structured Systems Analysis and Design Method, see SSADM
Structured text, 12
Structured walkthrough, 74, 221
Style guide, 237
Subroutine, 340, 345
Switching
 circuit, 366
 packet, 367
Synchronisation, 261
Synthesis by rule (artificial speech), 227
System
 'albatross', 2
 batch, 259
 deferred online, 236
 hierarchical view of, 4
 information, 4, 5
 online, 236
 operational, 5
 single, 3
 strategic, 4
 tactical, 5
 the organisation as a, 3-4
System boundary, 223-224
System changeover, 390-393
System development
 stages in, 6
 structured approaches, 24, 25, 27-28
 traditional approaches, 21-24, 25, 28
System R, 312
System Telecom
 company and systems background, 13-20
 customer file, 279
 and business analysis, 90
 and collecting data, 147
 and communicating with people, 60
 and data communication, 378
 and databases, 303-306
 and file organisation, 302
 and interpreting information collected, 195
 and physical data design, 334
 and project management, 105-106
 and quality, 81
 and recording information, 166
 and structured methods, 41
 and system protection, 222
 and terms of reference, 126
Systematic activity sampling, 143
Systems analysis, guidelines, 109

Table, 229, 306, 317, 327
Task analysis, 404
Taylor, F W, 409, 414
TCP (Transmission Control Protocol), 374
Teletypewriter, 262
Template, see Dialogue
Terms of reference, 117, 118, 126
Testing
 acceptance 386
 integration, 383, 385
 performance, 385
 regression, 385
 system, 383-385
 unit, 93, 382
 volume/soak, 385
Text design, 238
3GL (third generation language), 323, 327
Third normal form, see RDA
3 Ts model, 55
 see also Speaking, effective
TickIT, 61, 73
Time management, 109, 116
 guidelines, 122-123
Time sequence diagram, 255-257
TNF (Third Normal Form) analysis, see RDA
Token passing, 366, 369
 see also Network topology
Token ring, 369
Touch sensitive screen/tablet, 235-236
TQM (Total Quality Management), 71-72
 see also QMS
Training
 analyst and designer, 32
 user 28, 29, 190, 234, 240, 246, 387, 388, 404-405
Transaction file, see File
Transfer file, see File
Transmission control protocol, see TCP
Transport protocol, 373
Tree structure, and interviewing, 131-132
Tree, 366
 see also Network topology
Turnaround document, 230, 235

Unfreezing, 398, 407
Unit testing, see Testing
United States Government Accounting Office, 61
UNIX, 258, 260, 354, 355, 357, 363
Updating, 292, 293
User, 2
 and database design, 304-306
 and interface design, 223, 226, 227, 229 233-234, 238-240
 and ownership, 109, 165
 and requirements specification, 185
 approval by, 27
 as interviewee, 129-133
 experienced, 241-243
 involvement of, 27-28, 95, 154-155, 171, 322
 role of, in SSADM, 35-36
 role of, in Yourdon, 30
 views of, 109, 129, 180-181, 217-218
 see also Customer
User control, 245
'User-friendly', 199
User implementation model, 403
 see also Yourdon methodology
User interface, 403
User requirements, 165, 278
User options, 186-87
User privilege, 216-217, 219, 221
User support, 246
User training, see Training

Validation, 188, 234, 340, 341
Validation route, 188
VAX machine, 358
VDU (Visual Display Unit), 227, 230, 238, 247, 262, 354
VDU 90/270 (EC directive), 247
'Victorian novel', 197
Virtual terminal, 371
 see also Layering
Virus, 205, 206, 207-208, 214
Visual aids, 193
 see also Presentations
Volatility, 293, 302

Wait-states, 360
Wakefield, Tony, 9
WAN (Wide Area Network), 201, 368-369, 372, 374, 375, 377, 378
WBS (Work breakdown strucure), 95
Wealth of Nations, see Smith, Adam
Weinburg, G M, 8, 9
WIMP interface, 243-246, 388
Wiring, structured, 376
Work station, 223, 247-249
Workstation, computer, 351, 355, 357
Writing, effective, 50-51, 193, 238
 see also Communication
WYSIWYG (What You See Is What You Get), 245

Yourdon, Edward, 29
Yourdon methodology, for software development, 29 -31, 41, 196, 335

Zapping, of data areas, 203
Zero defects, 64
 see also Crosby, Philip